T0134935

Lecture Notes in Social Networks

More information about this series at http://www.springer.com/series/8768

Rokia Missaoui • Talel Abdessalem
Matthieu Latapy
Editors

Trends in Social Network Analysis

Information Propagation, User Behavior Modeling, Forecasting, and Vulnerability Assessment

Editors
Rokia Missaoui
Department of Computer Science
& Engineering
University of Quebec in Outaouais
Gatineau, QC, Canada

Talel Abdessalem
Department of Computer Science
and Networks
Telecom ParisTech
Paris, France

Matthieu Latapy
UPMC Univ Paris 06, CNRS
LIP6 UMR 7606
Sorbonne Universitès
Paris, France

ISSN 2190-5428 ISSN 2190-5436 (electronic)
Lecture Notes in Social Networks
ISBN 978-3-319-85149-5 ISBN 978-3-319-53420-6 (eBook)
DOI 10.1007/978-3-319-53420-6

Printed on acid-free paper

This Springer imprint is published by Springer Nature
The registered company is Springer International Publishing AG
The registered company address is: Gewerbestrasse 11, 6330 Cham, Switzerland

Preface

A rapidly growing number of Internet users participate in social networks, belong to communities, produce, broadcast and use media content in different ways. This tremendous growth of social networks has led to a multidisciplinary research on analyzing and mining such networks according to various directions and trends as shown in the present book.

The book contains recent studies in social network analysis and represent extended versions of a selected collection of articles presented at the 2015 IEEE/ACM international Conference on Advances in Social Network Analysis and Mining (ASONAM), which took place in Paris between August 25 and 28, 2015. The topics covered by this book are: assortativity, influence propagation and maximization, social user's profile and behavioral modeling, sarcasm analysis, delurking, social engineering and vulnerability assessment, link prediction and social media forecasting.

"The Perceived Assortativity of Social Networks: Methodological Problems and Solutions" by David N. Fisher, Matthew J. Silk and Daniel W. Franks studies assortativity (also called assortative mixing) in social networks and shows that while social networks are more assortative than non-social ones, they tend to be positively assortative only when they are built using group-based methods. To overcome this bias, a number of solutions based on advances in sociological and biological fields are exploited.

"A Parametric Study to Construct Time-aware Social Profiles" by Sirinya On-at, Arnaud Quirin, André Péninou, Nadine Baptiste-Jessel, Marie-Françoise Canut, Florence Sèdes focuses on using information shared on users' egocentric networks to extract their interests. To that end, a time-aware method is applied inside an existing social profile building process and aims at weighting user's interests in the social profile according to their temporal relevance (temporal score). An empirical study on scientific publication networks (DBLP/Mendeley) is then used to compare the effectiveness and accuracy of the proposed social profile construction technique against the time agnostic technique.

“Sarcasm Analysis on Twitter Data Using Machine Learning Approaches” by Santosh Kumar Bharti, Ramkrushna Pradhan, Korra Sathya Babu and Sanjay Kumar Jena tackles the new and challenging problem of sarcasm analysis in social networks by proposing four approaches that aim at extracting text features such as lexical, hyperbole, behavioral and universal facts. Then, the accuracy of the proposed solutions for Twitter data is computed and reveals a considerable improvement over existing sarcasm identification techniques.

The goal of the chapter, “The DEvOTION algorithm for Delurking in Social Networks” by Roberto Interdonato, Chiara Pulice and Andrea Tagarelli is to delurk silent members of a social network, i.e., to encourage them to get more involved in the network. To reach such a goal, the authors define a delurking-oriented targeted influence maximization problem under the linear threshold model and propose an approximate solution based on a greedy algorithm named DEvOTION. The superiority of the defined procedure over other existing delurking approaches is then given.

“Social Engineering Threat Assessment using a Multi-layered Graph-based Model” by Omar Jaafor and Babiga Birregah deals with a security issue related to social engineering vulnerability, i.e., evaluating the set of attacks that focus on deceiving humans into performing actions or disclosing information. The study proposes a graph-based multi-layered model built from layers that depict different attack scenarios, different states in an attack and multiple contexts that could be used. It highlights the interconnections between the different elements in an attack such as actions performed, users involved and resources used, and represents actions that do not necessarily leave traces in a monitored system.

In the chapter titled “Through The Grapevine: A Comparison of News in Microblogs and Traditional Media”, the authors Byungkyu Kang, Haleigh Wright, Tobias Höllerer, Ambuj K. Singh and John O’Donovan propose two novel algorithmic approaches, namely content similarity computation and graph analysis, to automatically capture the main differences in newsworthy content between microblogs and traditional news media.

“Prediction of Elevated Activity in Online Social Media Using Aggregated and Individualized Models” by Jimpei Harada, David Darmon, Michelle Girvan and William Rand deals with information propagation to a large set of social network members by focusing on the identification of the time periods when a large portion of a target population is active, which requires modeling users’ behavior. Three methods for behavior modeling are then proposed and validated on data collected from a set of users on Twitter in 2011 and 2012.

“Unsupervised Link Prediction based on time frames in Weighted-Directed Citation Networks” by Mehmet Kaya, Mujtaba Jawed, Ertan Bütün and Reda Alhajj proposes a time-frame based unsupervised link prediction method for directed and weighted networks. To that end, weighted temporal events are first defined. Then, a novel approach based on the common neighbor metric for computing the time-frame based node score is given. The empirical study exploits an unsupervised learning strategy on a weighted-directed citation network to show that the proposed method gives accurate prediction and promising results.

"An Approach to Maximize the Influence Spread in Social Networks" by Ibrahima Gaye, Gervais Mendy, Samuel Ouya and Djaraf Seck deals with the influence maximization problem by proposing a Spanning Connected Graph algorithm (with three variants) that computes the seeds from which the information propagation is initiated. The first variant builds the children of the nodes randomly while the second one uses the neighborhood for the identification of children. The third variant is a generalization of the first two ones and takes an arbitrary graph as input while the first two variants require a connected graph as input. These procedures are effective and have a polynomial time complexity.

Eva García Martín, Niklas Lavesson, and Håkan Grahn on "Energy Efficiency Analysis of the Very Fast Decision Tree Algorithm" addresses a general issue in data mining applications and could be useful for mining social networks. It introduces energy consumption and energy efficiency as important factors to consider during data mining algorithm analysis and evaluation. The impact of varying the parameters of the Very Fast Decision Tree (VFDT) algorithm on energy consumption and accuracy is empirically studied. The conclusion is that energy consumption is affected by such parameters and can be reduced significantly while maintaining accuracy.

We would like to conclude this preface by conveying our appreciation to all contributing authors and our warm thanks to Professor Reda Alhajj for giving us the opportunity to be the Guest Editors of this book. We also would like to express our gratitude to Christopher T. Coughlin and his team members from Springer US for their help in the preparation of this volume.

QC, Canada — Rokia Missaoui
Paris, France — Talel Abdessalem
Paris, France — Matthieu Latapy
December 2016

Contents

The Perceived Assortativity of Social Networks: Methodological Problems and Solutions 1
David N. Fisher, Matthew J. Silk, and Daniel W. Franks

A Parametric Study to Construct Time-Aware Social Profiles 21
Sirinya On-at, Arnaud Quirin, André Péninou, Nadine Baptiste-Jessel, Marie-Françoise Canut, and Florence Sèdes

Sarcasm Analysis on Twitter Data Using Machine Learning Approaches 51
Santosh Kumar Bharti, Ramkrushna Pradhan, Korra Sathya Babu, and Sanjay Kumar Jena

The DEvOTION Algorithm for Delurking in Social Networks 77
Roberto Interdonato, Chiara Pulice, and Andrea Tagarelli

Social Engineering Threat Assessment Using a Multi-Layered Graph-Based Model 107
Omar Jaafor and Babiga Birregah

Through the Grapevine: A Comparison of News in Microblogs and Traditional Media 135
Byungkyu Kang, Haleigh Wright, Tobias Höllerer, Ambuj K. Singh, and John O'Donovan

Prediction of Elevated Activity in Online Social Media Using Aggregated and Individualized Models 169
Jimpei Harada, David Darmon, Michelle Girvan, and William Rand

Unsupervised Link Prediction Based on Time Frames in Weighted –Directed Citation Networks 189
Mehmet Kaya, Mujtaba Jawed, Ertan Bütün, and Reda Alhajj

An Approach to Maximize the Influence Spread in the Social Networks .. 207
Ibrahima Gaye, Gervais Mendy, Samuel Ouya, and Djaraf Seck

Energy Efficiency Analysis of the Very Fast Decision Tree Algorithm 229
Eva Garcia-Martin, Niklas Lavesson, and Håkan Grahn

Glossary 253

Contributors

Reda Alhajj Department of Computer Science, University of Calgary, Calgary, AB, Canada

Korra Sathya Babu National Institute of Technology Rourkela, Rourkela, India

Nadine Baptiste-Jessel Toulouse Institute of Computer Science Research (IRIT), University of Toulouse, CNRS, INPT, UPS, UT1, UT2J, Toulouse, France

Santosh Kumar Bharti National Institute of Technology Rourkela, Rourkela, India

Babiga Birregah Charles Delaunay Institute, UMR CNRS 6281, University of Technology of Troyes, Troyes, France

Ertan Bütün Department of Computer Engineering, Fırat University, Elazığ, Turkey

Marie-Françoise Canut Toulouse Institute of Computer Science Research (IRIT), University of Toulouse, CNRS, INPT, UPS, UT1, UT2J, Toulouse, France

David Darmon Department of Mathematics, University of Maryland, College Park, MD, USA

David N. Fisher Department of Integrative Biology, University of Guelph, Guelph, ON, Canada

Daniel W. Franks Department of Biology, University of York, York, UK

Department of Computer Science, University of York, York, UK

Eva Garcia-Martin Blekinge Institute of Technology, Karlskrona, Sweden

Ibrahima Gaye UCAD-ESP, LIRT of Sénégal, Dakar, Fann, Senegal

Michelle Girvan Department of Physics, University of Maryland, College Park, MD, USA

Håkan Grahn Blekinge Institute of Technology, Karlskrona, Sweden

Jimpei Harada Center for Complexity in Business, University of Maryland, College Park, MD, USA

Tobias Höllerer Department of Computer Science, University of California, Santa Barbara, CA, USA

Roberto Interdonato DIMES, University of Calabria, Arcavacata di Rende (CS), Italy

Omar Jaafor Charles Delaunay Institute, UMR CNRS 6281, University of Technology of Troyes, Troyes, France

Mujtaba Jawed Department of Computer Engineering, Fırat University, Elazığ, Turkey

Sanjay Kumar Jena National Institute of Technology Rourkela, Rourkela, India

Byungkyu Kang Department of Computer Science, University of California, Santa Barbara, CA, USA

Mehmet Kaya Department of Computer Engineering, Fırat University, Elazığ, Turkey

Niklas Lavesson Blekinge Institute of Technology, Karlskrona, Sweden

Gervais Mendy UCAD-ESP, LIRT of Sénégal, Dakar, Fann, Senegal

John O'Donovan Department of Computer Science, University of California, Santa Barbara, CA, USA

Sirinya On-at Toulouse Institute of Computer Science Research (IRIT), University of Toulouse, CNRS, INPT, UPS, UT1, UT2J, Toulouse, France

Samuel Ouya UCAD-ESP, LIRT of Sénégal, Dakar, Fann, Senegal

André Péninou Toulouse Institute of Computer Science Research (IRIT), University of Toulouse, CNRS, INPT, UPS, UT1, UT2J, Toulouse, France

Ramkrushna Pradhan National Institute of Technology Rourkela, Rourkela, India

Chiara Pulice DIMES, University of Calabria, Arcavacata di Rende (CS), Italy

Arnaud Quirin Toulouse Institute of Computer Science Research (IRIT), University of Toulouse, CNRS, INPT, UPS, UT1, UT2J, Toulouse, France

William Rand Poole College of Management, North Carolina State University, Raleigh, NC, USA

Djaraf Seck UCAD-FST, LMDAN of Sénégal, Dakar, Fann, Senegal

Florence Sèdes Toulouse Institute of Computer Science Research (IRIT), University of Toulouse, CNRS, INPT, UPS, UT1, UT2J, Toulouse, France

Matthew J. Silk Environment and Sustainability Institute, University of Exeter, Penryn, UK

Ambuj K. Singh Department of Computer Science, University of California, Santa Barbara, CA, USA

Andrea Tagarelli DIMES, University of Calabria, Arcavacata di Rende (CS), Italy

Haleigh Wright Department of Computer Science, University of California, Santa Barbara, CA, USA

The Perceived Assortativity of Social Networks: Methodological Problems and Solutions

David N. Fisher, Matthew J. Silk, and Daniel W. Franks

1 Introduction

Network theory is a useful tool that can help us explain a range of social, biological and technical phenomena [1–3]. For instance, network approaches have been used to investigate diverse topics such as the global political and social system [4, 5] to the formation of coalitions among individuals [6–9]. Networks can be described using a number of local (related to the individual) and global (related to the whole network) measures. One important global measure is degree correlation or assortativity (we use the latter term for brevity), which was formally defined by Newman [10], although Pastor-Satorras et al. [3] had calculated an analogous measure previously. The assortativity of a network measures how the probability of a connection between two nodes (individuals) in a network depends on the degrees of those two nodes (the degree being the number of connections each node possesses). The measure quantifies whether those with many connections associate with others with many connections (assortative networks or networks with assortativity), or if "hubs" form where well-connected individuals are connected to many individuals with few other connections (disassortative networks or networks with dissassortativity). If the

D.N. Fisher (✉)
Department of Integrative Biology, University of Guelph, Guelph, ON, N1G 2W1, Canada
e-mail: davidnfisher@hotmail.com

M.J. Silk (✉)
Environment and Sustainability Institute, University of Exeter, Penryn Campus, Penryn, TR10 9FE, UK
e-mail: matthewsilk@outlook.com

D.W. Franks (✉)
Department of Biology, University of York, York, YO10 5DD, UK

Department of Computer Science, University of York, York, YO10 5DD, UK
e-mail: daniel.franks@york.ac.uk

R. Missaoui et al. (eds.), *Trends in Social Network Analysis*, Lecture Notes in Social Networks, DOI 10.1007/978-3-319-53420-6_1

tendency for nodes to be connected is independent of each other's degrees a network has neutral assortativity. Assortativity is calculated as the Pearson's correlation coefficient between the degrees of all pairs of connected nodes, and ranges from -1 to 1 [10]. The Pastor-Satorras method involves plotting the degree of each node against the mean degree of its neighbours, and judging the network assortative if the slope is positive and disassortative if the slope is negative [3]. In this article we will focus on the Newman measure, as it has been more commonly used by the scientific community, gives a coefficient bounded between -1 and 1 rather than a slope and typically is supported by a statistical test, something general absent from the reporting of the Pastor-Satorras method.

Assortativity is a key property to consider when understanding how networks function, especially when considering social networks. A network's robustness to attacks is increased if it is assortative [11, 12]. However, the speed of information transfer and ability to act in synchrony is increased in disassortative networks [13, 14]. Thus the assortativity of a social network in which an individual is embedded can have a substantial impact on that individual.

Networks are typically found to be neutrally or negatively assortative [10, 15, 16]. When considering all networks comprised of eight nodes, Estrada [17] observed that only 8% of over 11,000 possible networks were assortative. Despite this general trend, social networks are often said to differ from other networks by being assortative [15, 18]. This has led to those finding disassortativity in networks of online interactions [19], mythical stories [20] or networks of dolphin interactions [21] to suggest they are different to typical human social networks. While the generality of assortativities in social networks has been questioned [16, 22], a wide variety of recent research still states that this is a property typical of social networks (e.g. [17, 23–28]). With assortativity being a key network property and subject of interest in a range of fields, this topic requires clarification.

In this paper we review assortativity in the networks literature, with emphasis on social networks. We assess the generality of the hypothesis that social networks tend to be distinct from other kinds of networks in their assortativities and explore whether the precise method of social network construction influences this metric. We go on to show how particular methodologies of social network construction could result in falsely assortative networks, and present a number of solutions to this by drawing on advances in sociological and biological fields.

2 Assortativity in Social and Other Networks

Random networks should be neutrally assortative [10]. However, simulations by Franks et al. [29] showed that random social networks constructed using group-based methods are assortative unless extensive sampling is carried out. Group-based methods are where links are formed between individuals not when they directly interact, but when they both are found in the same group, or contribute to a joint piece of work, e.g. co-author a paper or appear in a band together. We hypothesised that the suggestion that social networks possess assortativity was due to a preponderance of social networks constructed using group-based methods in the early literature. We thus conducted a literature search, recording the (dis)assortativity of networks, network types (social or non-social) and for the social networks, method of construction (direct interactions of group-based). We expect that social networks built using group-based methods would be more assortative than the other two classes of network, which would be similarly assortative.

2.1 *Literature Search: Method*

Assortativity has been calculated for a wide range of networks in the last decade. We compiled a list of the assortativity of published networks based on the table in Whitney and Alderson [16], literature searches with the terms "degree correlation" and "assortativity", and examining the articles citing Newman [10]. If it could not be determined how the network was constructed, or how large it was, the network was excluded. We did not include the average assortativity when reported from a range of similar networks when the individual scores were not reported. We also did not include assortativities of networks from studies re-analysing existing datasets, to avoid pseudo-replication. Only undirected networks were considered; see Piraveenan et al. [30] for a review of assortativities in directed networks. We then classified these networks as non-social networks, social networks constructed using direct interactions or social networks constructed using group-based methods. We then compared assortativity across network classes using a Kruskal–Wallis test as assortativity scores were not normally distributed. If this revealed significant differences among network classes, we then compared network classes to each other using Wilcoxon rank sum tests and to the a neutral assortativity of zero with Wilcoxon signed rank tests.

2.2 *Literature Search: Results*

In published papers we found assortativities for 88 networks that met our criteria for inclusion, see Table 1.52 of these were social networks, of which 25 were constructed using group-based methods. The assortativities for the network classes are shown as boxplots in Fig. 1. The Kruskal–Wallis tests indicated that there were differences among groups (Kruskal–Wallis $\chi^2 = 26.8$, d.f. $= 2$, $p < 0.001$). All networks types were different from each other, with the group-based social networks being more assortative than both other classes of network, and the direct social networks being more assortative than the non-social networks (all Wilcoxon rank sum tests, group-based social networks vs. direct social networks, $W = 214$, $n(\text{direct}) = 27$, $n(\text{group-based}) = 25$, $p = 0.024$; group-based social networks vs. non-social networks, $W = 783$, $n(\text{group-based social networks}) = 25$, $n(\text{non-social network}) = 36$, $p < 0.001$; direct social networks vs. non-social network, $W = 716$,

Table 1 88 networks of various types and methods of construction

Network	Size	Type	Assortativity	Method of construction	Source
Beowulf (myth)	74	Social	−0.1	Direct	[20]
Cyworld (online)	12,048,186	Social	−0.13	Direct	[22]
Email address books	16,881	Social	0.092	Direct	[18]
Epinions neg (online)	131,828	Social	−0.022	Direct	[31]
Epinions pos (online)	131,828	Social	0.217	Direct	[31]
Facebook (online)	721,000,000	Social	0.226	Direct	[32]
Flickr (online)	1,846,198	Social	0.202	Direct	[22]
Gnutella P2P (online)	191,679	Social	−0.109	Direct	[22]
Ground squirrels	65	Social	0.82	Direct	[33]
Iliad (myth)	716	Social	−0.08	Direct	[20]
LiveJournal (online)	5,284,457	Social	0.179	Direct	[22]
Mixi (online)	360,802	Social	0.122	Direct	[22]
MySpace (online)	100,000	Social	0.02	Direct	[22]
Nioki (online)	20,259	Social	−0.13	Direct	[19]
Orkut (online)	100,000	Social	0.31	Direct	[22]
Pussokram (online)	29,341	Social	−0.048	Direct	[19]
Slashdot neg (online)	82,144	Social	−0.114	Direct	[31]
Slashdot pos (online)	82,144	Social	0.162	Direct	[31]
Student relationships	573	Social	−0.029	Direct	[18]
Táin (myth)	404	Social	−0.33	Direct	[20]
Twitter (online)	4,317,000	Social	−0.025	Direct	[34]
Whisper (online)	690,000	Social	−0.011	Direct	[34]
Xiaonei (online)	396,836	Social	−0.0036	Direct	[22]
YouTube (online)	1,157,827	Social	−0.033	Direct	[22]
Chinese science citations	81	Social	−0.036	Direct	[35]
Barbary macaque (grooming)	141	Social	0.351	Direct	[36]
GitHub (online)	671,751	Social	−0.0386	Direct	[37]
Australian dolphins	117	Social	0.003	Group	[38]
Biology co-authors	1,520,251	Social	0.13	Group	[10]
Birds (mixed species)	93	Social	0.29	Group	[39]
Company directors	7,673	Social	0.28	Group	[10]
Condensed matter co-authors	36,458	Social	0.18	Group	[40]
Condensed matter co-authors 1995–9	16,729	Social	0.18	Group	[41]
Film actors	449,913	Social	0.21	Group	[18]
Killer whales	7	Social	−0.48	Group	[42]
New Zealand dolphins	64	Social	−0.044	Group	[21]
Maths co-authors	253,339	Social	0.12	Group	[10]
Physics co-authors	52,909	Social	0.36	Group	[10]

(continued)

Table 1 (continued)

Physics co-author2	16,264	Social	0.18	Group	[10]
Scottish dolphins	124	Social	0.17	Group	[43]
Sticklebacks	94	Social	0.66	Group	[44]
TV series actor collaboration	79,663	Social	0.53	Group	[22]
Grant proposals (accepted)	24,181	Social	−0.1018	Group	[45]
Grant proposals (rejected)	46,567	Social	−0.1145	Group	[45]
Brazilian co-authorship (Humanities)	74,490	Social	0.3737	Group	[46]
Brazilian co-authorship (Linguistics, letters and arts)	15,375	Social	0.3761	Group	[46]
Brazilian co-authorship (Engineering)	15,375	Social	0.0273	Group	[46]
Brazilian co-authorship (Agricultural science)	55,695	Social	0.0769	Group	[46]
Brazilian co-authorship (Biological science)	75,304	Social	0.1404	Group	[46]
Brazilian co-authorship (Exact and earth sciences)	65,221	Social	0.1173	Group	[46]
Brazilian co-authorship (applied social sciences)	48,340	Social	0.1373	Group	[46]
Brazilian co-authorship (health sciences)	114,169	Social	0.1301	Group	[46]
Airports in Pakistan	35	Transport	−0.47	NA	[47]
Berlin U & S Bahn	75	Transport	0.096	NA	[16]
Bike	131	Mechanical	−0.2	NA	[16]
Brain connections	160	Biological	0.058	NA	[48]
Brain connections, Polymicrogyria	160	Biological	0.044	NA	[48]
Car door	649	Mechanical	−0.16	NA	[16]
Domain network (online)	11,174	Technological	−0.17	NA	[49]
Film references	40,008	Commercial	−0.057	NA	[50]
Fresh water food web	92	Biological	−0.33	NA	[18]
Grand piano action key 1	71	Mechanical	−0.32	NA	[16]
Internet	10,697	Technical	−0.19	NA	[18]
Jet engine	60	Mechanical	−0.13	NA	[16]
London underground	92	Transport	0.01	NA	[16]
Marine food web	134	Biological	−0.26	NA	[18]
Metabolic pathways	765	Biological	−0.24	NA	[18]
Moscow subway	51	Transport	0.18	NA	[16]
Moscow subway and regional rail	129	Transport	0.26	NA	[16]
Neural pathways	307	Biological	−0.23	NA	[18]
Power grid	4941	Technical	−0.003	NA	[18]
Proteins	2115	Biological	−0.16	NA	[18]
Pseudomonas strains	37	Biological	−0.553	NA	[51]
Router network	228,298	Technological	−0.01	NA	[49]

(continued)

Table 1 (continued)

Network	Size	Type	Assortativity	Method of construction	Source
Six speed transmission	143	Mechanical	−0.18	NA	[16]
Software	3162	Technical	−0.016	NA	[18]
Tokyo regional rail	147	Transport	−0.09	NA	[16]
Tokyo regional rail and subway	191	Transport	0.043	NA	[16]
V8 engine	243	Mechanical	−0.27	NA	[16]
World Wide Web	269,504	Technical	−0.067	NA	[18]
Yeast genes	333	Biological	−0.15	NA	[49]
Yeast proteins	1066	Biological	−0.12	NA	[49]
Copenhagen streets	1637	Transport	−0.07	NA	[52]
London streets	3010	Transport	−0.06	NA	[52]
Paris streets	4501	Transport	−0.06	NA	[52]
Manhattan streets	1046	Transport	−0.26	NA	[52]
San Francisco streets	3110	Transport	−0.01	NA	[52]
Toronto streets	2599	Transport	−0.06	NA	[52]

Type indicates the nature of the links between nodes, social are networks based on social interactions, technical are networks of interacting technology systems, biological are networks of some kind of biological process, transport networks are networks of a mode of transport in a particular area, mechanical networks are based on connections between parts in a defined object and links in commercial networks are formed when an employee moved from one company to another. Method of construction for social networks can be either "direct" where one to one interactions are used to build the network, or "group" where interactions are inferred based on shared use of physical space or contribution to a piece of work. All quoted degree correlations are from the Newman [10] method of calculating assortativity

n(direct social networks $= 27$), n(non-social network) $= 36$, $p = 0.001$). The direct social networks possessed neutral assortativities (mean $= 0.054$, Wilcoxon signed rank test, $V = 213$, $n = 27$, $p = 0.572$), while the group-based social networks were assortative (mean $= 0.157$, Wilcoxon signed rank test, $V = 288$, $n = 25$, $p < 0.001$) and the non-social networks were disassortative (mean $= -0.117$, Wilcoxon signed rank test, $V = 94.5$, $n = 36$, $p < 0.001$).

2.3 *Literature Search: Conclusions*

To confirm our hypothesis, that a preponderance of group-based methods makes it appear as if social networks typically possess assortativity, we found that group-based social networks were more assortative than direct social networks and non-social networks. However, direct social networks were still more assortative than non-social networks, which showed disassortativity on average. This therefore indicates that social networks only tend to be assortative if they are constructed with group-based methods, and non-social networks are typically disassortative. The

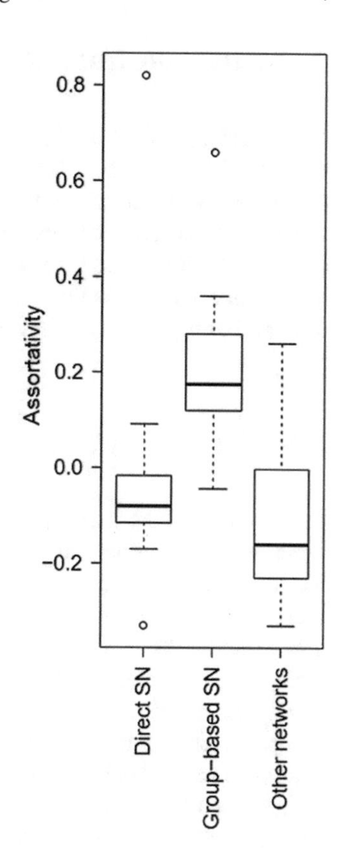

Fig. 1 Boxplots of assortativities in each class of network. Differences between all classes were statistically significant (all Wilcoxon rank sum tests, $p < 0.025$ in all cases). Direct social networks did not possess assortativities different from zero (mean $= 0.054$, $p = 0.572$), while the group-based social networks were assortative (mean $= 0.157$, $p < 0.001$) and the non-social networks were dissortative (mean $= -0.117$, $p < 0.001$; all Wilcoxon signed rank tests)

processes that create dissortative networks is a large and active research area (e.g. [53, 54]) and so we will not investigate this result further here. We also note that many of the direct social networks in our table were based on interactions online. This is perhaps problematic if interactions online are fundamentally different to human interactions in the real world. Our choice of networks was simply based on what was available. Therefore, to confirm whether the assortativities of offline and online human social networks made using the same method (i.e. with direct interactions) are different, more offline human networks using direct interactions need to be constructed and published.

The bias towards assortativity when using group-based methods has been used has previously been noted [18, 55], but this appears to have escaped the notice of much of the research community, who continue to state that assortativity is a characteristic quality of a social network (e.g. [17, 23–28] but see [16, 22]). This may be because no clear rationale for this has been presented, nor solutions offered to combat it. We now go on to present why we think this assortativity was being discovered with group-based methods, if they are erroneous, and how to avoid incorrect estimates of assortativity in the future.

3 Methodological Pitfalls and False Assortativity

In this section we will discuss the issues that exist when using assortativity to describe the structure of social networks. We explore how methodologies used to sample networks could influence their assortativities, and whether changing how assortativity is calculated may also be important. Understanding the consequences of the method used to sample networks on properties such as assortativity could have important implications for our understanding of networks in general, as well as contributing to the further development of relevant analytical techniques.

3.1 Group-Based Networks and Assortativity

In situations where it is not possible or feasible to directly observe social interactions, social networks are constructed using group-based methods. This approach has been applied both in constructing collaboration networks in humans, e.g. jazz musicians, scientific co-authorship networks and film actors [56–58] and also in networks based on co-occurrence in a social group in animals, e.g. song birds, dolphins and sharks [59–61]. Networks built using this method assume that every member of the group is associating with every other member of the group at each sampling census. This seems perfectly reasonable, hence this method for constructing social networks has been used by many studies [55, 62]. The assumption that meaningful social networks can be constructed based on this type of co-occurrence data has been termed "Gambit of the group" in the animal behaviour literature [63], which hints at the risk involved. This broad usage of group-based methods makes it very important that we fully understand the implications of using this method on the network and individual-level metrics calculated.

One study has previously investigated the influence of variation in sampling effort of group-based networks on various network-level metrics, including assortativity [29]. Franks et al. [29] investigated the impact of different group-based sampling regimes, changing both the total number of censuses (the number of times a population is sampled using a group-based approach) and the proportions of individuals sampled at each census, in random networks. The assortativity of a random network, which should on average be zero [10], was *always positive* if an insufficient number of censuses were completed before network construction ([29]; our emphasis). Crucially, any network not sampled intensively enough could show assortativity, either due to too few censuses or by not sampling enough of the population. Therefore, the use of group-based methods can produce a sampling bias, and requires further statistical analysis to determine the importance of the results obtained.

3.2 Modeling Group-Based Sampling

We wanted to extend the findings of Franks et al. [29] to demonstrate how assortativity changes as individuals are recorded in increasing numbers of groups. The aim was to show that in a system where freely moving individuals form social groups randomly (as opposed to a system based on random social networks), assortativity would decline as the effect of two individuals being seen in the same group diminished. We used a simulation-based approach where individuals associated randomly to construct networks using a group-based approach in a simple population of 100. Individuals were allocated randomly between 20 possible groups during each "census" using a symmetric Dirichlet distribution (with a uniform shape parameter) to define the size of each group. This enabled us to generate variation around a fixed expected group size, whilst maintaining a fairly consistent number of groups in each census (occasionally the size of a group could be zero). Any remaining individuals were allocated to a random group, meaning that all individuals were sampled in each census. Association data was collected over 20 censuses, with this process repeated for 10 different repeats of the simulation. After each census cumulative association data was recorded and the networks were dichotomised to create binary networks. In these networks individuals are either connected if they were observed in the same group at least once, or not connected if never observed together. This is necessary as assortativity is an unweighted measure, i.e. only the number of different associations an individual had is counted, not the frequency with which it associated with them. We discuss issues related to this method later. We then calculated assortativity for each set of cumulative association data. Simulations were carried out and degree correlations calculated in R 3.0.1 [64] (http://www.R-project.org).

We plot the results of these simulations in Fig. 2. From this plot it is clear that networks possess assortativity at a low number of censuses, and become gradually more neutrally assortative as the number of censuses increases (Fig. 2). This pattern emerges because of how group-based approaches are used in network construction. At low numbers of censuses, individuals that have been found in the same groups will both be connected and have similar degree, thus giving the network assortativity [15, 18]. As the number of censuses increases; however, this connection will gradually break down. Therefore, our simple simulation model shows that the assortativity found using group-based sampling approaches is highly dependent on the number of censuses completed. Indeed a low number of censuses can lead to many network measures being distorted [29, 65]. This is therefore an important consideration when deciding the sampling regimes used when constructing social networks using this method. The significance of the assortativity can only be established by comparison to appropriate null networks that truly highlight what aspects of the real network are interesting. This is achieved by randomly resampling observations using the correct group-size distribution [66], a method that will be outlined in more detail in Sect. 3.2.

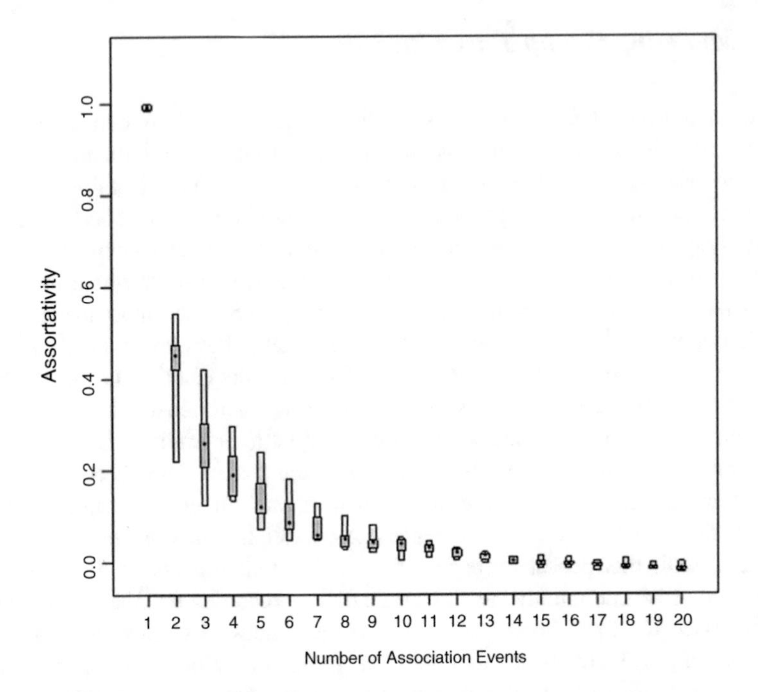

Fig. 2 The relationship between sampling effort (in terms of number of association events observed) and assortativity as a result of random social interactions in a simulated population. For each census the *black point* represents the median value, the *dark grey box* the interquartile range and the *light grey box* the range from ten runs of the simulation

3.3 Filtering Networks

Filtering networks involves taking a weighted (or valued) network, which contains information about how strong associations are as well as whether they are present or absent, and removing edges below a certain weight, i.e. infrequent or unimportant associations. This results in a binary network, with only edges above a certain weight present. The effect of dichotomising networks to transform them from weighted to binary has been studied for a number of network-level metrics [29]. The effect of this on assortativity is striking. While there was a limited effect of filtering at a low threshold (and therefore removing few edges), filtering at a high threshold (removing relatively many edges) had a considerable impact on the assortativity even when a high number of censuses on the simulated networks were completed. While for unfiltered, weighted random networks the assortativity reduced to zero as expected when a sufficient number of censuses were completed, this was not the case for networks filtered with a high threshold. These networks typically remained assortative at all levels of sampling (Figure 3 of [29]), despite being based on networks that were originally random and would be expected to neutrally assortative. For example, removing all edges with a weight of less than 0.5 (out of 1) meant that the assortativity reached 0.4 after 10 censuses, compared to 0.1 after 10 censuses if the threshold weight was set at 0.2. Many studies that have used group-based approaches to construct social networks have filtered the social networks produced before analysis [55], and it seems likely that the use of this approach may have played an additional role in inflating the degree correlation calculated for social networks.

4 Solutions

There are a number of solutions available to reduce the occurrence of erroneous assortativity that address each of the major issues outlined above. These methods have been developed in different parts of the social network literature, and if used collectively can greatly improve our understanding of the true variation in assortativity in all kinds of social networks.

4.1 Increased Sampling

When using group-based approaches, increasing the number of censuses above a threshold should produce a more accurate measure of the network's true assortativity [29]. Thus a suitable minimum number of censuses completed is required whenever using group-based approaches. If 80% of the population of interest can be sampled, then 10 censuses should be sufficient information on social interactions to accurately calculate network metrics. If only 40% of the population can be sampled, then 15–20 censuses may be required [29]. Collaboration networks, a key example of social networks that possess assortativity, typically do not have sampling periods. Instead all papers published over an agreed time period are looked at. To allow better comparison with networks and techniques that do have sampling periods, collaborations could be examined over particular, regular time periods, then each time period used as an "observation" from which to generate associations and so networks. However, if altering the sampling regime is not feasible, ensuring erroneous conclusions are not reached requires more rigorous statistical analysis.

4.2 Use of Null Models

Socio-biologists typically use resampling procedures to account for group-based sampling (and other, random) effects in their animal social networks [63, 66]. Resampling the observations used to construct the observed network using the original group-size data, and potentially additional biologically meaningful constraints, provides a large number of null networks against which to compare the observed network and test relevant hypotheses [66]. This method has been continually revised to take into account the non-independence of group sightings [67] and the risk of producing biased group compositions [68]. This approach has been used in a number of studies (e.g. [69, 70]) and means that assortativity would only be considered interesting if the null model does not also display assortativity. We found one example of this approach being used in human social networks, in a study performed on a network of a board of directors [15]. However, no statistical tests were used for this comparison, and the method does not seem to have been universally adopted by the wider social network community. The importance of adopting this method is further highlighted the results of Franks et al. [29] which show that assortativity can be expected in networks where individuals randomly interact.

Random permutations are used by other network analysis techniques such as quadratic assignment procedure (QAP) to overcome problems with structural autocorrelation [71], but the Bejder et al. [66] method and extensions go further by randomly permuting the raw data, rather than the network matrix, to account for the distribution of group sizes and any other biases in sampling. Appropriate null models like these are the most effective way to make statistical inferences about networks constructed using group-based approaches, where controlling for effects such as group size is of considerable importance. There have been some recent examples of appropriate null models being used to test various hypotheses in the wider social networks literature (e.g. [72, 73]) and we add to their calls for this to be more widely embraced.

4.3 Analysing Weighted Networks

Weak interactions are potentially important (e.g. [74, 75]), and removing them can further increase error in the degree correlations calculated [29, 41]. If a population is observed for an extended period of time, and a large number of censuses are performed, the probability that two individuals are never observed to associate approaches zero. In a binary network, this would result in all individuals having the same, maximum, degree (which would be the population size -1), and therefore the assortativity would increase before becoming undefined. Therefore, it is preferable to use weighted, unfiltered networks. For example, if collaboration networks are based on some measure of the information contained in an email [76], or the relative roles of those involved [77], you could then produce weighted networks that would be more informative to study and be less likely to be falsely assortative. The analysis of weighted networks is becoming increasingly manageable (e.g. [78]), reducing the need to filter networks and analyse them as binary data. As such, this approach should only be used when absolutely required [79], preferably with filtering using low thresholds (few edges removed). The filtering of some networks is unavoidable, for example, in brain imaging data which is likely to be very noisy [80], but we still recommend the lowest amount of filtering possible while accounting for edge uncertainty. Additionally, it would also be highly beneficial to continue to develop social network analysis in weighted networks to further reduce the requirement for binary network data, for example, by continuing the development of exponential random graph models that can be used in weighted networks [81, 82].

4.4 Using Diadic Over Group-Based Approaches

The use of group-based methods is instrumental in creating some of the problems we suggest solutions for in previous subsections. If possible, alternative methods to group-based approaches should be used when constructing a network. For example, instead of paper co-authorship being used to define interactions between scientists, direct, diadic interactions based on correspondence such as emails could be used [76, 77]. Human social networks based on communication data such as with mobile phones or online are also examples of diadic interactions, and are becoming increasingly common [22, 76, 83–85]. In animal networks, this can be achieved by focussing on observation of suitable behavioural interactions [33, 86] or using reality mining methodologies [87], which are more fully outlined in Sect. 4.5. Social networks constructed using direct interactions on average possess neutral assortativity (see above), indicating they are not prone to the same problems with false assortativities that networks built using group-based approaches are.

4.5 Modern Technology

Missing associations, and therefore edges in the network, is a problem largely unique to social networks. This can be prevented in human social networks, especially collaboration networks, by examining information-rich communications data. Information-rich interpersonal communications such as texts, tweets and emails can be used to construct networks on a very large scale with a high degree of accuracy [83]. In animal social networks a solution is to use reality mining [87], the concept of which has been borrowed from the sociology social network literature [88]. In these studies, modern technologies such as GPS trackers or proximity loggers are used to track animal movements and monitor interactions or accurately infer associations. For example, proximity loggers have been used to automatically record associations based on spatial and temporal overlap between individuals in animals such as cows, Tasmanian devils, crows and badgers [89–91], although see [92, 93] for discussion of the potential problems associated with using these devices.

These methods substantially reduce the number of missing edges in the network. Additionally, even when they don't allow us to move towards constructing networks using direct interactions, they can make it much more feasible to complete a large number of censuses if a group-based approach is still required. Furthermore, with careful thought technologies such as these could be used to complement the study of human social networks, potentially combined with data gathered using other methods. For instance, in the workplace, are those an employee physically associates with, e.g. at the water cooler or on a cigarette break, the same as those they communicate with electronically? What about different types of electronic communication, e.g. Facebook messages compared to work emails? Such application of modern technologies will only enhance the ability of scientists to measure social interactions in a wide range of networks.

4.6 Alternatives to the Newman Degree Correlation Measure

Recent research has suggested that calculating the correlation coefficient proposed by Newman [10] may not be appropriate for large networks, as it tends to mis-estimate the assortativity, especially in disassortative networks [25]. Litvak and van der Hofstad [25] suggested that using a method that ranks the degree of nodes, rather than using their absolute values (like a Spearman's rank correlation coefficient), may produce more valid results. Using this measure, the assortativity was consistent across a wide range of network sizes and typically consistently different from those calculated using Newman's assortativity coefficient (Figure 1 of [25]). It may be that changing the way that assortativity is calculated could reduce the differences between group-based networks, other social networks and different network types. This can be combined with the Bejder et al. [66] method to control for both group-size variation and biases associated with whole network size.

Alternatively, several authors have proposed individual-orientated metrics to quantify the tendency for well-connected individuals to connect to other individuals. The "Rich-clubs" of Zhou and Mondragon [94, 95], and the number of "differences" in node degree between neighbours of Thedchanamoorthy et al. [28] may both be robust to these pitfalls. Understanding the consequences of using different statistical approaches for calculating degree correlations should be the subject of further modeling work, in order to determine whether this can have an important influence to the properties with which a network is prescribed.

5 Conclusions

Assortativity has often been suggested to be a property of social networks that makes them different from non-social networks. We show that it is likely that this phenomenon may be driven by the data collection methods, resulting in previous studies overestimating the extent to which social networks possess assortativity. With alterations to methods or analyses, however, this problem with erroneous assortativities being described can be mitigated.

Through the use of methods initially developed in biology and sociology, we suggest that careful application of sampling and statistics can afford us increased confidence that finding assortativity in social networks is genuine, and not an artefact of the methodology used to construct them. It may well be that social networks are unique amongst types of network; there are certainly good reasons why they could have developed to be more assortative than non-social networks, e.g. increased robustness [11, 12]. However, given that information transfer and the ability to act in synchrony are greater in networks that show disassortativity [13, 14], it would perhaps be surprising if this property is as universal as has previously been described [10, 15].

Many of the methods discussed have been developed in one part of the social networks literature, and would benefit researchers in other fields greatly. Furthermore, following the integration of methods used in these different fields, it is likely that new insights will become available and further progress made in our ability to make inferences about assortativity in networks more generally.

References

1. Girvan, M., Newman, M.E.J.: Community structure in social and biological networks. Proc. Natl. Acad. Sci. U. S. A. **99**, 7821–7826 (2002)
2. Krause, J., Croft, D.P., James, R.: Social network theory in the behavioural sciences: potential applications. Behav. Ecol. Sociobiol. **62**, 15–27 (2007)
3. Pastor-Satorras, R., Vázquez, A., Vespignani, A.: Dynamical and correlation properties of the Internet. Phys. Rev. Lett. **87**, 258701 (2001)
4. Nemeth, R.J., Smith, D.A.: International trade and world-system structure: a multiple network analysis. Rev. (Fernand Braudel Cent). **8**, 517–560 (2010)
5. Snyder, D., Kick, E.L.: Structural position in the world system and economic growth, 1955–1970: a multiple-network analysis of transactional interactions. Am. J. Sociol. **84**, 1096–1126 (1979)
6. Kapferer, B.: In: Boissevain, J., Mitchell, J.C. (eds.) Norms and the Manipulation of Relationships in a Work Setting, pp. 83–110. Netw. Anal. Stud. Hum. Interact. Mouton, Paris (1969)
7. Thurman, B.: In the office: networks and coalitions. Soc. Networks. **2**, 47–63 (1979)
8. Voelkl, B., Kasper, C.: Social structure of primate interaction networks facilitates the emergence of cooperation. Biotechnol. Lett. **5**, 462–464 (2009)
9. Zachary, W.W.: An information flow model for conflict and fission in small groups. J. Anthropol. Res. **473**, (1977)
10. Newman, M.: Assortative mixing in networks. Phys. Rev. Lett. **2**, 1–5 (2002)
11. Hasegawa, T., Konno, K., Nemoto, K.: Robustness of correlated networks against propagating attacks. Eur. Phys. J. B. **85**, 1–9 (2012)
12. Jing, Z., Lin, T., Hong, Y., et al.: The effects of degree correlations on network topologies and robustness. Chin. Phys. **16**, 3571–3580 (2007)
13. Di Bernado, M., Garofalo, F., Sorrentino, F.: Effects of degree correlation on the sychronization of networks of oscillators. Int. J. Bifurcation Chaos. **17**, 3499–3506 (2007)
14. Gallos, L., Song, C., Makse, H.: Scaling of degree correlations and its influence on diffusion in scale-free networks. Phys. Rev. Lett. **100**, 248701 (2008)
15. Newman, M., Park, J.: Why social networks are different from other types of networks. Phys. Rev. E. **68**, 036122 (2003)
16. Whitney, D., Alderson, D.: Are technological and social networks really different? Unifying Themes Complex Syst. **6**, 74–81 (2008)
17. Estrada, E.: Combinatorial study of degree assortativity in networks. Phys. Rev. E. **84**, 047101 (2011)
18. Newman, M.: Mixing patterns in networks. Phys. Rev. E. **67**, 026126 (2003)
19. Holme, P., Edling, C.R., Liljeros, F.: Structure and time evolution of an Internet dating community. Soc. Networks. **26**, 155–174 (2004)
20. Mac Carron, P., Kenna, R.: Universal properties of mythological networks. EPL Europhys. Lett. **99**, 28002 (2012)
21. Lusseau, D., Newman, M.: Identifying the role that animals play in their social networks. Proc. R. Soc. B Biol. Sci. **271**, S477–S481 (2004)
22. Hu, H.-B., Wang, X.-F.: Disassortative mixing in online social networks. EPL Europhys. Lett. **86**, 18003 (2009)

23. Araújo, E.B., Moreira, A.A., Furtado, V., et al.: Collaboration networks from a large CV database: dynamics, topology and bonus impact. PLoS One. **9**, e90537 (2014)
24. Furtenbacher, T., Arendás, P., Mellau, G., Császár, A.G.: Simple molecules as complex systems. Sci. Rep. **4**, 4654 (2014)
25. Litvak, N., van der Hofstad, R.: Uncovering disassortativity in large scale-free networks. Phys. Rev. E. **87**, 022801 (2013)
26. Mac Carron P, Kenna R. A quantitative approach to comparative mythology. nestor.coventry.ac.uk (2013)
27. Palathingal, B., Chirayath, J.: Clustering similar questions in social question answering services. In: Shan, L.P., Cao, T.H. (eds.) The 16th Pacific Asia Conference on Information Systems (PACIS), USA, 13–15 July 2012 (2012)
28. Thedchanamoorthy, G., Piraveenan, M., Kasthuriratna, D., Senanayake, U.: Node assortativity in complex networks: an alternative approach. Proc. Comput. Sci. **29**, 2449–2461 (2014)
29. Franks, D.W., Ruxton, G.D., James, R.: Sampling animal association networks with the gambit of the group. Behav. Ecol. Sociobiol. **64**, 493–503 (2009)
30. Piraveenan, M., Prokopenko, M., Zomaya, A.: Assortative mixing in directed biological networks. IEEE/ACM Trans. Comput. Biol. Bioinf. **9**, 66–78 (2012)
31. Ciotti, V., Bianconi, G., Capocci, A., et al.: Degree correlations in signed social networks. Phys. A Stat. Mech. Appl. **422**, 25–39 (2015)
32. Ugander J, Karrer B, Backstrom L, Marlow C. The anatomy of the facebook social graph. arXiv Prepr arXiv. 1–17 (2011)
33. Manno, T.G.: Social networking in the Columbian ground squirrel, *Spermophilus columbianus*. Anim. Behav. **75**, 1221–1228 (2008)
34. Wang G, Wang B, Wang T, et al.. Whispers in the dark. In: Proceedings of the 2014 Conference on Internet Measurement Conference—IMC '14, ACM Press, New York, NY, pp. 137–150 (2014)
35. Shan, W., Liu, C., Yu, J.: Features of the discipline knowledge network: evidence from China. Technol. Econ. Dev. Econ. **20**, 45–64 (2014)
36. Sosa, S.: Structural architecture of the social network of a non-human primate (*Macaca sylvanus*): a study of its topology in La Forêt des Singes, Rocamadour. Folia Primatol. (Basel). **85**, 154–163 (2014)
37. Lima A, Rossi L, Musolesi M. Coding Together at Scale: GitHub as a Collaborative Social Network. In: Proceedings of 8th AAAI International Conference on Weblogs and Social Media (ICWSM) (2014)
38. Wiszniewski, J., Lusseau, D., Möller, L.M.: Female bisexual kinship ties maintain social cohesion in a dolphin network. Anim. Behav. **80**, 895–904 (2010)
39. Farine, D.R., Aplin, L.M., Sheldon, B.C., Hoppitt, W.: Interspecific social networks promote information transmission in wild songbirds. Proc. R. Soc. B. **282**, 20142804 (2015)
40. Illenberger, J., Flötteröd, G.: Estimating network properties from snowball sampled data. Soc. Networks. **34**(4), 701–711 (2012)
41. Kossinets, G.: Effects of missing data in social networks. Soc. Networks. **28**, 247–268 (2006)
42. Whitehead, H.: Analysing Animal Societies: Quantatitive Methods for Vertebrate Social Analysis. The University Chigaco Press, Chicago (2008)
43. Lusseau, D., Wilson, B., Hammond, P.S., et al.: Quantifying the influence of sociality on population structure in bottlenose dolphins. J. Anim. Ecol. **75**, 14–24 (2006)
44. Croft, D.P., James, R., Thomas, P.O.R., et al.: Social structure and co-operative interactions in a wild population of guppies (*Poecilia reticulata*). Behav. Ecol. Sociobiol. **59**, 644–650 (2006)
45. Tsouchnika, M., Argyrakis, P.: Network of participants in European research: accepted versus rejected proposals. Eur. Phys. J. B. **87**, 292 (2014)
46. Mena-Chalco, J.P., Digiampietri, L.A., Lopes, F.M., Cesar, R.M.: Brazilian bibliometric coauthorship networks. J. Assoc. Inf. Sci. Technol. **65**, 1424–1445 (2014)
47. Mohman, Y.T., Wang, A., Chen, H.: Statistical analysis of the airport network of Pakistan. Pramana. **85**, 173–183 (2015)

48. Im, K., Paldino, M.J., Poduri, A., et al.: Altered white matter connectivity and network organization in polymicrogyria revealed by individual gyral topology-based analysis. Neuroimage. **86**, 182–193 (2014)
49. Lee, I., Kim, E., Marcotte, E.M.: Modes of interaction between individuals dominate the topologies of real world networks. PLoS One. **10**, e0121248 (2015)
50. Spitz, A., Horvát, E.-Á.: Measuring long-term impact based on network centrality: unraveling cinematic citations. PLoS One. **9**, e108857 (2014)
51. Aguirre-von-Wobeser, E., Soberón-Chávez, G., Eguiarte, L.E., et al.: Two-role model of an interaction network of free-living γ-proteobacteria from an oligotrophic environment. Environ. Microbiol. **16**, 1366–1377 (2014)
52. Jiang, B., Duan, Y., Lu, F., et al.: Topological structure of urban street networks from the perspective of degree correlations. Environ. Plan. B Plan. Des. **41**, 813–828 (2014)
53. Mussmann S, Moore J, Pfeiffer JJ, Neville J. Assortativity in Chung Lu random graph models. In: Proceedings of the 8th Workshop on Social Network Mining and Analysis—SNAKDD'14, ACM Press, New York, NY, pp. 1–8 (2014)
54. Yang R.. Modifying network assortativity with degree preservation. In: 29th International Conference on Computers and Their Applications. CATA 2014, International Society for Computers and Their Applications, Winona, MN, pp. 35–40 (2014)
55. Croft, D.P., James, R., Krause, J.: Exploring Animal Social Networks. Princeton University Press, Oxford (2008)
56. Gleiser, P.M., Danon, L.: Community structure in jazz. Adv. Complex Syst. **06**, 565–573 (2003)
57. Newman, M.E.: The structure of scientific collaboration networks. Proc. Natl. Acad. Sci. U. S. A. **98**, 404–409 (2001)
58. Watts, D.J., Strogatz, S.H.: Collective dynamics of "small-world" networks. Nature. **393**, 440–442 (1998)
59. Aplin, L.M., Farine, D.R., Morand-Ferron, J., Sheldon, B.C.: Social networks predict patch discovery in a wild population of songbirds. Proc. Biol. Sci. **279**, 4199–4205 (2012)
60. Lusseau, D.: The emergent properties of a dolphin social network. Proc. Biol. Sci. **270**, 186–188 (2003)
61. Mourier, J., Vercelloni, J., Planes, S.: Evidence of social communities in a spatially structured network of a free-ranging shark species. Anim. Behav. **83**, 389–401 (2012)
62. Newman, M.: Coauthorship networks and patterns of scientific collaboration. Proc. Natl. Acad. Sci. U. S. A. **101**, 5200–5205 (2004)
63. Whitehead, H., Dufault, S.: Techniques for analyzing vertebrate social structure using identified individuals: review and recommendations. Adv. Study Behav. **28**, 33–73 (1999)
64. R Core Team. R: A language and environment for statistical computing. R Foundation for Statistical Computing. Vienna (2013)
65. Perreault, C.: A note on reconstructing animal social networks from independent small-group observations. Anim. Behav. **80**, 551–562 (2010)
66. Bejder, L., Fletcher, D., Bräger, S.: A method for testing association patterns of social animals. Anim. Behav. **56**, 719–725 (1998)
67. Sundaresan, S.R., Fischhoff, I.R., Dushoff, J.: Avoiding spurious findings of nonrandom social structure in association data. Anim. Behav. **77**, 1381–1385 (2009)
68. Krause, S., Mattner, L., James, R., et al.: Social network analysis and valid Markov chain Monte Carlo tests of null models. Behav. Ecol. Sociobiol. **63**, 1089–1096 (2009)
69. Aplin, L.M., Farine, D.R., Morand-Ferron, J., et al.: Individual personalities predict social behaviour in wild networks of great tits (*Parus major*). Ecol. Lett. **16**, 1365–1372 (2013)
70. Wey, T.W., Burger, J.R., Ebensperger, L.A., Hayes, L.D.: Reproductive correlates of social network variation in plurally breeding degus (*Octodon degus*). Anim. Behav. **85**, 1407–1414 (2013)
71. Krackhardt, D.: Predicting with networks: nonparametric multiple regression analysis of dyadic data. Soc. Networks. **10**, 359–381 (1988)

72. Hanhijarvi S, Garriga GC, Puolmakai K. Randomization techniques for graphs. In: Proceedings of the 9th SIAM International Conference on Data Mining (SDM '09), pp. 780–791 (2009)
73. La Fond T, Neville J. Randomization tests for distinguishing social influence and homophily effects. In: Proceedings of the 19th international conference on World wide web—WWW '10, ACM Press, New York, NY, p. 601 (2010)
74. Farine, D.R.: Measuring phenotypic assortment in animal social networks: weighted associations are more robust than binary edges. Anim. Behav. **89**, 141–153 (2014)
75. Granovetter, M.: The strength of weak ties. Am. J. Sociol. **78**, 1360–1380 (1973)
76. Garton, L., Haythornthwaite, C., Wellman, B.: Studying online social networks. J. Comput. Mediated Commun. (2006). doi:10.1111/j.1083-6101.1997.tb00062.x
77. Rowe R, Creamer G, Hershkop S, Stolfo SJ. Automated social hierarchy detection through email network analysis. In: Proceedings of the 9th WebKDD and 1st SNA-KDD 2007 workshop on Web mining and social network analysis–WebKDD/SNA-KDD '07, ACM Press, New York, NY, pp. 109–117 (2007)
78. Opsahl, T., Agneessens, F., Skvoretz, J.: Node centrality in weighted networks: Generalizing degree and shortest paths. Soc. Networks. **32**, 245–251 (2010)
79. Noldus, R., Van Mieghem, P.: Assortativity in complex networks. J. Complex Networks. **3**, 507–542 (2015)
80. Iturria-Medina, Y., Canales-Rodríguez, E.J., Melie-García, L., et al.: Characterizing brain anatomical connections using diffusion weighted MRI and graph theory. Neuroimage. **36**, 645–660 (2007)
81. Krivitsky, P.: Exponential-family random graph models for valued networks. Electron. J. Stat. **6**, 1100–1128 (2012)
82. Krivitsky P., ergm.count: Fit, simulate and diagnose exponential-family models for networks with count edges. The Statnet Project (2015). http://www.statnet.org. R package version 3.2.2. http://CRAN.R-project.org/package=ergm.count
83. De Choudhury M, Mason W. Inferring relevant social networks from interpersonal communication. In: Proceedings of the 19th International Conference on World wide web, pp. 301–310 (2010)
84. Expert, P., Evans, T.S., Blondel, V.D., Lambiotte, R.: Uncovering space-independent communities in spatial networks. Proc. Natl. Acad. Sci. U. S. A. 7663–7668 (2011)
85. Peruani, F., Tabourier, L.: Directedness of information flow in mobile phone communication networks. PLoS One. **6**, e28860 (2011)
86. Wey, T.W., Blumstein, D.T.: Social cohesion in yellow-bellied marmots is established through age and kin structuring. Anim. Behav. **79**, 1343–1352 (2010)
87. Krause, J., Krause, S., Arlinghaus, R., et al.: Reality mining of animal social systems. Trends Ecol. Evol. **28**, 1–11 (2013)
88. Eagle, N., Pentland, A.: Reality mining: sensing complex social systems. Pers. Ubiquitous Comput. **10**, 255–268 (2005)
89. Böhm, M., Hutchings, M.R., White, P.C.L.: Contact networks in a wildlife-livestock host community: identifying high-risk individuals in the transmission of bovine TB among badgers and cattle. PLoS One. **4**, e5016 (2009)
90. Hamede, R.K., Bashford, J., McCallum, H., Jones, M.: Contact networks in a wild Tasmanian devil (*Sarcophilus harrisii*) population: using social network analysis to reveal seasonal variability in social behaviour and its implications for transmission of devil facial tumour disease. Ecol. Lett. **12**, 1147–1157 (2009)
91. Rutz, C., Burns, Z.T., James, R., et al.: Automated mapping of social networks in wild birds. Curr. Biol. **22**, R669–R671 (2012)
92. Boyland, N.K., James, R., Mlynski, D.T., et al.: Spatial proximity loggers for recording animal social networks: consequences of inter-logger variation in performance. Behav. Ecol. Sociobiol. **67**, 1877–1890 (2013)

93. Drewe, J.A., Weber, N., Carter, S.P., et al.: Performance of proximity loggers in recording intra- and inter-species interactions: a laboratory and field-based validation study. PLoS One. **7**, e39068 (2012)
94. Zhou, S., Mondragón, R.: The rich-club phenomenon in the Internet topology. IEEE Commun. Lett. 1–3 (2004)
95. Colizza, V., Flammini, A., Serrano, M.A., Vespignani, A.: Detecting rich-club ordering in complex networks. Nat. Phys. **2**, 1–18 (2006)

A Parametric Study to Construct Time-Aware Social Profiles

Sirinya On-at, Arnaud Quirin, André Péninou, Nadine Baptiste-Jessel, Marie-Françoise Canut, and Florence Sèdes

1 Introduction

In information systems, user profile is an essential element for adaptive information mechanisms (e.g., personalization, information access, recommendation). These mechanisms rely on users' profiles to propose relevant content according to the user specific needs. Various models of user profile have been proposed (e.g., vector of weighted keywords, vector of weighted topics, semantic network, etc.), as presented in [1]. In this work, we model the user profile by using the keyword-based approach. Each keyword represents user's interest to which we associate a numerical weight according to its importance in the profile.

Users' profiles can be built by using the user's own information or by gathering information from external sources such as user's social network. In this paper, we build user profile by extracting user's interests from the information and relationships of the user social network, so-called social profile. We focus on user "egocentric network" which only considers the user's directed contacts. Social network-based user profiling is useful, on one hand, to provide additional information that can improve the performance of existing user profile and, on the other hand, to overcome the cold start problem or to complete non-existing/missing profile of inactive users.

We observe that in Online Social Networks (OSN), users are encouraged to contribute, broadcast contents, and connect to those who share the same interests and/or activities. As the behavior of online user evolves quickly over time, the

S. On-at (✉) • A. Quirin • A. Péninou (✉) • N. Baptiste-Jessel • M.-F. Canut • F. Sèdes
Toulouse Institute of Computer Science Research (IRIT), University of Toulouse, CNRS, INPT, UPS, UT1, UT2J, 118 route de narbonne, 31062 Toulouse, Cedex 9, France
e-mail: sirinya.on-at@irit.fr; andre.peninou@irit.fr

R. Missaoui et al. (eds.), *Trends in Social Network Analysis*, Lecture Notes in Social Networks, DOI 10.1007/978-3-319-53420-6_2

information and relationships in an OSN can rapidly become obsolete [2, 3]. Thus, building a social profile from this kind of data without taking into account its evolving characteristic may lead to misinterpretation of user long-term interests.

As a motivating example, given a user u is not football fan but supports occasionally his national team during the world cup, the well-known sport event that takes place every 4 years. He follows the official account of his favorite team and/or players in social media (e.g., Facebook, Twitter), to follow updated information about this team and the players. After the world cup, the information shared from these accounts become less meaningful for him. If this fact is not considered, we may find "football," "world cup," "France national team" as interests in his profile for a long time. As a result, the adaptive mechanisms that exploit this profile (e.g., personalized news recommendation) will continue to propose him the up-to-date news about football while this information becomes actually useless to him. This example can show that: for a user, his/her relationships between his/her social networks friends may drift overtime. The link between two users can be relevant for a period of time and may become less meaningful for a later period of time. Furthermore, information sharing behavior in social network has a dynamic and non-persistent characteristic. Some information shared by a user may not always reflect his/her interests for a long-time period.

To overcome this problem, we focus on taking into account the evolution of user's interests in social network-based user profiling process in order to build the more relevant and up-to-date user profile. We try to answer to the following questions: how to select the relevant individuals in the user social network as meaningful information sources? And then how to select relevant information from these selected individuals to extract the relevant and fresh interests to build the user social profile?

In our previous work [4], published in Asonam' 2015, we applied a time-aware method to improve a time-agnostic approach previously proposed in our team. We integrated a "temporal score" to each interest according to its relevance and freshness. The time-aware weight of each interest is computed by combining two time-aware weights: (1) the time-aware individual weight that weights the relevance of the information sources (individuals), computed by applying a time-aware link prediction technique; (2) the time-aware semantic weight that weights the relevance of information used to extract the user's interests. This weight is computed by applying a term frequency method and a time-weighted function.

In this paper, we extend this work in many ways. We conduct first a parametric study in order to find out the best value and the impact of different parameters used to calculate the temporal score. The parametric study shows the impact of the individual score compared to the information score in the temporal score calculation. We describe the improvements made in calculating the temporal score compared to the corresponding one of our previous work. At last, we validate our new method using (DBLP/Mendeley) dataset with providing new experiments showing the relevance and the compatibility of two data sources.

The rest of the paper is organized as follows. Section 2 describes the related work with presenting our previous user profiling method and existing techniques to mine user interests' evolution. In Sect. 3, we present our time-aware social profiling method. In Sect. 4, we describe the results of various experiments conducted on two co-authorships networks: DBLP and Mendeley. Section 5 concludes and presents some future works.

2 Related Works

2.1 User Profile Building Process

In the literature, different user profiling techniques are discussed. According to [1], user's interests can be extracted from information explicitly given by users or explicitly gathered when they interact with the system. Several works [5–7] also proposed to extract the user's interests from their social networks in order to build his/her social profile. In that case, user's interests are extracted by using the information shared by his/her social network members. We are particularly interested by extracting user's interests from their egocentric network. An egocentric network is a specific social network which takes into account only direct connections to a user called "ego." An egocentric network is thus composed of individuals having a direct relationship with the "ego" user and relationships between these individuals. This decision is motivated by the work of [8] which assumes that a user tends to connect with the people who share common interests with him/her. To build a social profile, user's interests can be extracted either from individual people, as adopted in [5, 6], or by extracting this information from their communities [7]. The effectiveness of social profiles has been proved with empirical results [5–7]. Tchuente et al. [7] also showed the effectiveness of the proposed community-based approach compared to an individual-based approach. However, we observe that user's interests and social networks evolution have not been widely taken into consideration in the proposed social profile building process. So, in the next section, we provide an overview about existing techniques that deal with the user's interests evolution.

2.2 Incorporating Dynamic Interests in the Profile

User's interests evolution is considered as one of the concept drift examples [9–11]. "In supervised learning context, concept drift refers to an online supervised learning scenario when the relation between the input data and the target variable changes over time. The real concept drift refers to changes in the conditional distribution of the output (i.e., target variable) given the input (input features), while the distribution of the input may stay unchanged. A typical example of the real concept drift is a change in user's interests when following an online news stream. Whilst the distribution of the incoming news documents often remains the same, the conditional distribution of the interesting (and thus not interesting) news" [9].

In our work context, we are interested in extracting user's interests from his/her social network to build his/her social profile. Dynamic social network properties allow us to consider that user's relations and shared information within this kind of network cannot be considered relevant for a long-time period. To build an effective user social dimension in our work context, it is necessary to take into account the evolution characteristic of online social network. It means that we should prevent user's interests drift regarding the relevance of social relations in user social network and also the relevance of shared information in user social network.

We can distinguish two main axes for incorporating dynamic user interests. The first axis consists in taking into account dynamic characteristic of the information sources during the interest extraction process. The second axis consists in updating an existing profile after completing the construction process [12, 13]. In this paper, we focus on the first axis only to build a relevant and up-to-date social profile. Then, as a future work, we can extend the work to include the study of user profile updating process to maintain the relevant of the profile over time.

To deal with the user's interest evolution, we can apply different techniques used to handle the concept drift issue [10, 14]. The first technique is based on a time window or time forgetting technique which selects only the information from the latest time periods [15]. In the same approach, outdated information outside a time window is completely ignored. In the information retrieval literature, the same principle is applied to build a short-term user profile [16]. However, in some cases, the ignored information could eventually be valuable [17]. Thus, this might lead to the loss of useful knowledge.

The second technique, called instance weighting, consists in weighting different time periods according to their relevance. To perform this computation, time decay functions, which assign the higher weights to the most recent information, are widely used [18, 19]. This technique enables the use of all available information in a restricted way. In fact, the relevant information is selected by weighting their importance using some temporal factors. To extract user interests for a recommendation system, [20] apply an exponential temporal function to score the tags before using them. This concept is also adopted in personalized information retrieval literature. For example, [21] applied a temporal function to weight the user interests according to the freshness of the information sources.

The third technique relies on ensemble learning to generate a family of predictors [14]. Each predictor is weighted by its relevance according to the present time point (e.g., predictors that were more successful on recent instances get higher weights). Koren [14] modeled the evolution of user behavior during the whole time period for collaborative filtering, with showing the effectiveness of his contribution for movies recommendation. Unfortunately, this kind of techniques requires a training dataset to achieve relevant families of predictors.

In our case, user's interests are extracted from the information shared by the connected individuals in the user social network. To build an effective social profile, it is important to take into account the dynamic of social networks.

2.3 Social Network Evolution

In social network analysis literature, social network evolution is related to network structural dynamics (existence, creation, and persistence of social relationships among social actors) and information flows (information sharing and diffusion between social actors) [22, 23].

In the OSNs context, the advent of social media incites a rapid evolution of social networks in terms of network structure (users' relationships) and information flow [24]. Since links are quickly established in OSNs, two users creating a relationship are not required to know each other in real life. Thus, the links persistency is not always maintained in this case. Thus, the links persistency is not always an evidence of a real relationship. Arnaboldi et al. [3] analyzed the records of a Twitter communication to study the dynamics that govern the preservation of online social relationships and discovered that Twitter users tend to keep weak social relationships rather than strong ones, with a high turnover of contacts in their networks. In terms of information flow, social events or viral marketing (buzz) increase information sharing, which in turn enhance online social content sharing. Often, this social phenomenon occurs for a short period, then disappears and may reappear in another period. For example, during the World Cup 2014, 672 million tweets were posted related to the tournament. The Ice Bucket Challenge campaign that became viral on social media during July–August 2014 generated more than 2.2 million hashtag mentions on Twitter [25]. Thus, for a user, the relationships and information in his/her social network can drift overtime. Information can be relevant for him/her for some time and becomes obsolete later.

According to [26], the dynamic characteristic of social networks depends on users' behaviors. Some users are very active and tend to frequently share information. In this case, the relationships and/or information in their social networks quickly evolve. Conversely, some users are less active and their social networks change slowly. Moreover, as previously mentioned, users of social networks can also evolve chaotically regarding social events or buzz. Thus, to apply a time-aware approach, we should choose a technique which can fit to different types of social networks characteristics.

The time window technique can lead to lose meaningful information in case of gradually evolving networks. The ensemble learning technique shows good performance in terms of accuracy. However, this technique requires a training dataset as well as it is not appropriate for new or less active users who have poor or empty relationships or information in the network. Consequently, we adopt in our contribution an instance weighting technique which enables us to use all available information in a differentiated manner by applying temporal factors. We describe this contribution in the next section.

3 Proposition: Temporal Scores to Construct Social Profiles

In this section, we present our proposal of time-aware social profile building process. We first present the agnostic social profile building process (CoBSP) developed in our team [7]. Then, we introduce the temporal score that we integrate to the CoBSP in order to take into account user interests evolution. Finally, the algorithm is described showing the whole process.

A global view of the profiling process is presented as follows:

For a given user (ego), for whom we desire to build social profile, we use his egocentric networks as information sources to extract the interests and build then his social profile. We collect the information shared by each member of user's egocentric network.

Then, we extract the keywords from all collected information and aggregate them by using a scoring function. This step is customizable so we can apply additional features or techniques to calculate the score of each extracted element.

We derive all calculated elements (interests) to the social profile according to their score. Finally, the derived social profile is represented in the form of a vector of weighted user interests.

Based on the existing social dimension derivation process (CoBSP) proposed by Tchuente et al. [7], we integrate a temporal score to the interests weighting step of the social profile construction process. We first describe hereafter the CoBSP process followed by our proposal time-aware social profile construction process called CoBSPT.[1]

3.1 Notations

In the social profile construction process, we denote u as the user (ego) for whom we desire to build social profile. The egocentric network of a user is defined as follows: for each user (u) we consider the undirected graph $G(u) = (V, E)$ where V is the set of nodes directly connected to u, and E is the set of relationships between each node pair of V. We emphasize that u is not included in V. We use the term individuals (SetIndiv) to represent the set of user's egocentric network members. For each individual Indiv $\in$SetIndiv, his shared information is called $\text{Info}_{\text{Indiv}}$. Each $\text{Info}_{\text{Indiv}}$ contains elements called e, extracted by using classical text analysis techniques [7]. Note that an element e is a keyword that represents a user interest.

[1] T for "Temporal".

3.2 Community-Based Social Profile Construction Process

The social dimension derivation process (CoBSP) proposed by [7] takes the following as inputs: (1) a given user u (ego), (2) the egocentric network of u, and (3) all information shared by all individuals in the egocentric network of u. The output is a social profile of u in the form of a vector of weighted elements (keywords). The algorithm consists of four steps described as follows:

- Step 1: Extraction of the communities from the user egocentric network. This step uses the iLCD algorithm [27] which performs very well with overlapping communities. The choice of algorithm is motivated by [7] where the iLCD is compared to the InfoMap algorithm [28], the CFinder algorithm [29], and the social cohesion-based algorithm [30].
- Step 2: Building the profile of each community found in the first step. The profile of a community is computed by analyzing the behavior of all members of this community. For each community c_i, we extract the elements e from the Info of all members. We use the term frequency tf measure to compute the score of the extracted elements, called S_{tf}, standing for tf score. The score of an element found in the community c_i is represented by $S_{tf}(e, c_i)$.
- Step 3: Computing the score of each element found in the profile of each community. The score of an element e from a community c_i is the combination of the structural score (S_{struct}) of c_i and the semantic (S_{sem}) score of e. The structural score (Eq. (1)) applied to an element e is the centrality value of c_i in the egocentric network compared to other communities.

$$S_{Struct}\left(c_i\right) = \text{centrality\_score}\left(c_i, C\right) \tag{1}$$

The semantic score of an element e of a community c_i depends on the weight of this element for all members of this community. This score can be computed by using tf or tf-idf measure [31].

The combination of the structural and the semantic scores is performed using Eq. (2).

$$S_{Sem,Struct}\ \left(e, c_i\right) = \text{combination}\left(S_{struct}\left(e, c_i\right), S_{sem}\left(e, c_i\right), \alpha\right) \tag{2}$$

The combination is a linear function: combination $(A, B, p) = p \times A + (1-p) \times \mathrm{B}$, where $p \in [0,1]$ represents the proportion between A and B. In Eq. (3), α is the proportion which varies the importance of the structural score compared to the semantic one. We describe in Sect. 5 the parametric study which enabled us to find out the fittest values of α.

- Step 4: Deriving the extracted interests for each community into the social dimension (social profile) according to the weights computed in the third step. At the end of step 3, a same element e may have different weights in different communities in the user egocentric network. In order to obtain a single weight for the element e, the function CombMNZ, proposed by [32], is adopted to combine different weights of this element from different communities.

3.3 Community-Based Social Profile Construction Process with Temporal Score

The main difference between the proposed CoBSPT and the existing CoBSP algorithms lies in the integration of temporal factors to the community profiling (step 2 of CoBSP). In the step 2, we propose to assign a temporal score to each element from each community. The input of step 2 is the information shared by the individuals in the user's egocentric network. While introducing temporal factors in our context, it is important to note that we are in an egocentric network where people know and interact with each other, in particular with the user (*ego*). These relationships exist because ego and these individuals share some common interests. We use this specific characteristic in the interests weight calculation with assuming that the temporal factors are also related to the interactions between individuals and *ego*.

3.3.1 Temporal Score Calculation

When considering any individual *Indiv* of the user's egocentric network and the information Infoshared by *indiv*, we introduce two different temporal factors for the information.

First, as described in Sect. 2.2, we use a time decay function in order to evaluate the freshness of information. This allows us to give a temporal score to each information in a classical way. Nevertheless, rather than considering the freshness with respect to a particular timestamp (current date), we consider the freshness with respect to the timestamp of the last interaction between *Indiv* and *ego*. We thus assume that the information shared by *Indiv* at a timestamp close to their last interactions with *ego* is more relevant for *ego* than a more distant one.

Second, we assume that ego may be more influenced by the individuals having a strong relationship with him. So, another way of computing the temporal weight of the information $\text{Info}_{\text{Indiv}}$ extracted from the individual *Indiv* is to assume that it is proportional to the relationship strength between *Indiv* and *ego*.

Finally, these two temporal factors will be combined and their possible combinations will be studied to decide which element is more important to compute the social profile: information weight only, relationship weight only, or some combination of the two weights. In the following section, we explain in more details how to compute the two score and their combination method.

Existing Temporal Function

To weight any information *Info* using temporal factors, we are interested in the time exponential function, which is widely used in many applications to gradually decay the importance of past behavior as time goes by. Based on the demonstrated performance of the time exponential function (3) proposed in [33], we decided to adopt this function to calculate the information temporal score.

$$f(t) = e^{-\lambda t} \tag{3}$$

The value of t represents the elapsed time (e.g., days, weeks, months, years, etc.) between the information timestamp and a given timestamp. Thus, $t = 0$ represents the value of the most recent period. For example, if we use day as a time unit, $t = 0$ for the current date, $t = 1$ for yesterday, and so on. $\lambda \in [0,1]$ represents the time decay rate. The higher the λ is, the less important the old information is. The value t of an information *Info* is computed by considering the temporal distance between the *Info* timestamp and the current time, as shown in Eq. (4).

$$t = |\text{Current date–timestamp (info)}| \tag{4}$$

Information Temporal Score Calculation (calculInfoTempScore)

We present in this section, the calculInfoTempScore function of the CoBSPT algorithm. In this function, we assign a temporal score to each information $\text{Info}_{\text{Indiv}}$ shared by an individual *Indiv* belonging to the user *u* egocentric network. The information temporal score is computed by applying the temporal function $f(t)$ of Eq. (3).

In our case, we do not use the value of t as a simple elapsed time between the information timestamp and the current time. We suggest that the *Info* shared close to the last interaction between *Indiv* and the central user *u* is considered more relevant to *u* than the distant one. So, we modified Eq. (4) by changing the current date by the timestamp of the last interaction between *u* and *Indiv* as shown in Eq. (5). The relevance of information shared by *Indiv* before and after the interaction is reduced according to their temporal distance from the last interaction between *Indiv* and *u*. For example, if the last interaction between *Indiv* and *u* is in 2013, $t_{\text{interaction}}$ of the information shared in 2012 or 2014 is equal to 1, and t of the information shared in 2011 or 2015 is equal to 2 and so on (see Fig. 1).

$$t_{\text{Info}_{\text{Indiv}}} \; |(\text{last interaction timestamp } (\text{Indiv}, u)) - (\text{Info}_{\text{Indiv}} \text{ timestamp})| \tag{5}$$

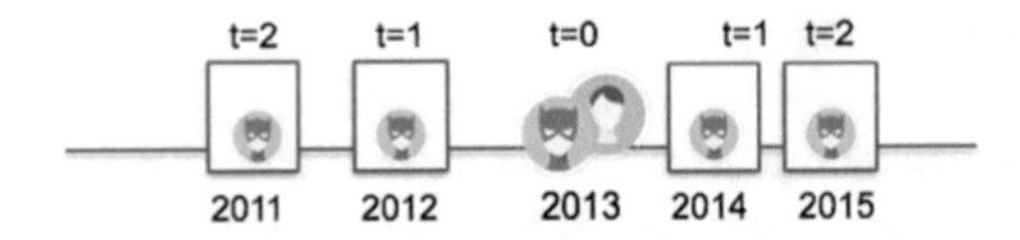

Fig. 1 The temporal relevance of information based on the last interaction of the user and a given individual

Finally, the information temporal score of a given information *Info* shared by an individual *Indiv* that will be assigned to each element *e* extracted from *Info* by using Eq. (6):

$$S_{\mathrm{Temp}}^{\mathrm{Info_{Indiv}}} = \mathrm{calculInfoTempScore}\left(\mathrm{Info_{Indiv}}\right) = f\left(t_{\mathrm{Info_{Indiv}}}\right) \tag{6}$$

Relationships Strength Techniques

There are several techniques to compute the relationship strength between two individuals. The interaction frequency is one of the widely used techniques. In the bibliometric field, we can use the co-publication frequency to identify the relationship strength between co-authors. We can also apply an advanced method by assigning the higher weight to the publication that has fewest co-authors [5]. In the case of social networks, we can use topology-based metrics (e.g., common neighbors, shortest paths, etc.) to compute the relationship strength between two nodes. These metrics are also applied for link prediction [34]. In this work, we propose to apply a link prediction metric to compute a similarity score between two connected nodes. This similarity score will represent the relationship persistency between two nodes and approximate their relationship strength. We assume that evaluating the possible persistency of a relationship in a future is more relevant than evaluating its current relationship strength.

Based on the link prediction metrics comparative study [34], we decide to adopt the Adamic/Adar metric which, despite its simplicity, outperforms the other metrics in terms of accuracy. With this metric, a pair of nodes having a common neighbor that is not common to several other nodes is considered more important. The similarity score is computed by the following Eq. (7).

$$\mathrm{AdamicAdar}\,(x, y) = \sum_{z \epsilon \{\Gamma(x) \cap \Gamma(y)\}} \frac{1}{\log|\Gamma(z)|} \tag{7}$$

where $\Gamma(x)$ represents the set of the neighbors of x.

To take into account the temporal factor, we adopt a time-aware link prediction technique introduced by [35] which proposes the integration of the temporal score into the existing Adamic/Adar metric. Based on this work, we use the following "temporal Adamic/Adar" metric:

$$\text{TemporalAdamicAdar}\,(x,y) = \sum_{z\epsilon\{\Gamma(x)\cap\Gamma(y)\}} \frac{w\,(x,z) \bullet w\,(z,y)}{\log|\Gamma(z)|} \tag{8}$$

where $\Gamma(x)$ represents the set of the neighbors of x. The $w(x_1, x_2)$ function represents the temporal relevance score of two given nodes x_1 and x_2. This function can be customized according to the required techniques to compute the temporal score for a given dataset.

Temporal Individuals Score Calculation (calculIndivTempScore)

We present in this section the function calculIndivTempScore of the CoBSPT algorithm. We adopt the temporal Adamic/Adar metric from Eq. (8) to compute the relationship strength between the central user u and a given individual *Indiv*. We suggest that individuals that have the most recent relationships with u are the ones that have the highest probability to share up-to-date common interests with him/her. In order to introduce it in Adamic/Adar metric, we apply the temporal function from Eq. (3) to compute the temporal relevance score (for Adamic/Adar metric) between two individuals x_1 and x_2 as follows (9):

$$w'\,(x_1, x_2) = f\Big(\text{CurrentTimeStamp} - \text{LastInteractionTimeStamp}\;\,(x_1, x_2) \tag{9}$$

Finally, the temporal individual relevance score of Indiv that will be assigned to each element e extracted from any information Info shared by Indiv is computed by using the temporal relevance score (Eq. (9)) for two individuals as follows (10):

$$\begin{aligned} S_{\text{Temp}}^{\text{Indiv}} &= \text{calculIndivTempScore (Indiv)} = \text{TempAdamicAdar}\;\,(u, \text{Indiv}) \\ &= \sum_{z\epsilon\{\Gamma(u)\cap\Gamma(\text{indiv})\}} \frac{w'\,(u,z) \bullet w\,(z, \text{indiv})}{\log|\Gamma(z)|} \end{aligned} \tag{10}$$

We note that this temporal score is the same for each element e extracted from any *Info* shared by the individual *Indiv*.

Final Temporal Score Calculation

The final temporal score of an element e extracted from the information $\text{Info}_{\text{Indiv}}$ shared by an individual *Indiv* is computed by linearly combining the temporal information weight and the temporal individual weight with a parameter γ as follows (11):

$$P_{Temp}^{Indiv,Info}(e) = combination\left(P_{Temp}^{Info,Indiv}, P_{Temp}^{Indiv}, \gamma\right)\Big|\,\text{e} \in \text{Info} \tag{11}$$

We present in Sect. 5, the parametric study to determine the best value of γ.

3.3.2 Temporal Score Integration

Once the temporal scores $\left(S_{\text{Temp}}^{\text{Comb}}(e)\right)$of all elements for each community are computed, we aggregate them by computing their weight compared to the total weight of all elements found in the community (the AggregationScore function of the CoBSPT algorithm, computed by Eq. (12)).

$$S_{\text{Temp}}^{C_i}(e) = \text{AgregationScore}\left(S_{\text{Temp}}^{\text{comb}}(e, c_i)\right) = \frac{\sum S_{\text{Temp}}^{\text{Comb}}(e)}{\sum_{f \in E(C_i)} S_{\text{Temp}}^{\text{Comb}}(f)} \tag{12}$$

$E(c_i)$ is the set of extracted elements from the information shared by all individuals in the community c_i.

In step 3, we replace the term frequency (S_{tf}) used in CoBSP by the temporal score $\left(S_{\text{Temp}}^{c_i}\right)$ calculated from the second step to compute their semantic score as follows (13):

$$S_{sem}(e, c_i) = S_{Temp}^{c_i}(e) \tag{13}$$

We keep the same step 4 as the CoBSP algorithm to derive the social profile. Finally, we return the social profile as a vector of weighted user interests.

We present below the proposed time-aware social profile building process based on the CoBSP algorithm.

Used notations:

u is the given user (ego) for whom we desire to compute social profile

k is the egocentric network of the user u.

c_i is a community extracted from the egocentric network.

$C = \{ c_i \}$ is the set of communities extracted from the egocentric network.

e is an element (term or keyword) extracted from a piece of information.

$E(c_i)$ is the set of extracted elements from the information shared by all individuals in the community c_i.

Notations of the form S_{Temp}^{Xxxx} are temporal weights calculations detailed in previous sections.

Algorithm CoBSPT input: u, k output: Social profile of user u

Begin

// 1st step: detect communities from the egocentric network k with iLCD algorithm

C: = iLCD(k)

// 2nd step: community profiling

For each community $c_i \in C$ *do*

For each individual *Indiv* $\in c_i$ do

// formula (10), section 3.3.1.4

$S_{Temp}^{Indiv} = calculIndivTempScore(Indiv)$;

For each information *Info* shared by *Indiv* do

//formula (6) section 3.3.1.2

$S_{Temp}^{Info_{Indiv}} = calculInfoTempScore\ (Info_{Indiv})$;
//formula (11) section 3.3.1.5

$$S_{Temp}^{Comb}(e) = combination\left(S_{Temp}^{Indiv}, S_{Temp}^{Info,Indiv}, \gamma\right)$$

End For
End For
// Update the score of each element *e* found in the community c_i by aggregating $S_{Temp}^{Comb}(e)$ from the set of elements found in the community
For each element $e \in E(c_i)$ do
//formula (12), section 3.3.2
$S_{Temp}^{C_i}(e) = AggregationScore\left(S_{Temp}^{Comb}(e), c_i\right)$;
End For
End For
// 3rd step: community structural and semantic score calculation
For each community $c_i \in C$ do
$S_{Struct}(c_i) = \text{calculCentralityScore}\ (c_i)$;
For each $e \in E(c_i)$ do
//formula (13) section 3.3.2
$S_{sem}(e, c_i) = S_{Temp}^{c_i}(e)$;
//formula (2) section 3.2
$S_{Sem,Struct}(e, c_i) = combination(S_{struct}(c_i), S_{sem}(e, c_i), \alpha)$;
End For
End For
// 4th step: Social profile derivation
For each element $e \in E(c_i)$, $\forall c_i \in C$, do

$$S_{Social}(e) = CombMNZ\left(S_{sem,struct}(e, c_i)\right), \forall c_i \in C\Big);$$

End For
Return social profile of *u*;
End

4 Experiments

To validate our proposition, we compare the relevance of our time-aware social profile building technique (CoBSPT) against the existing time-agnostic technique (CoBSP) as explained in Fig. 2. The strategy is to find out which one can provide social profiles close to the real user profile. We use the precision and recall metrics to assess the comparison. Note that [7] yet showed the effectiveness of the community-based approach (CoBSP) compared to a classical individual-based approach.

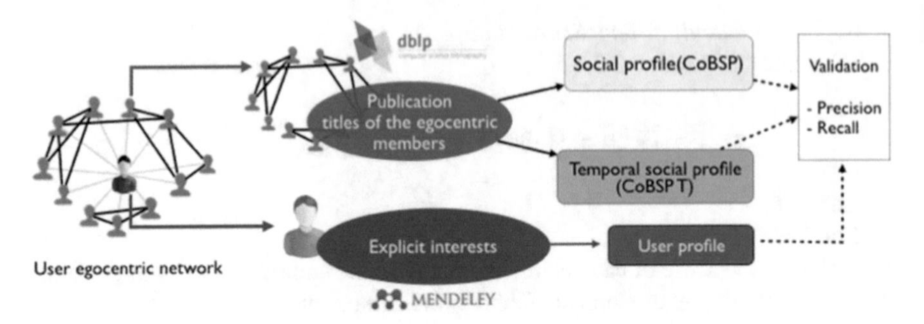

Fig. 2 User profile and social profile building process

4.1 Dataset Description

We conducted the experiments on two distinct and not connected co-authorships networks: DBLP and Mendeley. In the DBLP network, the nodes represent the authors. Two authors are connected if they published together at least one article. Two authors are connected as often as they published together and each link is labeled by the publication date. We used DBLP to build the social profile using the two approaches (CoBSP and CoBSPT). Using this data, we extract the user's interests from the publication titles. On the other hand, the real profile is built by using the interests indicated explicitly by the user in his/her Mendeley profile. This allows to use two distinct data sources and thus avoiding many biases in the results obtained and associated interpretations.

In our experiment, we consider the authors who exist in both DBLP and Mendeley databases and have enough interests in their Mendeley profile to significant and valuable results. Subsequently, we included authors that have at least 5 interests in their Mendeley profile. Our dataset contains 112 users (authors) distributed between 24 and 495 co-authors, with in average 85 of co-authors.

4.2 Analysis of Common Keywords Between DBLP and Mendeley

As we used distinct data sources to build both user and social profiles, we analyzed the number of common keywords found on the Mendeley user's profile and the publication titles of the target user in DBLP. We aimed to demonstrate the relevance and the compatibility of the two data sources and even to dismiss incompatible data. Figure 3 presents the number of users for a given number of keywords in the Mendeley profile. The 112 users of our dataset have between 5 and 24 interests in Mendeley.

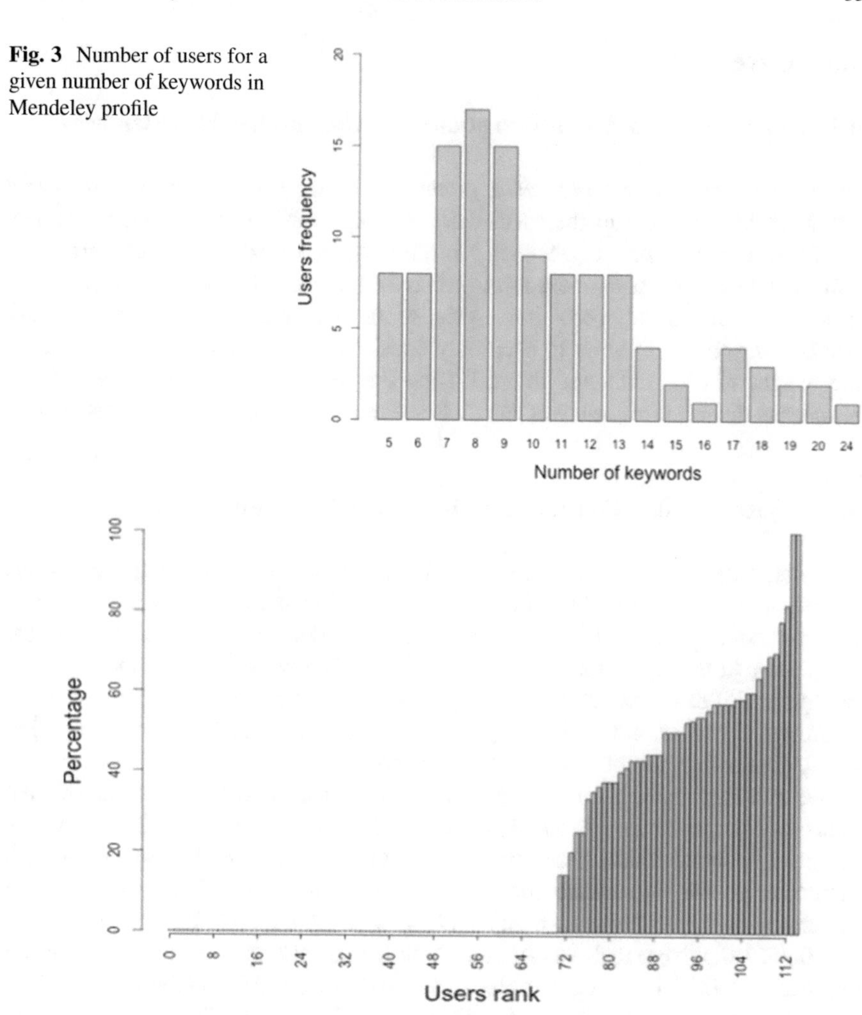

Fig. 3 Number of users for a given number of keywords in Mendeley profile

Fig. 4 Percentage of Mendeley keywords found in the DBLP publication titles

Figure 4 shows the percentage of Mendeley keywords found in the DBLP publication titles for all users, sorted by this percentage. The average percentage of common keywords for all users is 19.3%. In the sub-group of 42 users that have at least one keyword of Mendeley in their DBLP publication titles, we observe that they have between 5 and 20 keywords in their Mendeley profile. For this sub-group, the average percentage of common keywords is 50.5%, the minimum value is 14.3%, and the maximum value is 100%.

4.3 *Case Study*

4.3.1 Ground Truth: Extraction of the Real User Profile from Mendeley

To build the real user profile as a ground truth, we use the interests indicated explicitly by the users in their Mendeley profile. First, for each user, we collect the keywords from his/her Mendeley profile. Then we extract interests from the collected keywords using text-mining classical tools: we used dictionaries and thesaurus to merge keywords having the same meaning and we removed empty words using filters, in order to keep only consistent interests. Finally, we compute the weight of each extracted interest using the term frequency (tf) measure that represents the term frequency of each interest in the set of all found interests [7].

4.3.2 Social Profiles Construction and Parametric Study

The first step of the social profile building process consists in gathering the co-authors of each user to build his/her egocentric network. Then, we extract the communities from the user egocentric networks. For each community, the publication titles of the members are collected. Finally, we analyze the collected publication titles to extract the meaningful keywords using the text-mining engine as presented in Sect. 4.3.1. We apply the temporal score to each keyword according to Eq. (11) before computing their semantic score.

We build the social profiles with a parametric study in order to infer the suitable values for parameters α, γ, and λ, represented in Eqs. (2), (3), and (11), that give the most accurate results. Furthermore, our aim is to find out the impact of each parameter to the time-aware social profile building process. The value of each parameter is ranged between 0 and 1 (λ, γ, $\alpha \in \{0.01, 0.05, 01, 0.25, 0.5, 0.75, 0.9, 0.95, 1.0\}$). Note that we can build both the social profile of the time-agnostic approach (CoBSP) by fixing the value of $\gamma = 0.0$ and $\lambda = 0.0$, and the social profiles of our time-aware approach with the different combination of the parameters.

4.4 *Results*

This section presents the results of our experiments by comparing the effectiveness of the existing time-agnostic approach CoBSP and our proposed time-aware approach CoBSPT. We only consider the most relevant interests in the social profile to compute the precision and recall (the top n of all interests sorted in descending order). Note that all following results shown in this section are computed by using the top 5 interests. Results with top 10 interests are comparable but precision is less meaningful with real user profiles having less than ten keywords.

It is important to mention that, all results presented hereafter, we compare CoBSP and our CoBSPT results with respect to λ, γ, and α. Nevertheless, λ, γ are not implied in CoBSP and thus CoBSP results will never vary whatever the values of λ and γ (the representation is a straight line).

4.4.1 All Users Results

We first present the results of our parametric study for all users (112). The global results of each approach are presented by the average of the precision and recall for all users.

For the CoBSP approach, the best precision (0.1821) is observed when $\alpha \in \{0.01, 0.05, 0.1, 0.25, 0.5\}$. The best recall (0.099) is observed when $\alpha = 0.05$. For our proposed approach CoBSPT, the best precision (0.225) is observed when the parameters (α, γ, λ) equal to (0.05, 0.95, 1) and (0.01, 0.01, 0.01). The best recall (0.1218) is observed when $\alpha = 0.05$, $\gamma = 0.95$, and $\lambda = 1$. We present a complete parametric study on this dataset in terms of precision in Fig. 11 in the appendix. The results in terms of recall are presented in Fig. 12 in the appendix.

Figure 5 presents the comparison of the best precision and recall of the social profiles built by the CoBSP and CoBSPT approaches. As the best results for both approaches are obtained with $\alpha = 0.05$, we fixed this value for the precision graph (see left side of Fig. 6). The red curve represents the precision of the CoBSPT approach for different value of λ. The precision is computed by fixing $\gamma = 0.95$ (as we obtain the best precision by this value and when $\alpha = 0.05$). The blue dot represents the best precision of the CoBSP approach and thus it does not vary whatever the values of λ and γ. The recall graph (right part of Fig. 5) represents

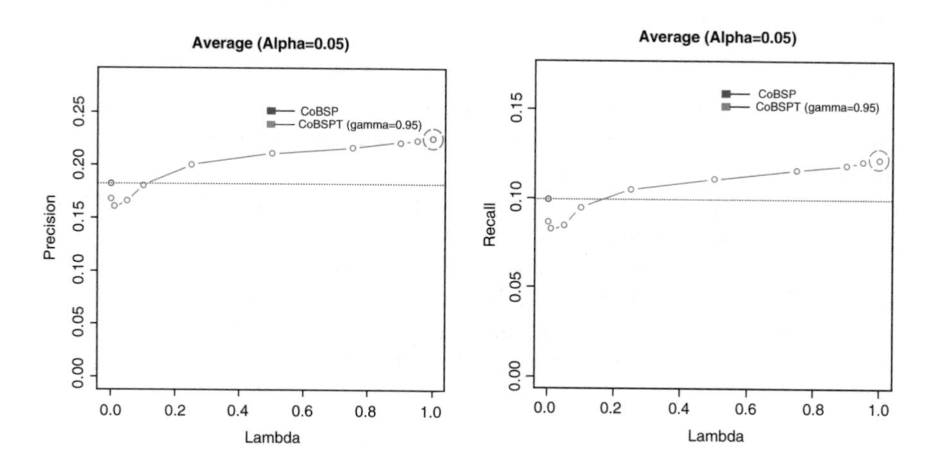

Fig. 5 *Left*: comparison of the average precision for all users with the best parameters for each approach for $\alpha = 0.05$ ($\gamma = 0.95$ for CoBSPT). *Right*: comparison of the average recall for all users with the best parameters for each approach for $\alpha = 0.05$ ($\gamma = 0.95$ for CoBSPT)

the same information as for the precision graph with the fixed value of $\gamma = 0.95$ (as we obtain the best recall with this value when $\alpha = 0.05$).

We found that the best results of CoBSPT outperform the obtained ones by CoBSP of 23.5% and 23% in terms of precision and recall, respectively. This improvement shows the effectiveness of our temporal method compared to the time-agnostic one. Nevertheless, any improvement is achieved only if $\lambda > 0.2$.

4.4.2 Results for Selected Users

Based on the analysis of common keywords between DBLP and Mendeley, previously explained in the Sect. 4.2, we observe that only 42 users have at least a common keyword between their Mendeley profile and their publication titles in DBLP. This means that Mendeley keywords have very few chances to be retrieved by CoBSP or CoBSPT algorithms, and thus can affect the experimentation results. Thus, we performed another study by taking into account only those users who have more than 50% of common keywords between DBLP and Mendeley. This new dataset contains 25 users.

For the CoBSP approach, the best precision (0.226) is observed when $\alpha \in \{0.01, 0.05, 0.1, 0.25, 0.5\}$. The best recall (0.1154) is observed when $\alpha = 0.95$. For our proposed approach CoBSPT, the best precision (0.2869) is observed when $\alpha = 0.01$, $\gamma = 1.0$, and $\lambda = 1.0$. The best recall (0.1472) is observed when $\alpha = 0.01$, $\gamma = 1.0$, and $\lambda = 1.0$. We present a complete parametric study on this dataset in terms of precision shown in Fig. 13 in the appendix. The results in terms of recall are presented in Fig. 14 in the appendix.

Figure 6 presents the comparison of the best precision and recall of the social profiles built by the CoBSP and CoBSPT approaches for the selected users. As the best results for both approaches are obtained when $\alpha = 0.01$, we fixed this value for the precision graph (see left side of Fig. 6). The red curve represents the precision of the CoBSPT approach for different values of λ. As we obtained the best precision when $\alpha = 0.01$ and $\gamma = 1.0$, we fixed these values. The blue dot represents the best precision for the CoBSP approach. Note that the precision of CoBSP is only computed when $\lambda = 0.0$ and $\gamma = 0.0$.

The recall graph (see right side of Fig. 6) presents the same information as the precision graph when the values of $\gamma = 1.0$ and $\alpha = 0.01$.

We can observe that the best result of the CoBSPT algorithm also outperforms the result of CoBSP of 23.89 and 27.8% in terms of precision and recall, respectively. Furthermore, the precision and recall of both approaches are better than those compared to the results for all users. Nevertheless, any improvement is now achieved only if $\lambda > 0.2$.

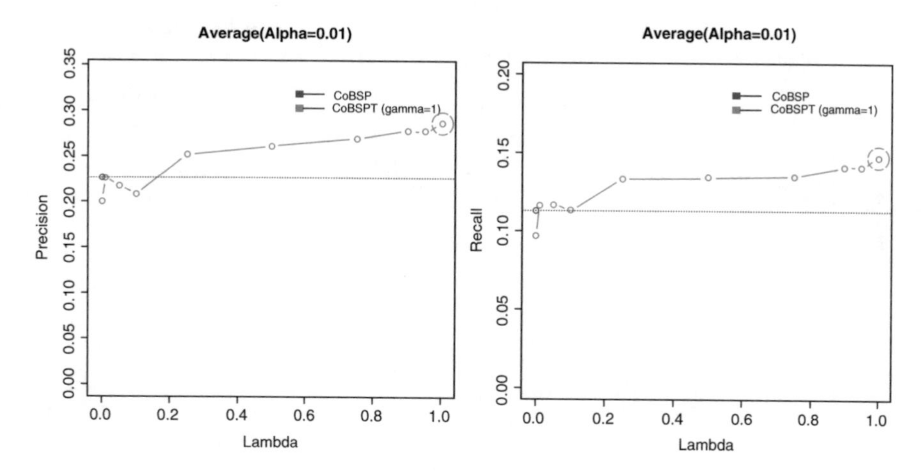

Fig. 6 *Left*: comparison of the average precision for selected users with the best parameters for each approach with $\alpha = 0.01$ ($\gamma = 1$ for CoBSPT). *Right*: comparison of the average recall for selected users with the best parameters for each approach with $\alpha = 0.01$ ($\gamma = 1$ for CoBSPT)

4.4.3 Different Time Decay Rate for the Relationships and the Information

In CoBSPT algorithm, to compute the temporal score of each interest, we apply a time decay function for the individual and information temporal score calculation. This decay function uses a time decay rate (λ) set in previous experiments to the same values for both calculations. With this technique, we obtain better results than the existing approach as shown in Sects. 4.4.1 and 4.4.2.

In this section, we hypothesis that the time decay rate of the relationships and the information could be different. In fact, it could be possible that in a social network, to have people tend to create/share information more than creating new contacts or interact with each other, and vice versa. Thus, in this section we study the impact of the relationship dynamic (individual temporal score) compared to the information dynamic (information temporal score). Note that we use the "all users" dataset of Sect. 4.4.1 for this experiment.

In order to study the impact of the individual temporal score, we focus on the precision and recall values for $\gamma = 1$. In fact, this value represents the results of the social profile computed by considering only the individual temporal score. We obtained the best precision (0.214) when $\lambda = 0.75$ and $\alpha \in \{0.25, 0.5\}$. For the recall, we got the best value (0.114) when $\lambda = 0.75$ and $\alpha = 0.25$. With these settings, the result of CoBSPT outperforms the result of CoBSP of 17.5 and 23.9% in terms of precision and recall (see Fig. 7). These values ensure the benefits of using a temporal link selection method in building the user social profile.

Similarly, we focus on the precision and recall values computed when fixing $\gamma = 0$ to study the impact of the information temporal score (ignoring individual temporal score). We obtained the best precision (0.221) when $\lambda = 0.01$ and

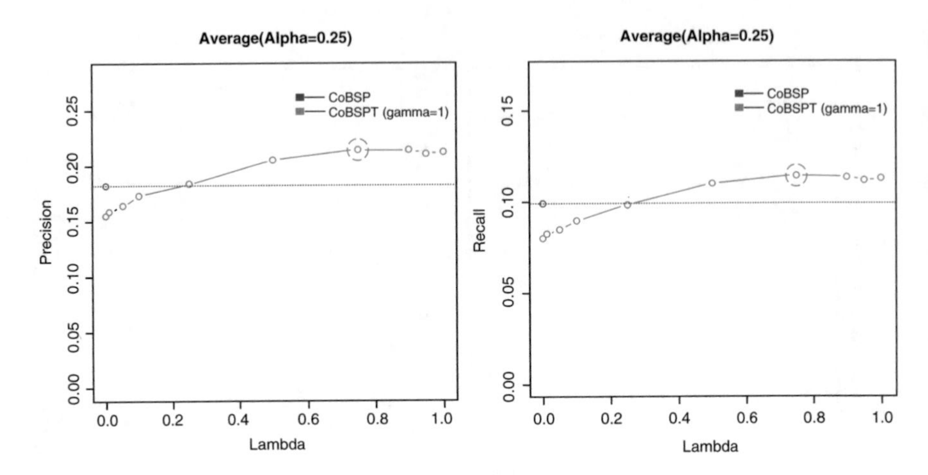

Fig. 7 Comparison of the average precision and recall for all users with the best parameters for each approach for $\alpha = 0.25$ ($\gamma = 1$ for CoBSPT)

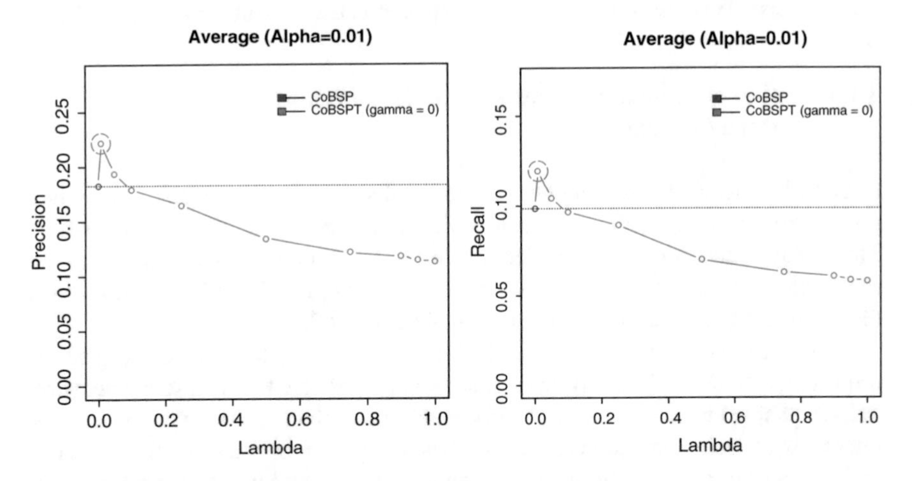

Fig. 8 Comparison of the average precision recall average for all users with the best parameters for each approach with $\alpha = 0.01$ ($\gamma = 0$ for CoBSPT)

$\alpha = 0.01$. For the recall, we got the best value (0.1195) when setting $\lambda = 0.01$ and $\alpha = 0.01$. With these settings, the result of CoBSPT outperforms the result of CoBSP of 21.3 and 29.89% in terms of precision and recall (see Fig. 8). These results show the benefit of exploiting the temporal characteristic of the information to extract meaningful user's interests to build relevant user social profile. This ensures also the fact that the diffusion of information plays also an important role in the evolution of the user's interests in online social networks, as the social network structure does. However, we observe that the value $\lambda = 0.01$, corresponding to the best precision and recall for the CoBSPT, is different from the individual temporal score which is $\lambda = 0.75$.

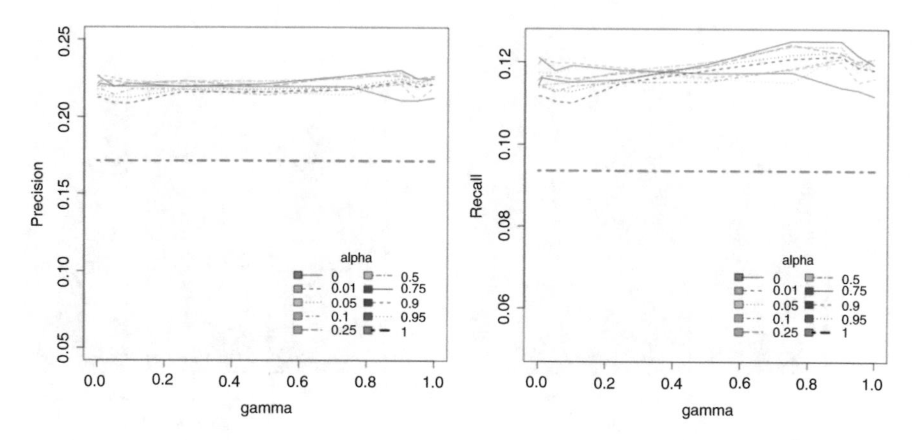

Fig. 9 Comparison of the precision average of all users on the basis of γ and α approach by fixing λ to 0.75 for the individual temporal score and by λ to 0.01 for the information temporal score

We then build the social profile using CoBSPT approach with fixing the parameters λ to 0.75 of the individual temporal score, and λ to 0.01 of the information temporal score. We check the results on the basis of γ and α. According to the left part of Fig. 9, we got the best precision (0.23) when $\gamma = 0.9$ and $\alpha = 0.75$. For the recall on the right part of Fig. 8, we obtained the best value (0.125) when $\gamma = 0.75$ and $\alpha = 0.75$.

Based on Fig. 9, we observe that the curves are almost flat even if temporal information and the individual scores clearly have different properties (as underlined by Figs. 7 and 8). This behavior can be interpreted in two ways: (1) the two scores are correlated, and therefore the same set of profiles is correctly recognized when varying γ is; or (2) the two scores are not correlated, but when changing γ a certain amount of profiles is not correctly recognized anymore and thus a comparable amount of profiles is now correctly recognized (the loss is compensated by the gain).

We obtained a better improvement compared to the resulted one by the previous approach, in which we fixed the same value for the temporal individual score and the temporal information score (see Fig. 10). The CoBSPT algorithm outperforms the CoBSP with an improvement of 26.5% and 26.3% in terms of precision and recall, respectively.

4.4.4 Discussion

Obviously, the experiments shown the effectiveness of our proposition compared to the existing time-agnostic egocentric network-based user profiling process in terms of precision and recall. This proves how much is beneficial of leveraging the evolution of user's interests (by considering the social network evolution) in the social network-based user profiling process. One can be surprised and skeptical

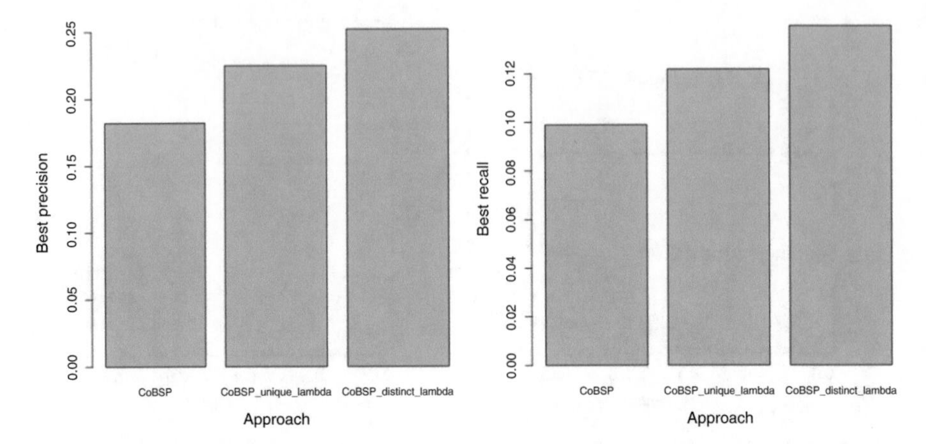

Fig. 10 Comparison of the best result from each technique (CoBSP, CoBSPT_unique_lambda), obtained by using the same value of λ for the individual temporal score and the information temporal score (CoBSPT_distinct_lambda), obtained by fixing distinct values of λ ($\lambda = 0.75$ for the individual temporal score and $\lambda = 0.01$ for the information temporal score)

considering the low scoring results. We emphasize that the results are compared to optional data from users' Mendeley profiles which therefore might be incompatible and/or outdated.

We have assumed that a time-aware social profiling method proposed answers to the two research questions stated in the introduction (how to select the relevant individuals and how to select the relevant information). Choosing both relevant information and individuals is a crucial issue. The selection criteria and features can vary according to the social network type and users' behaviors. Our time-aware method is thus more generic and it seems to give interesting results.

The results have been improved on the 25 selected users having more than 50% of common keywords between DBLP. This demonstrates that the matching step of keywords has also an impact on the precision and recall of the two algorithms in the evaluation process. We suggest that the keywords-based approach can lead to have wide range of words having the same meaning (synonym) and sometimes the keywords are too specific to the users. For example, a user may be interested in machine learning in general but when we extract keywords from publication titles of his friends we can found "svm" which is a specific machine learning method. So, it is not obvious to link "svm" to "machine learning." As a result, the extracted interests are too specific to the user and we can miss the fact that the user is also interested in other machine learning methods. This problem can be more crucial on other types of social network which are more general in terms of users and shared information. For example, on Twitter or Instagram, users tend to share information about their daily life or social events, buzz, etc. In this case, it would be appropriate to apply other techniques to extract the user interests or other models to represent the social profile. For example, we can adopt the concept-based model that represents concepts rather than words or bags of words and the concepts can be also set on different hierarchical level. We can also apply the semantic network model that represents the profile by a weighted semantic network in which each node represents a concept [1].

To obtain the best results, the time decay rates applied to relationships and information must be defined differently. The best decay rate applied to relationships and used to compute the individual temporal score is 0.75 and that of the information is 0.01. We can conclude that the relationships between users evolve more quickly than the information in the co-author network context. Thus, taking into account the freshness of relationships is very important meanwhile taking into account the freshness of information is less important. Furthermore, the best proportion between the individual temporal score and that of the temporal information score is 0.9. This shows the importance of the individual temporal score compared to that of the information in the context of co-author networks. These observations can be related with the characteristic of our dataset. We can observe that co-authorships networks do not exhibit a rapid evolution of information compared to other OSNs (Facebook, Twitter, Reddit). The authors can collaborate with different co-authors overtime, but their research field generally remains the same for a long period. So, their publication titles could remain related to the same topic domains whereas the relationships with their co-authors change more rapidly over time. Thus, we should take into consideration the dynamic characteristic of the relationships between users compared to the information dynamics on co-authorships networks.

5 Conclusion and Future Works

In this paper, we proposed to take into account the social network dynamic in user social profile building process. Considering the importance of the individuals of user social networks and their information, we defined temporal criteria (temporal scores) to weight both individuals and information in order to extract relevant and up-to-date user interests. The experiments shown the effectiveness of this method compared to the existing time-agnostic one. This demonstrates the benefit of taking into account the evolution of user interests (by considering the social network evolution) in the social network-based user profiling process. We observed that in the co-authorship network context, the time decay rate of user relationship is required to be higher than the one of the shared information. We also observed that the individual temporal score has a larger preference over the information temporal score in our final temporal score calculation.

We plan to apply our approach on other social networks with even a higher dynamic characteristic than co-authorships networks in order to evaluate our proposition for large-scale data and also to study the impact of link dynamics and information dynamics on these social networks (such as Facebook, Twitter, or Reddit). We plan also to use other link prediction algorithms and other time-weight functions to enhance the performance of our approach. Other models of social profile will be also studied and adopted for our future work to overcome the limit of granularity and synonym of the bags of words model.

Our long-term perspective consists in the proposition of a generic platform that extracts the information and builds the user social profile according to the type and the specific characteristics of the underlying social network. Such a platform would be parameterized by the characteristics of the targeted social network using a machine learning approach.

Appendix

Figure 11 represents the result of the parametric study in terms of precision for all users. We plot a graph for each value of the parameter α. For a given graph, each curve represents the precision for a given value of the parameter γ, for all values of the parameter λ, shown in the X-axis. Figure 12 represents the same information in terms of recall. We recall that α represents the proportion of the structural score

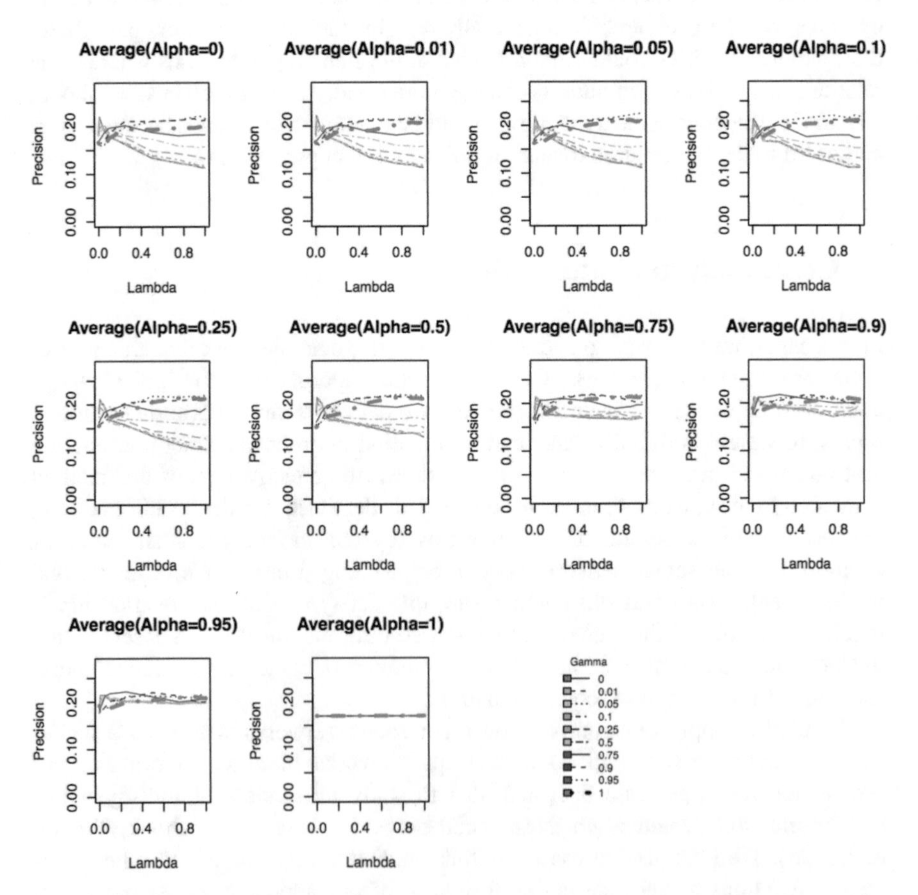

Fig. 11 The average precision for all users with the parametric study for parameters α, γ, and λ

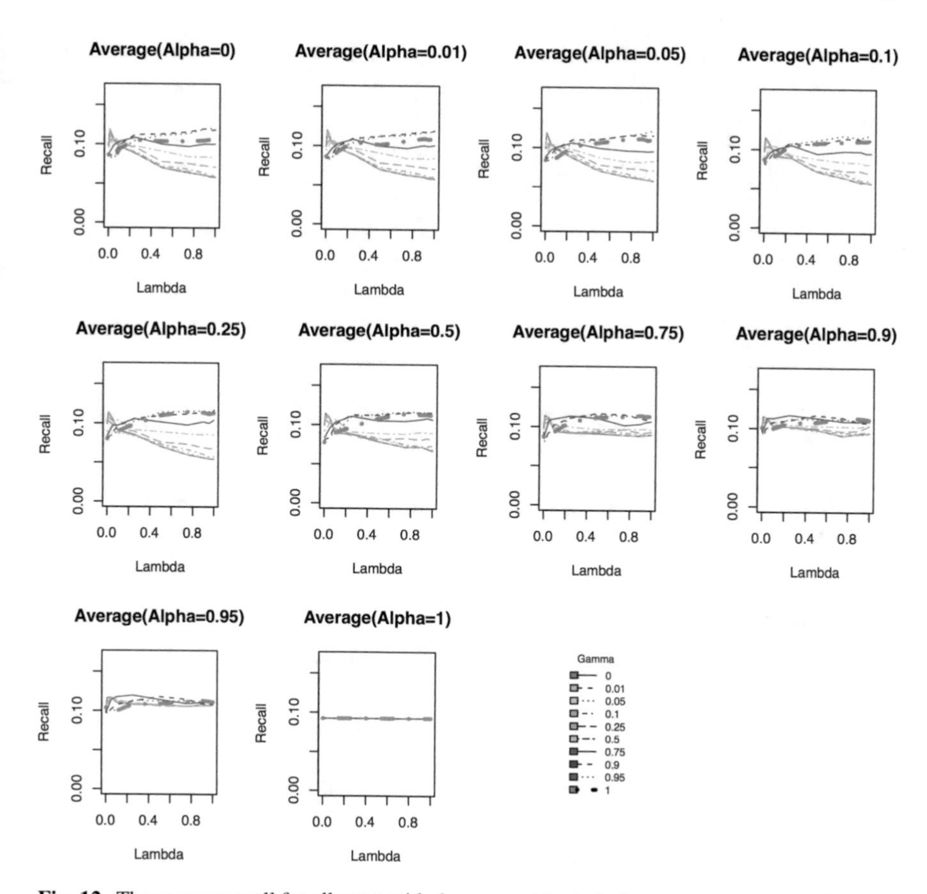

Fig. 12 The average recall for all users with the parametric study for parameters α, γ, and λ

compared to the semantic score as presented in Eq. (1), λ represents the time decay rate as presented in Eq. (2), and γ represents the proportion of the individual temporal score compared to the information temporal score as presented in Eq. (9). As noted above, the set of points corresponding to $\lambda = 0.0$ and $\gamma = 0.0$ represents the CoBSP results. Figures 13 and 14 represent the result of the parametric study in terms of precision and recall for 25 selected users that have at least 50% of common keyword between Mendeley and DBLP.

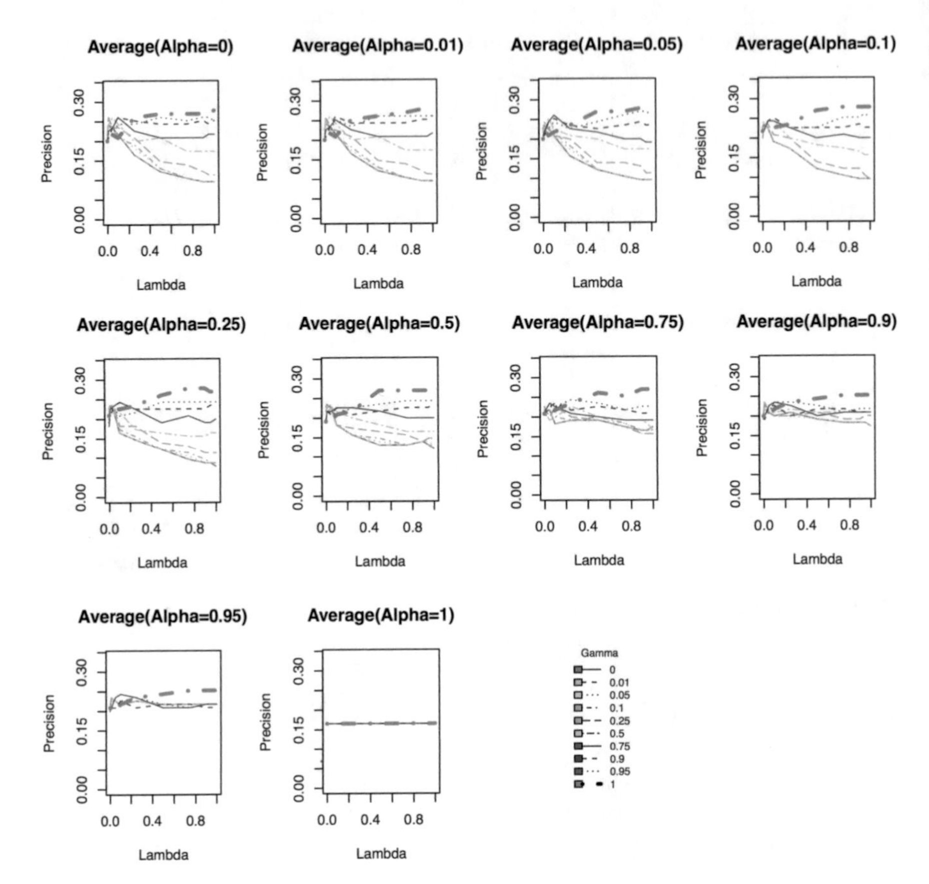

Fig. 13 The average precision for selected users with the parametric study for parameters α, γ, and λ

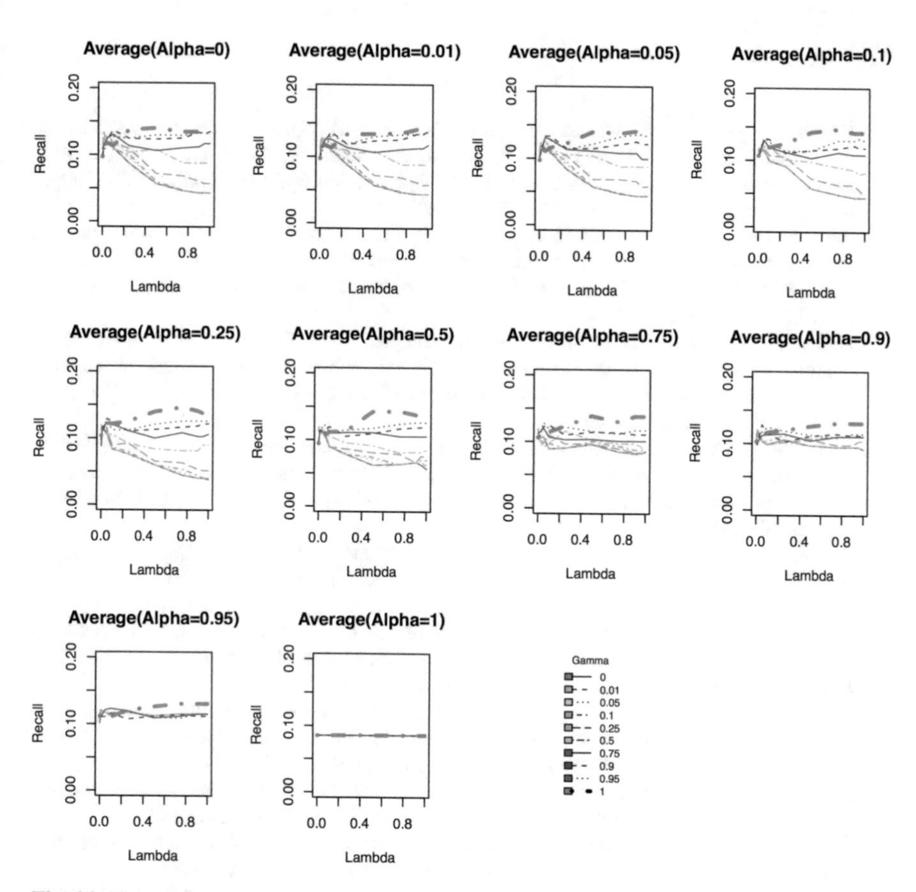

Fig. 14 The recall average for all users with the parametric study for parameters α, γ, and λ

References

1. Gauch, S., Speretta, M., Chandramouli, A., Micarelli, A.: User profiles for personalized information access. In: Brusilovsky, P., Kobsa, A., Nejdl, W. (eds.) The Adaptive Web. Lecture Notes in Computer Science, vol. 4321, pp. 54–89. Springer, Berlin (2007). http://link.springer.com/chapter/10.1007/978-3-540-72079-9_2
2. Abel, F., Gao, Q., Houben, G.-J., Tao, K.: Analyzing temporal dynamics in twitter profiles for personalized recommendations in the social web. In: Proceedings of the 3rd International Web Science Conference, pp. 2:1–2:8, WebSci '11, ACM, New York, NY (2011). doi:10.1145/2527031.2527040
3. Arnaboldi, V., Conti, M., Passarella, A., Dunbar, R.: 2013. Dynamics of personal social relationships in online social networks: a study on Twitter. In: Proceedings of the First ACM Conference on Online Social Networks, pp. 15–26. COSN '13, ACM, New York, NY. doi:10.1145/2512938.2512949
4. Canut, M.-F., On-At, S., Péninou, A., Sèdes, F.: 2015. Time-aware egocentric network-based user profiling. In: Proceedings of the 2015 IEEE/ACM International Conference on Advances in Social Networks Analysis and Mining 2015, pp. 569–572, ASONAM '15, ACM, New York, NY. doi:10.1145/2808797.2809415
5. Cabanac, G.: Accuracy of inter-researcher similarity measures based on topical and social clues. Scientometrics. **87**(3), 597–620 (2011). doi:10.1007/s11192-011-0358-1
6. David, C., Zwerdling, N., Guy, I., Ofek-Koifman, S., N. Har'el, I. Ronen, E. Uziel, S. Yogev, S. Chernov. 2009. Personalized social search based on the user's social network. In: Proceedings of the 18th ACM Conference on Information and Knowledge Management, pp. 1227–1236, CIKM '09, ACM, New York, NY. doi:10.1145/1645953.1646109
7. Tchuente, D., Canut, M.-F., Jessel, N., Peninou, A., Sèdes, F.: A community-based algorithm for deriving users' profiles from egocentrics networks: experiment on Facebook and DBLP. Soc. Network Anal. Min. **3**(3), 667–683 (2013). doi:10.1007/s13278-013-0113-0
8. Aral, S., Walker, D.: Tie strength, embeddedness, and social influence: a large-scale networked experiment. Manage. Sci. **60**(6), 1352–1370 (2014). doi:10.1287/mnsc.2014.1936
9. Gama, J., I. Žliobaitė, A. Bifet, M. Pechenizkiy, A. Bouchachia. 2014. A survey on concept drift adaptation." ACM Comput. Surv. 46 (4): 44:1–44:37. doi:10.1145/2523813.
10. Tsymbal, A.. 2004. The problem of concept drift: definitions and related work. Technical Report, Department of Computer Science, Trinity College, Dublin
11. Widmer, G., M. Kubat. 1993. Effective learning in dynamic environments by explicit context tracking. In: P.B. Brazdil (ed.) Machine Learning: ECML-93. Lecture Notes in Computer Science, vol. 667, pp. 227–243, Springer, Berlin. http://link.springer.com/chapter/10.1007/3-540-56602-3_139
12. Mianowska, B., Nguyen, N.T.: Tuning user profiles based on analyzing dynamic preference in document retrieval systems. Multimedia Tools Appl. **65**(1), 93–118 (2013). doi:10.1007/s11042-012-1145-6
13. Sugiyama, K., K. Hatano, M. Yoshikawa. 2004. Adaptive web search based on user profile constructed without any effort from users. In: Proceedings of the 13th International Conference on World Wide Web, pp. 675–684, WWW '04, ACM, New York, NY. doi:10.1145/988672.988764
14. Koren, Y.. 2009. Collaborative filtering with temporal dynamics. In Proceedings of the 15th ACM SIGKDD International Conference on Knowledge Discovery and Data Mining, pp. 447–456, KDD '09, ACM, New York, NY. doi:10.1145/1557019.1557072
15. Maloof, M.A., Michalski, R.S.: Selecting examples for partial memory learning. Mach. Learn. **41**(1), 27–52 (2000). doi:10.1023/A:1007661119649
16. Bennett, P.N., R.W. White, W. Chu, S.T. Dumais, P. Bailey, F. Borisyuk, X. Cui. 2012. Modeling the impact of short- and long-term behavior on search personalization. In: Proceedings of the 35th International ACM SIGIR Conference on Research and Development in Information Retrieval, pp. 185–194, SIGIR '12, ACM, New York, NY. doi:10.1145/2348283.2348312

17. Tan, B., X. Shen, and C. Zhai. 2006. Mining long-term search history to improve search accuracy. In: Proceedings of the 12th ACM SIGKDD International Conference on Knowledge Discovery and Data Mining, pp. 718–723, KDD '06, ACM, New York, NY. doi:10.1145/1150402.1150493
18. Li, D., Cao, P., Guo, Y., Lei, M.: Time weight update model based on the memory principle in collaborative filtering. J. Comput. **8**(11), 2763–2767 (2013). doi:10.4304/jcp.8.11.2763-2767
19. Li, L., Zheng, L., Yang, F., Li, T.: Modeling and broadening temporal user interest in personalized news recommendation. Expert Syst. Appl. **41**(7), 3168–3177 (2014). doi:10.1016/j.eswa.2013.11.020
20. Zheng, N., Li, Q.: A recommender system based on tag and time information for social tagging systems. Expert Syst. Appl. **38**(4), 4575–4587 (2011). doi:10.1016/j.eswa.2010.09.131
21. Kacem, A., M. Boughanem, R. Faiz. 2014. Time-sensitive user profile for optimizing search personlization. In: V. Dimitrova, T. Kuflik, D. Chin, F. Ricci, P. Dolog, G.-J. Houben (eds.) User eModeling, Adaptation, and Personalization. Lecture Notes in Computer Science, vol. 8538, pp. 111–121, Springer International Publishing, Basel. http://link.springer.com/chapter/10.1007/978-3-319-08786-3_10
22. Stattner, E., Collard, M., Vidot, N.: D2SNet: dynamics of diffusion and dynamic human behaviour in social networks. Comput. Hum. Behav. Adv. Hum. Comput. Interact. **29**(2), 496–509 (2013). doi:10.1016/j.chb.2012.06.004
23. Weng, L., J. Ratkiewicz, N. Perra, B. Gonçalves, C. Castillo, F. Bonchi, R. Schifanella, F. Menczer, A. Flammini. 2013. The role of information diffusion in the evolution of social networks. In: Proceedings of the 19th ACM SIGKDD International Conference on Knowledge Discovery and Data Mining, pp. 356–364, KDD '13, ACM, New York, NY. doi:10.1145/2487575.2487607
24. Gomez Rodriguez M., J. Leskovec, B. Schölkopf. 2013. Structure and dynamics of information pathways in online media. In: Proceedings of the Sixth ACM International Conference on Web Search and Data Mining, pp. 23–32, WSDM '13, ACM, New York, NY. doi:10.1145/2433396.2433402
25. Koohy, H., Koohy, B.: A lesson from the ice bucket challenge: using social networks to publicize science. Front. Genet. **5**, (2014). doi:10.3389/fgene.2014.00430
26. Kumar, R., J. Novak, A. Tomkins. 2006. Structure and evolution of online social networks. In: Proceedings of the 12th ACM SIGKDD International Conference on Knowledge Discovery and Data Mining, pp. 611–617, KDD '06, ACM, New York, NY. doi:10.1145/1150402.1150476
27. Cazabet, R., M. Leguistin, F. Amblard. 2012. Automated community detection on social networks: useful?Efficient?Asking the users. In: Proceedings of the 4th International Workshop on Web Intelligence; Communities, pp. 6:1–6:8, ACM, New York, NY. doi:10.1145/2189736.2189745
28. Rosvall, M., Bergstrom, C.T.: An information-theoretic framework for resolving community structure in complex networks. Proc. Natl. Acad. Sci. U. S. A. **104**(18), 7327–7331 (2007). doi:10.1073/pnas.0611034104
29. Pollner, P., Palla, G., Vicsek, T.: Parallel clustering with cfinder. Parallel Process. Lett. **22**(1), 1240001 (2012). doi:10.1142/S0129626412400014
30. Friggeri, A., G. Chelius, E. Fleury. 2011. Triangles to capture social cohesion. arXiv:1107.3231 [Physics]. http://arxiv.org/abs/1107.3231
31. Salton, G., Waldstein, R.K.: Term relevance weights in on-line information retrieval. Info. Process. Manage. **14**(1), 29–35 (1978). doi:10.1016/0306-4573(78)90055-9
32. Shaw, J.A., E.A. Fox, J.A. Shaw, E.A. Fox. 1994. Combination of multiple searches. In: The Second Text REtrieval Conference (TREC-2), pp. 243–252
33. Ding, Y., Jacob, E.K., Caverlee, J., Fried, M., Zhang, Z.: Profiling social networks: a social tagging perspective. D-Lib Mag. **15**(3/4), (2009). doi:10.1045/march2009-ding

34. Liben-Nowell, D., J. Kleinberg. 2003. The link prediction problem for social networks. In: Proceedings of the Twelfth International Conference on Information and Knowledge Management, pp. 556–559, CIKM '03, ACM, New York, NY. doi:10.1145/956863.956972
35. Tylenda, T., R. Angelova, S. Bedathur. 2009. Towards time-aware link prediction in evolving social networks. In: Proceedings of the 3rd Workshop on Social Network Mining and Analysis, pp. 9:1–9:10, SNA-KDD '09, ACM, New York, NY. doi:10.1145/1731011.1731020.

Sarcasm Analysis on Twitter Data Using Machine Learning Approaches

Santosh Kumar Bharti, Ramkrushna Pradhan, Korra Sathya Babu, and Sanjay Kumar Jena

1 Introduction

Sentiment analysis is a process that "aims to determine the attitude of a speaker or a writer on some topic" [24]. Social media has fuelled the spread of user sentiments in the online space as ratings, reviews, comments, etc. The need for precise and reliable information about consumer preferences has led to increased interest towards analysis of social media content. For many businesses, online opinion has turned into a kind of virtual currency that can make or break a product in the marketplace. Sentiment analysis means monitoring social media posts and discussions, then figuring out how participants are reacting to it.

Sarcasm is a particular type of sentiment which derived from the French word "Sarcasmor" that means "tear flesh" or "grind the teeth." In simple words, it means to speak bitterly. The literal meaning is different than what the speaker intends to say through sarcasm. The Random House dictionary defines sarcasm as "a harsh or bitter derision or irony" or "a sharply ironical taunt; sneering or cutting remark." Sarcasm can also be defined as a "contrast between a positive sentiment and negative situation" [29] and vice versa. For example, "I love working on holidays." In this tweet "love" gives a positive opinion but "working on holidays" is referring to a negative situation as people do not work on holidays.

For sarcasm analysis in text, it is of paramount importance to have a rudimentary knowledge of natural language processing (NLP) that aims to acquire, understand, and generate human languages such as English, Chinese, Hindi, etc. Part-of-speech (POS) tagging, parsing, tokenizations, etc. are the tasks performed in NLP, which are used for sarcasm detection.

S.K. Bharti (✉) • R. Pradhan • K.S. Babu • S.K. Jena
National Institute of Technology Rourkela, Rourkela, India
e-mail: sbharti1984@gmail.com; ramkrushna768@gmail.com; prof.ksb@gmail.com; skjena@nitrkl.ac.in

R. Missaoui et al. (eds.), *Trends in Social Network Analysis*, Lecture Notes in Social Networks, DOI 10.1007/978-3-319-53420-6_3

Sarcasm can be detected by considering lexical, pragmatic, hyperbole, or other such features of the statement. Some features can also be developed using certain patterns such as unigram, bigram, trigram, etc. There can be features based on verbal or gestural clues such as emoticons, onomatopoeic expressions in laughter, positive interjections, quotation marks, use of punctuation which can help in detecting sarcasm. But all these features are not enough to identify sarcasm in tweets until the context of the text is not known. The machine should be aware of the context of the text and relate it to general world knowledge to be able to identify sarcasm more accurately.

In this chapter, we focused on lexical (unigram, bigram, and trigram), hyperbole (intensifiers and interjections), behavioral (likes and dislikes), and universal facts as the text features for sarcasm detection in tweets. For the extraction of these features from the tweets, four algorithms were proposed, namely parsing-based lexical generation algorithm (PBLGA), likes and dislikes contradiction (LDC), tweet contradicting universal facts (TCUF), and tweet contradicting time-dependent facts (TCTDF). These algorithms produce the learned feature lists to identify sarcasm in different contexts of tweets such as a contradictory sentiment and situation, likes and dislikes contradiction, universal fact negation, and time-dependent fact negation. Various machine learning classifiers, namely support vector machine (SVM), maximum entropy (ME), Naive Bayes (NB), and decision tree (DT), were deployed to classify sarcastic tweets. These classifiers are trained using the extracted features lists.

This chapter is an extended version of our previous paper "Parsing-based Sarcasm Sentiment Recognition in Twitter Data" [3] published in the proceedings of ASONAM 2015. It discussed two algorithms, PBLGA and Interjection word start (IWS) for sarcasm detection in Twitter data. In this chapter, we include three additional algorithms to cover more sarcasm types. IWS is not considered in this extended version as the algorithm doesn't fit into the machine learning approach. IWS algorithm does not generate any feature list for training the classifiers mentioned earlier. The main differences lie in the experiments that have been evaluated through various machine learning techniques and have added a better comparison with existing methods, leading to new conclusions.

The rest of the chapter is organized as follows: Sect. 2 describes related work. Preliminary is given in Sect. 3. Data collection and preprocessing are discussed in Sect. 4. The proposed scheme and implementation details are given in Sect. 5. Section 6 depicts various machine learning classifiers. Experimental results are shown in Sect. 7. Finally, we conclude in Sect. 8.

2 Related Work

In recent times, research interest grew rapidly towards sarcasm detection in text [1, 3, 4, 8, 12, 15, 16, 18–21, 28, 29, 32, 35]. Many researchers have investigated sarcasm on the data collected from various sources such as tweets from Twitter, Amazon

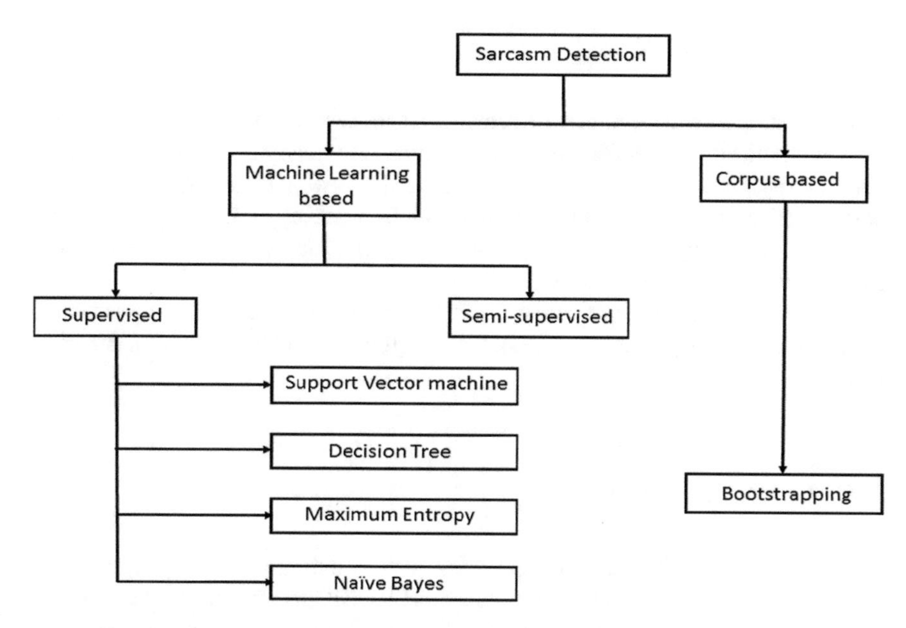

Fig. 1 Classification of sarcasm detection based on approaches used

product reviews, website comments, Google books, and online discussion forums on the basis of various features such as lexical, pragmatics, and hyperbole. Sarcasm detection can be classified into two categories based on classification approaches, namely machine learning-based approach and corpus-based approach as shown in Fig. 1.

2.1 Machine Learning-Based Approach

Machine learning-based approach requires a labelled dataset for training. It can be further classified into two categories, supervised and semi-supervised. In supervised method, the dataset should be fully labelled, while in semi-supervised, it can be partially labelled (combination of both labelled and unlabelled dataset).

Barbieri et al. [1] used machine learning-based approach to detect sarcasm through its inner structure. They haven't used any pattern of words as features. Liu et al. [19] used a supervised approach to classify movie reviews into two classes after performing subjective feature extractions. A semi-supervised approach was introduced by Davidov et al. [8] to detect sarcasm in tweets and Amazon product reviews. Irony and sarcasm have been approached as a computation problem recently by Carvalho et al. [6] using machine learning techniques. Lukin and Walker [20] used bootstrapping to improve the performance of sarcasm and nastiness classifiers for online dialogue. Liebrecht et al. [18] designed a machine learning model to detect sarcasm in Dutch tweets. Bhattacharyya and Verma [5] used machine learning-based methods for the classification of movie reviews.

2.2 Corpus-Based

Riloff et al. [29] built a model to detect sarcasm with a bootstrapping algorithm that automatically learns lists of positive sentiment phrases and negative situation phrases from a sarcastic tweet. Chaumartin [7] use sentiwordnet to find the polarity of newspaper headings. Verma and Bhattacharyya [37] used the same resource for developing both resource-based and machine learning-based methods for classification of movie reviews.

A similar classification of sarcasm identification based on their features can be made into three categories, namely lexical, pragmatics, and hyperbolic as shown in Fig. 2.

2.3 Lexical Features

The lexical features include text properties such as unigram, bigram, trigram, and *n*-gram. Kreuz and Roberts [16] introduced the lexical features that play a vital role in recognizing irony and sarcasm in text. Kreuz in his subsequent work along with Kreuz and Caucci [15] used these lexical and syntactic features to identify sarcasm in text. Barbieri et al. [1] considered seven lexical features to detect sarcasm through its inner structure such as unexpectedness, the intensity of the terms, or imbalance between registers. Davidov et al. [8] used two interesting lexical features, namely

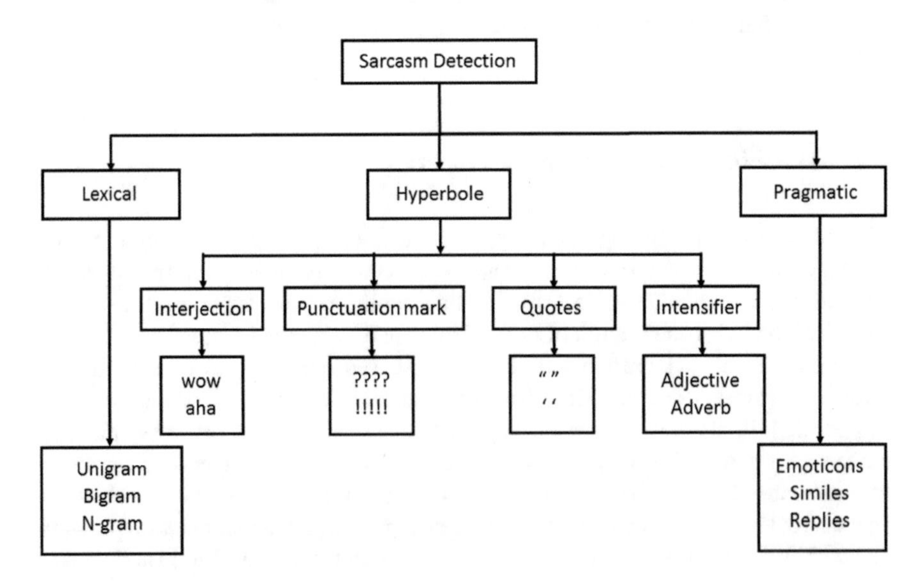

Fig. 2 Classification of sarcasm detection based on features used

pattern-based (high-frequency words and content words) and punctuation-based to build a weighted k-nearest neighbor based classification model to perform sarcasm detection. González-Ibánez et al. [12] explored numerous lexical features (derived from WordNet affect [31] and linguistic inquiry and word count (LWIC) [26]) to identify sarcasm. Riloff et al. [29] used a well-constructed lexicon-based approach to detect sarcasm, and for lexicon generation they used unigram, bigram and trigram features. Tsur et al. [34] observed that bigram, based features produce better results in detecting sarcasm in tweets and Amazon product reviews.

2.4 *Pragmatic Feature*

Pragmatic features include symbolic or figurative texts like smilies, emoticons, and replies. Kreuz and Roberts [16] introduced the concept of pragmatic features to identify sarcasm in textual data. González-Ibánez et al. [12] explored it further with emoticons, smileys and replies. They developed a system using the pragmatic features to identify sarcasm in tweets. Tayal et al. [32] used pragmatic features to identify sarcasm in political tweets. Carvalho et al. [6] created an automatic system for detecting irony relying on emoticons and special punctuations. They focused on detection of ironic style in newspaper articles. Rajadesingan et al. [28] used a systematic approach for sarcasm detection in tweets and used psychological and behavioral pragmatic features of a user with their present and past tweets.

2.5 *Hyperbolic Feature*

Hyperbole plays the most important role in identifying sarcasm in the text. A combination of the text properties such as intensifier, interjection, quotes, punctuation, etc. in tweets is called hyperbole. Lunando and Purwarianti [21] focused only on interjection words such as aha, bah, nah, wah, wow, yay, uh, etc. for sarcasm identification. They conclude that, if the text contains interjection words, it has more tendency to be classified as sarcastic. Liebrecht et al. [18] focused on hyperbole to detect sarcasm in tweets as utterance with a hyperbole. For example: "fantastic weather" when it rains is identified as sarcastic with more ease than the utterance without a hyperbole "the weather is good" when it rains. The utterance with the hyperbolic "fantastic" may be easier to interpret more sarcastic than the utterance with the non-hyperbolic "good." Utsumi [36] suggested that extreme adjectives and adverbs often provide an implicit way to display negative attitudes, i.e. sarcasm. Kreuz and Caucci [15] discussed the role of hyperbole features, such as an interjection (e.g., "gee" or "gosh") and punctuation symbols (e.g., "?"). Filatova [10] focused on hyperbole features to identify sarcasm in document level text

Table 1 Approaches, types, and features of sarcasm detection

Types of approaches used in sarcasm detection	
A1	Machine learning based
A11	Supervised
A12	Semi-supervised
A2	Corpus based
Types of sarcasm occur in text	
T1	Contrast between positive sentiment and negative situation
T2	Contrast between negative sentiment and positive situation
T3	Tweet start with interjection word
T4	Likes and dislikes contradiction—behavior based
T5	Tweet contradicting universal facts
T6	Tweet carries positive sentiment with an antonym pair
T7	Tweet contradicting time-dependent facts
Types of features	
F1	Lexical: unigram, bi-gram, tri-gram, *n*-gram, #hashtag
F2	Pragmatic: smileys, emoticons, replies
F3	Hyperbole: interjection, intensifier, punctuation mark, quotes
F31	Interjection: yay, oh, wow, yeah, nah, aha, etc.
F32	Intensifier: adverb, adjectives
F33	Punctuation mark: !!!!!, ????
F34	Quotes: " ", " "

because a phrase or a sentence level may not be sufficient for sarcasm detection. They considered the context of a sentence and the surrounding sentences to improve the accuracy of the detection. Tungthamthiti et al. [35] explored concept level knowledge to identify an indirect contradiction between sentiment and situation using the hyperbole feature as an intensifier. According to them words like "raining," "bad weather" are conceptually same. So, if "raining" is present in tweet, then consider it as a negative situation. They also focus on coherency if tweet contains multiple sentences.

Based on the classifications mentioned earlier, we have identified various approaches (denoted as A1, A2), types of sarcasm occurs in tweets (denoted as T1, T2, T3, T4, T5, T6, T7), and text features set (denoted as F1, F2, F3) which are used to detect sarcasm in Twitter data. The details are shown in Table 1. With the help of Table 1, a consolidated summary of previous studies in sarcasm identification is shown in Table 2.

Table 2 Previous studies in sarcasm detection in text

	Approaches			Types of sarcasm							Types of feature					
	A1		A2	T1	T2	T3	T4	T5	T6	T7	F1	F2	F3			
Study	A11	A12											F31	F32	F33	F34
Kreuz and Roberts [16]	✓			✓							✓	✓		✓		
Utsumi [36]	✓			✓							✓			✓		
Verma and Bhattacharyya [37]	✓		✓	✓							✓					
Bhattacharyya and Verma [5]	✓		✓	✓							✓					
Kreuz and Caucci [15]	✓			✓							✓	✓		✓		
Chaumartin [7]			✓	✓							✓					
Carvalho et al. [6]	✓			✓							✓					
Tsur et al. [34]		✓		✓							✓					
Davidov et al. [8]		✓		✓							✓				✓	
González-Ibánez et al. [12]	✓			✓							✓	✓				
Filatova [10]	✓		✓	✓							✓			✓		
Riloff et al. [29]	✓		✓	✓							✓					
Lunando and Purwarianti [21]	✓					✓					✓		✓			
Liebrecht et al. [18]	✓			✓							✓			✓		
Lukin and Walker [20]	✓		✓	✓							✓					
Tungthamthiti et al. [35]	✓			✓						✓	✓			✓		
Liu et al. [19]	✓				✓						✓	✓				✓
Justo et al. [14]	✓			✓							✓					✓
Kunneman et al. [17]	✓			✓							✓	✓	✓	✓		
Barbieri et al. [1]	✓			✓							✓					
Tayal et al. [32]	✓			✓							✓	✓				
Ingle et al. [13]	✓						✓	✓		✓	✓					
Pielage [27]	✓			✓							✓					
Rajadesingan et al. [28]	✓			✓			✓				✓	✓				
Bharti et al. [3]			✓	✓	✓	✓					✓		✓			

3 Preliminaries

Opinions posted in social media can be classified as positive, negative, and neutral. A positive sentiment can be further classified as actual positive or sarcastic sentiment. Similarly, a negative sentiment can be actual negative or sarcastic sentiment.

3.1 System Model

With the explosive growth of social media on the web, individuals and organizations are increasingly using this content in decision making. A system model that shows how the social media is useful for a user in their decision making proposed by us [3] is shown in Fig. 3. There are three model entities.

1. *Social media (SM)*: It may be a social networking website (Facebook, Twitter, Amazon, etc.), where people use to write an opinion, reviews, and post some blogs, or micro-blogs about any social media entities.
2. *Social media entities (SME)*: An entity about which users want to post and retrieve comments, tweets and give reviews on social media.

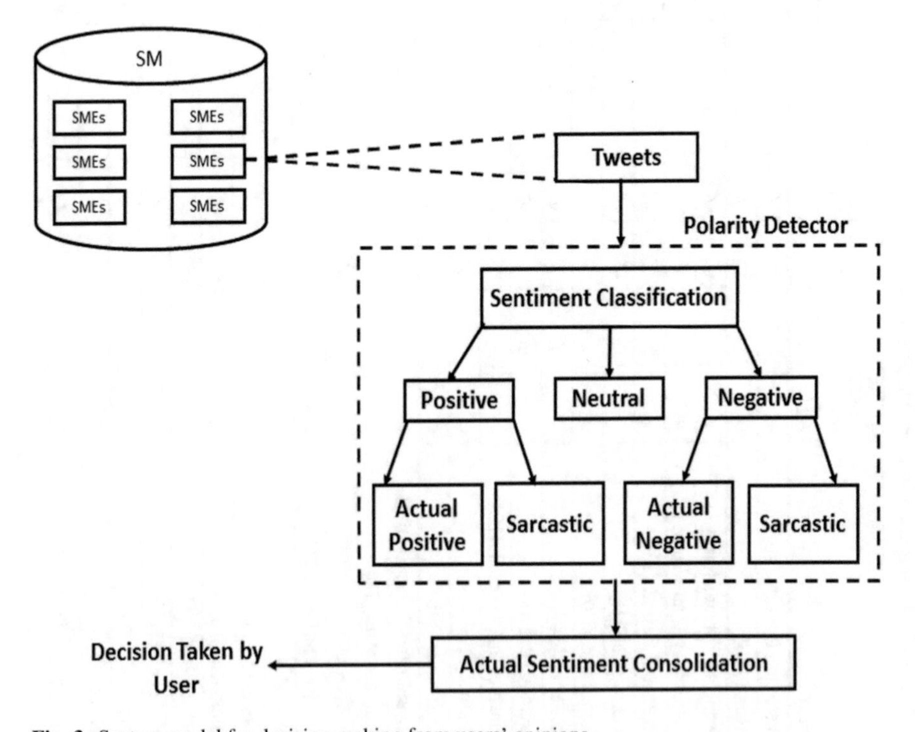

Fig. 3 System model for decision making from users' opinions

3. *Polarity detector (PD)*: An automated system which is capable of identifying the actual sentiment or sarcastic sentiment from text.

Whenever an event starts, or product is launched people start tweeting, writing reviews, posting comments, etc. on social media. People rely very much on these social media to get the perfect review about any product before they buy. Organizations are also dependent on these sites to know the marketing status of their product and subsequently improve their product quality. However, finding and monitoring correct opinions or reviews about SMEs remains a formidable task. It makes difficult for humans to read all the reviews and decode the sarcastic opinions. Also, the average human reader will have difficulty in recognition of sarcasm in twitter data, product review, etc., which misleads the individuals or organizations. To avoid this, there is a need for an automated sarcasm recognition system. Figure 3 shows how to find opinions automatically from SMEs. Tweet is one of the SMEs which passes through a polarity detector to get the opinions. Polarity detector classifies it as either negative, positive, or neutral. If the tweet is classified as either positive or negative, then further checks are required to see whether it has actual positive/negative sentiment or sarcastic opinion. Based on the result of the polarity detector, users can take the decision.

3.2 Part-of-Speech (POS) Tagging

It is a process of taking a word from a text (corpus) as input and assign corresponding part-of-speech information to each word as output based on its definition and context, i.e. relationship with adjacent and related words in a phrase, sentence, or paragraph. In this chapter, TEXTBLOB is used to identify POS tagging of the given text. It is a Python based package working on NLTK (Natural Language Toolkit), and freely available on the internet. Penn Treebank tag set notations [22] such as JJ-adjective, NN-noun, RB-adverb, VB-verb, and UH-interjection, etc. were used.

3.3 Parse Tree Generation

Parsing is the process of analyzing the grammatical structure of a language. Given a sequence of words, a parser forms units like subject, verb, object and determines the relations between these units, according to the rules of a formal grammar and generate a parse tree. An example of parse tree generation of the tweets "I love waiting forever for my doctor" and "I hate Australia in cricket because they always win" are shown in Fig. 4.

The notations S, NP, VP, ADJP, ADVP denote statement, noun phrase, verb phrase, adjective phrase, and adverb phrase, respectively. To generate a parse tree, one requires the parsed data. Here, TEXTBLOB was used to find parse data of

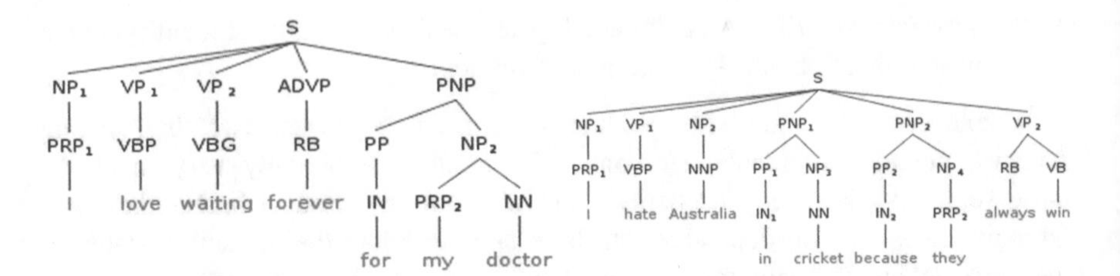

Fig. 4 Parse tree for tweet (*left*) "I love waiting forever for my doctor" and (*right*) "I hate Australia in cricket, because they always win"

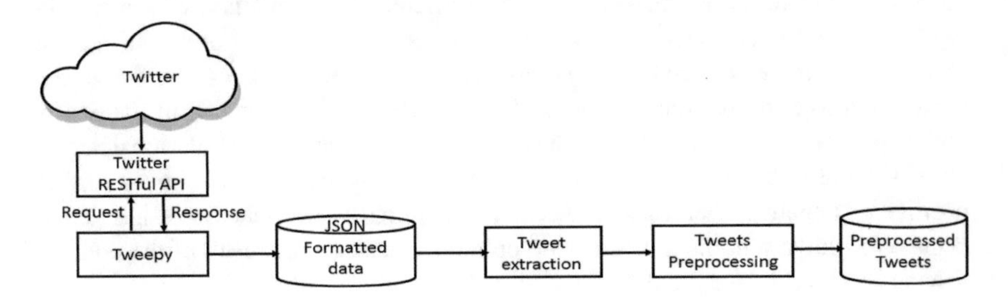

Fig. 5 Pipeline for tweets collection and preprocessing

tweets. The output is given by TEXTBLOB for the above tweet (I love waiting forever for my doctor) is I/PRP/B-NP/O, love/NN/I-NP/O, waiting/VBG/B-VP/O, forever/RB/B-ADVP/O, for/IN/B-PP/B-PNP, my/PRP$/B-NP/I-PNP, doctor/NN/I-NP/I-PNP. A parse tree as shown in Fig. 4(left) was generated by these tags. It shows the tweet type (T1) mentioned in Table 2 as "I love" indicating positive sentiment phrase and "waiting forever" showing adverse situation phrase, whereas Fig. 4(right) shows the tweet type (T2) mentioned in Table 2 as "hate" indicating negative sentiment and "always win" indicating a positive situation.

4 Data Collection and Preprocessing

RESTful is a Twitter developer API which provides an easy access of tweets stored in Twitter server. Tweepy is an open-sourced tool that enables Python to communicate with Twitter platform and use its API. In this work, we used Tweepy which provides an easy access to twitter's RESTful API for tweets collection from Twitter server as shown in Fig. 5.

4.1 Data Collection

We collected four sets of tweets to train the classifiers which are as follows:

1. To get the learned list of negative sentiment and positive situation and vice versa using PBLGA, a corpus of 200,000 tweets were collected from Twitter. Out of these 200,000 tweets, 100,000 tweets are sarcastic and remaining 100,000 tweets are regular. Sarcastic tweets were collected using keywords # sarcasm, # sarcastic, love, amazing, good, hate, sad, happy, bad, hurt, awesome, excited, nice, great, sick, etc.
2. Similarly, to get the likes and dislikes learned list of different users using LDC, we collected tweets from 50 Twitter users' accounts using the name of the user.
3. To get the (key, value) pair using TCUF and TCTDF for universal facts and temporary facts respectively, a corpus of 10,000 universal fact sentences and a corpus of 25,000 temporary facts sentences were collected. Temporary facts were gathered using some events, sports, festivals, etc.

For testing, a separate set of tweets for each proposed algorithm were collected as follows.

1. For PBLGA, totally 3000 tweets are collected, among these 1500 tweets with the sarcastic hashtag and 1500 random tweets.
2. For LDC, 100 random tweets were collected from each user's accounts.
3. For TCUF, 5000 random tweets are collected.
4. For TCTDF, 5000 random tweets (with timestamp) were collected.

4.2 Preprocessing

All the collected tweets were initially in JavaScript Object Notation (JSON) format, a language-independent open data format that uses human-readable text to express data objects consisting of attribute-value pairs. JSON data can be easily parsed with a wide variety of programming languages including JavaScript, PHP, Python, Ruby, and Java. In this work, we have used Python script to extract text from JSON tweets and removed RT (re-tweets), hashtags, URL, @user, uppercase word to lowercase from the respective tweet corpus as shown in Algorithm 1.

In Algorithm 1, Steps 1–4 extracts text part from JSON formatted tweets. We used function json.load() to extracts text from tweets. Next, Steps 5–10 is used to extract re-tweets, URL, hashtag, and @user from tweets using Python regular expression library and replaces with empty string.

Algorithm 1: `JSON_tweet_parsing_and_preprocessing`

Input: Tweet corpus in JSON format.
Output: Corpus of preprocessed tweets.
Notation: *re*: regular expression python library, *T*: tweets, *C*: corpus, *RRT*: remove re-tweet, *RHT*: remove hash tag, *RUN*: remove username, *RURL*: remove URL, *PTF*: parse tweet file
Initialisation : *PTF* = { ϕ }

```
1: for T in C do
2:     X = json · load (T)
3:     PTF ← X['text']
4: end for
5: for T in PTF do
6:     RRT = re.findall(T, "RT @\ w+")
7:     RURL = re.findall(T, "http (s)?://( · || [a-z] || [A-Z] || [0-9])+")
8:     RHT = re.findall(T, "#\ w+")
9:     RUN = re.findall(T, "@\ w+")
10:    replace RRT, RURL, RHT, RUN with empty strings
11: end for
```

5 Proposed Scheme

This section deals with the design goals, tweets collection, and the proposed algorithms. In all of the previous works on identifying sarcasm, efforts were made to design solutions that meet the few requirements of various types of sarcasm occurring in Twitter data as mentioned in Table 1. We observed that the authors have not yet worked on tweet types T2, T3, T5, T6, and T7 which may be due to lack of sufficient datasets availability. Most of the sarcastic tweets found in Twitter cover type T1 and T3. Therefore, majority of researchers considered T1 type tweets for their research. Regarding type T3, Lunando and Purwarianti [21] only discussed for Indonesian social media. If any tweet contains interjection word as wow, nah, oh, etc., then the tweet is probably sarcastic. For remaining tweet types (T4, T5, T6, and T7), finding sufficient dataset is difficult. In previous work [3], we proposed two algorithms namely PBLGA and IWS to cover tweets type T1, T2, and T3 for sarcasm detection. In this work, three additional algorithms were proposed to cover tweets type T4, T5, and T7 for sarcasm identification. The proposed algorithms are given below:

1. A `Parsing_based_lexical_generation_algorithm`(PBLGA) for sarcasm identification in Twitter data to recognize sarcasm in both types of tweets: the contradiction of negative sentiment and positive situation (T1) and contradiction of positive sentiment and negative situation (T2).
2. A behavioral based algorithm, `Likes_dislikes_contradiction`(LDC), is proposed to detect sarcasm, in particular, user's tweet that meets the requirements of type T4.

3. An algorithm, `Tweet_contradict_universal_facts`(TCUF), is proposed to detect sarcasm in tweets that intentionally negate the universal facts which serve the requirement of type T5.
4. An algorithm, `Tweet_contradict_time_dependent_facts`(TCTDF), that detects sarcasm in tweets which intentionally negate the time-dependent facts will serve the requirement of type T7.

5.1 PBLGA

PBLGA generates the lexicon to identify sarcasm in tweets in which sentiment contradicts with the tweet's situation. Generated lexicon contains the phrase in four categories, namely positive sentiment, negative situation, negative sentiment, and positive situation. PBLGA is given in Algorithm 2.

According to Algorithm 2, a parse tree divides a text into a set of a noun phrase (NP), verb phrase (VP), adjective phrase (ADJP), and adverb phrase (ADVP). According to Penn Treebank, verb (V) can be in any form as VB, VBD, VBG, VBN, VBP, and VBZ. Adverb (ADV) can be in any form as RB, RBR, WRB, and RBS. An adjective can take any form as JJ, JJR, and JJS. Noun (N) can take any form as NN, NNS, NNP, and NNPS. In the proposed algorithm, Step 2 does parsing of tweets in corpus and stores it tweet-wise (parse of one tweet in one line and the next tweet in another line) in a parse file. Step 6 finds the subset of each tweet in parse file. If a subset of parse contains either noun phrase followed by verb phrase or adjective phrase or only noun phrase, then add the phrase into the sentiment file. In the way of learning sentiment phrase, singular proper noun (NNS) and plural proper noun (NNPS) are not considered as it doesn't contain any sentiment. Similarly, in the verb phrase, gerund or present participle verb (VBG) are not considered. To learn situation phrase, there is no such restriction. If subset of parse tree contains either verb phrase or adverb phrase followed by verb phrase or verb phrase followed by adverb phrase or adjective phrase followed by verb phrase or verb phrase followed by noun phrase or verb phrase followed by adverb phrase followed by adjective phrase or verb phrase followed by adjective phrase followed by noun phrase or adverb phrase followed by adjective phrase followed by noun phrase, then the algorithm appends the phrase into situation file as shown in Algorithm 2.

To identify sentiment in given phrase, we used pre-defined lists of positive and negative words such as sentiwordnet [9]. It is a standard list of positive and negative English words. Using sentiword list, one can find the total number of positive and negative words in given phrase or sentence. Finally, sentiment score of given phrase is calculated as follows:

$$\mathrm{PR} = \frac{\mathrm{PWP}}{\mathrm{TWP}} \tag{1}$$

Algorithm 2: `Parsing_based_lexical_generation_algorithm` (PBLGA)

Input: Tweet corpus (C)
Output: Lexicon for negative sentiment, positive sentiment, negative situation and positive situation.
Notation: *VP*: verb phrase, *NP*: noun phrase, *ADJP*: adjective phrase, *ADVP*: adverb phrase, *PF*: parse file, *TWP*: tweet wise parse, *SF*: sentiment file, *sf*:situation file, *PSF*: positive sentiment file, *NSF*: Negative sentiment file, *psf*: positive situation file, *nsf*: negative situation file, *SC*: sentiment score, *P*: phrase, *T*: Tweet.
Initialisation : $SF = \{ \phi \}$, $sf = \{ \phi \}$, $PSF = \{ \phi \}$, $NSF = \{ \phi \}$, $psf = \{ \phi \}$, $nsf = \{ \phi \}$,

```
1:  for T in C do
2:      k = find_parse (T)
3:      PF ← TF ∪ k
4:  end for
5:  for TWP in PF do
6:      k = find_subset (TWP)
7:      if k = NP || ADJP || (NP + VP) then
8:          SF ← SF ∪ k
9:      else if k = VP || (ADVP + VP) || (VP + ADVP) || (ADJP + VP) || (VP + NP) ||
        (VP + ADVP + ADJP) || (VP + ADJP + NP) || (ADVP + ADJP + NP) then
10:         sf ← sf ∪ k
11:     end if
12: end for
13: for P in SF do
14:     SC = sentiment_score (P)
15:     if SC >0.0 then
16:         PSF ← PSF ∪ P
17:     else if SC <0.0 then
18:         NSF ← NSF ∪ P
19:     else
20:         Neutral Sentiment Phrase
21:     end if
22: end for
23: for P in sf do
24:     SC = sentiment_score (P)
25:     if SC >0.0 then
26:         psf ← psf ∪ P
27:     else if SC <0.0 then
28:         nsf ← nsf ∪ P
29:     else
30:         Neutral Situation Phrase
31:     end if
32: end for
```

Table 3 Sample learned phrases by PBLGA

Positive sentiment phrase: Great, lucky, cute, good, nice, happy, glad, delicious, best, awesome, perfect, joy, strong, pretty, proud, hilarious, better, gorgeous, honest, innocent, talented, excellent, lovely, outstanding, bright, elegant, easy, supportive, attractive, huge, interesting, successful, healthy, famous, pleasant, beloved, enjoy, appreciate, etc.
Negative sentiment phrase: terrible, little bit, ugly, excuse, dirty, little, expensive, half, cold food, tight jeans, troubled, least, hard, rude, wrong, dramatic, abusive, unhealthy, crap, bad, a few days, sick, dumb, pathetic, victim, useless, fluffy, violent, Poor, brutal, worst, crazy, etc.
Positive situation phrase: no regret, love, love seeing, just love discovering, just love, effectively making, just love living, absolutely LOVE, feel so loved, wanna be loved, winning, love falling, honestly tell, will fly, now enjoying managing, most entertaining, enjoying, thrilled, delighted, really enjoying, really is thrilled, are winning, most enjoy, most loved, enjoyed looking, feel so satisfied, proudly displayed, have liked, so impressed, wisely locked, is absolutely delighted, just loved getting being brilliantly led, feel loved, greatly appreciate, gladly appreciate, most appreciate, sincerely appreciate, very excited, good news, etc.
Negative situation phrase: Clarifying, are pumped, will be released, are arriving, babysitting, only run, kicking, gets stuck, is losing, crashing, destroyed, attacking, criticizing, is lying, biting, confused, dividing, exhausted, keep arguing, shouting, awkwardly standing, are limping, needs help, forgotten, can barely walk, already upset, crying, get dropped, feel tired, banning, stole, so upset, etc.

$$\mathrm{NR} = \frac{\mathrm{NWP}}{\mathrm{TWP}} \tag{2}$$

$$\text{Sentiment Score} = \mathrm{PR} - \mathrm{NR} \tag{3}$$

where

PR = positive ratio, NR = negative ratio, PWP = number of positive words in a given phrase, NWP = number of negative words in a given phrase, TWP = total words in a given phrase.

Further, to separate negative sentiment and positive sentiment from sentiment file and negative situation and positive situation from situation file, sentiment score criteria is applied. Equations (1) and (2) give a positive ratio and negative ratio for the given phrase, respectively. Finally, Eq. (3) gives the sentiment score of the phrase. If the sentiment score is greater than 0.0 in sentiment file, the algorithm appends it to a positive sentiment file, otherwise append in negative sentiment file. Similarly, if the sentiment score is higher than 0.0 in situation file, the algorithm appends it to the positive situation file, otherwise append in the negative situation file. If the sentiment score is exactly 0.0, then we decide manually based on text context for both sentiment and situation file. Hence, Algorithm 2 gives the learned list of phrases in four categories as shown in Table 3.

5.2 LDC

In this approach, behavioral based features are used for sarcasm identification such as likes and dislikes of individuals. Analyzing sarcasm in tweets, based on user's likes and dislikes contradiction (T4), one can analyze past tweets of particular user's Twitter account and observe his/her likes and dislikes habit. Based on these likes and dislikes lists of a particular user, one can easily identify whether the particular user's tweet is sarcastic or not. For example, if a fan of Sachin Tendulakar tweeting about Sachin and his Twitter account consists a tweet like "I love to see Sachin failure in batting," which contradicts his like's habit, then it's easily identifiable that the given tweet is sarcastic as shown in Algorithm 3. To get the learned lists of likes and dislikes, one can find all tweets, in particular, user's profile on Twitter and save as a corpus. Next, use the corpus as input to Algorithm 3. In the algorithm, steps 1–10 finds sentiment scores of all tweets in a corpus using Eqs. (1)–(3). Then, it checks the sentiment score for every tweet. If the sentiment score is positive, then append it to a positive sentiment tweet file. Similarly, if the sentiment score is negative, then append it to negative sentiment tweet file, otherwise discard the tweets. Now, there are two files, namely positive sentiment tweet file and negative sentiment tweet file.

Next, in steps 11–14, one can find POS tag for the positive sentiment tweet file and save it as a positive sentiment tag file (PSTF). In the steps 15–25, one can find a subset of every tweet in PSTF and obtain first tag (FT) value. If the first tag is a personal pronoun (like I, We ...), then find the object value of that tweet using Rusu_Triplets [30] method (it obtains subject, verb, and object of any sentence) and add into likes list (LL). Otherwise, find the subject value of that tweet using the same method and add it to likes list (LL). In the steps 26–40, one can apply the same rule as above for the negative sentiment tweet file, and the result is added to dislikes list (DLL). Hence, Algorithm 3 provides learned list of likes and dislikes of the particular Twitter user as shown in Table 4.

For testing a tweet from same user's profile, one can find sentiment scores of a particular tweet and then find a subject or object value according to Algorithm 3. If the sentiment score is positive, then check tweet's subject/object value in the user's likes list (LL). If present, then tweet is not sarcastic, otherwise, check it in the user's dislikes lists (DLL). If it is present in DLL, then tweet is sarcastic. Similarly, if the sentiment score is negative and subject/object value present in the user's likes lists (LL), then tweet is sarcastic, otherwise it is not sarcastic. If the sentiment score is zero, then discard the tweet. In the same way, one can form likes and dislikes list for many user profiles and test for sarcasm in their tweets.

Algorithm 3: `Likes_and_Dislikes_Contradiction` (LDC)

```
Input: Tweet corpus of any individual.
Output: List of likes and dislikes behaviour of same individual.
   Notation: N: Noun, T: Tweets, C: Corpus, PSTF: Positive
   sentiment tag file, NSTF: Negative sentiment tag file, t:
   tag, TWT: Tweet wise tag, FT: First tag, pstf: Positive
   sentiment tweet file, nstf: Negative sentiment tweet file, LL:
   Likes list, DLL: Dislikes list,.
   Initialisation : pstf = { ϕ } nstf = { ϕ } NSTF = { ϕ } PSTF = { ϕ } LF = { ϕ }
   DLF = { ϕ }
 1: for T in C do
 2:    SC = find_sentiment_score (T)
 3:    if SC >0.0 then
 4:       pstf ← pstf ∪ T
 5:    else if SC <0.0 then
 6:       nstf ← nstf ∪ T
 7:    else
 8:       Discard the tweet.
 9:    end if
10: end for
11: for T in pstf do
12:    k = find_postag (T)
13:    PSTF ← PSTF ∪ k
14: end for
15: for TWT in PSTF do
16:    t = find_subset (TWT)
17:    FT = find_first_tag (t)
18:    if (FT = PRP) then
19:       S = find_object (TWT)
20:       LL ← LL ∪ S
21:    else
22:       S = find_subject (TWT)
23:       LL ← LL ∪ S
24:    end if
25: end for
26: for T in nstf do
27:    k = find_postag (T)
28:    NSTF ← NSTF ∪ k
29: end for
30: for TWT in NSTF do
31:    t = find_subset (TWT)
32:    FT = find_first_tag (t)
33:    if (FT = PRP) then
34:       S = find_object (TWT)
35:       DLL ← DLL ∪ S
36:    else
37:       S = find_subject (TWT)
38:       DLL ← DLL ∪ S
39:    end if
40: end for
```

Table 4 Sample learned phrases by LDC

User	Likes	Dislikes
Santosh	Cricket	Swimming
Sumit	Sachin Tendulkar	Tennis
Rahul	Mango	Newspaper
Rakesh	Taj Mahal	Football
Rakesh	London	Shopping
Kranti	Movie	Watermelon
Kranti	Playing	Walking
Lakshmi	Traveling	Sweets
Deepak	Sleeping	TV serial

Algorithm 4: `Tweet_contradict_universal_facts`(TCUF)

Input: `Corpus of universal facts.`
Output: `Classification of tweets as sarcastic or not sarcastic.`
`Notation:` S: `Subject,` V: `verb,` O: `Object,` T: `tweets,` C: `corpus,` PF: `parse file,` TWP: `tweet wise parse phrase.`
Initialisation : $PF = \{ \phi \}$

1: **for** T in C **do**
2: p = `find_parsing` (T)
3: $PF \leftarrow PF \cup p$
4: **end for**
5: **for** TWP in PF **do**
6: S = `find_subject` (TWP)
7: V = `find_verb` (TWP)
8: O = `find_object` (TWP)
9: `forms` $\langle Key, Value \rangle$ `pair using S,V,O`
10: `Key` $\leftarrow$ `(S + V)`
11: `Value` $\leftarrow$ `O`
12: **end for**

5.3 TCUF

In this approach, facts are used to identify sarcasm in tweets that may be universal facts or time-dependent facts. A universal fact never change. For instance "Sun rises in the east" (T5) while time-dependent facts may change over a particular period such as "@MirzaSania becomes world number one. Great day for Indian tennis" (T7). In this TUCF algorithm, we used universal facts to identify sarcasm as shown in Algorithm 4. To identify sarcasm in tweets using universal facts, one can extract triplet value (subject, object, verb) from each tweet using Rusu_Triplets [30] method and forms $\langle key, value \rangle$ pair list for all the tweets which contains universal facts in the tweet corpus. Here, subject and verb are together called "key" and object is called "value." A sample $\langle key, value \rangle$ pair using TCUF is shown in Table 5. We train the system using this $\langle key, value \rangle$ pair. To test a tweet, extracts triplets of given tweet that forms $\langle key, value \rangle$ pair, and if key is matched with value in the trained system, the tweet is not sarcastic, otherwise, tweet is sarcastic.

Table 5 Sample learned phrases by TCUF

Key	Value
Mixture, Green, Orange	Blue
Sun, rises	East
Earth, revolves around	Sun

Table 6 Sample learned phrases by TCTDF

Key	Value
Holi, celebrates	March
India, won	World cup, 2011
India, won	World cup, 1983
Christmas, celebrates	December

In Algorithm 4, steps 1–4 finds parsing phrases such as noun phrase (NP), verb phrase (VP), adverb phrase (ADVP), etc. and append it to phrase file (PF). Steps 6–8 extract triplets $\langle subject, verb, object\rangle$ from parse phrase. Steps 9–11 forms $\langle key, value\rangle$ pair using these triplets where "key" contains subject (S) and verb (V) while "value" contains only objects (O). Algorithm 4 provides a learned list of universal facts in the form of $\langle key, value\rangle$ pair.

5.4 *TCTDP*

In this approach, the time-dependent facts are used to identify sarcasm in tweets. To identify sarcasm in tweets using time-dependent facts, one can extract triplets value (subject, object, verb) from each tweet using Rusu_Triplets [30] method and forms $\langle key, value\rangle$ pair list for all the tweets which contain time-dependent facts in the tweet corpus. Here, subject and verb are together called "key" and object and time-stamp are called "value." A sample $\langle key, value\rangle$ pair using TCUF is shown in Table 6. We train the system using the $\langle key, value\rangle$ pair. While testing a tweet, extract triplet of a given tweet and form $\langle key, value\rangle$ pair. If key is matched with the value in the trained system, the tweet is not sarcastic, otherwise it is sarcastic.

In Algorithm 5, steps 1–4 finds parsing phrases such as noun phrase (NP), verb phrase (VP), adverb phrase (ADVP), etc. and append it to phrase file (PF). Steps 6–8 extracts triplets (subject, verb, object) from parse phrase. Steps 9–11 forms $\langle key, value\rangle$ pair using these triplets where "key" contains the subject (S) and verb (V) while "value" contains an object (O) and time-stamp (TS). Hence, Algorithm 5 gives a learned list of time-dependent facts in the form of $\langle key, value\rangle$ pair.

6 Classifiers

Classifiers are used to classify the tweets into various possible classes. There are two approaches to classifying sarcastic sentiment, namely binary sarcastic

Algorithm 5: `Tweet_contradict_time_dependent_facts` (TCTDF)

```
Input: Corpus of temporary facts.
Output: Classification of tweets as sarcastic or not sarcastic.
   Notation: S: Subject, V: verb, O: Object, TS: Time-stamp, T:
   tweets, C: corpus, PF: parse file, TWP: tweet wise parse
   phrase.
   Initialisation : PF = { φ }
 1: for T in C do
 2:    p = find_parsing (T)
 3:    PF ← PF ∪ p
 4: end for
 5: for TWP in PF do
 6:    S = find_subject (TWP)
 7:    V = find_verb (TWP)
 8:    O = find_object (TWP)
 9:    forms ⟨Key, Value⟩ pair using S,V,O
10:    Key ← (S + V)
11:    Value ← (O + TS)
12: end for
```

sentiment classification and multi-class sarcastic sentiment classification. In binary sarcastic sentiment classification, each tweet of the corpus is classified into two classes, i.e., either as sarcastic or as non-sarcastic. In multi-class sarcastic sentiment classification, each tweet can be classified into more than two classes, i.e., as strong positive or sarcastic positive or neutral or sarcastic negative or strong negative [33]. In this work, implementation is done on binary sarcastic sentiment classification.

Here, four supervised machine learning algorithms, namely SVM, NB, ME, and DT are used for classification.

1. Support vector machine (SVM) Classifier: SVM is a non-probabilistic binary linear classifier [11]. Here, SVM model represents each tweet in vectorized form as a data point in space. This method is used to analyze the completely vectorized data and the fundamental idea behind the training of the model is to find a hyperplane represented by w. The set of textual data vectors are said to be optimally separated by hyperplane only when it is separated without error, and the distance between closest points of each class and hyperplane is maximum. After training of the model, the testing tweets are mapped into same space and predicted to belong to a class based on which side of the hyper-plane they fall on.
2. Naive Bayes (NB) Classifier: It is a probabilistic classifier which uses the properties of Bayes theorem assuming the substantial independence between the features [23]. One of the advantages of this classifier is that it requires a small amount of training data to calculate the parameters for prediction. Instead of calculating the complete co-variance matrix, the only variance of the feature is computed because of independence of features.

3. Maximum entropy (ME) Classifier: ME is an alternative technique which has proven useful in some natural language processing applications [2]. McCallum and Nigam [23] show that it sometimes, but not always, outperforms Naive Bayes for standard text classification.
4. Decision tree (DT) Classifier: DT is among the most widely used machine learning algorithms. They perform a general to the specific search of a feature space, adding the most informative features to a tree structure as the search proceeds [25]. The objective is to select a minimal set of features that efficiently partitions the feature space into classes of observations and assemble them into a tree. In this case, the observations are manually sense for Twitter data.

The supervised machine learning algorithms are applicable where the labelled dataset is available. To get the labelled dataset for these classifiers, the above proposed algorithms were used. The algorithms give learned list of phrase for different types of sarcasm occurring in tweets as presented in Table 2. The output of the above proposed algorithms is used to train these four classifiers as mentioned earlier. A sample learned phrases by PBLGA, LDC, TCUF, and TCTDF are shown in Tables 3, 4, 5, and 6, respectively. These phrases are directly used to train different classifiers to detect sarcasm in Twitter data.

7 Results and Discussion

This section describes the performances of proposed algorithms to extract various features like lexical, hyperbole, behavioral, and universal facts followed by classification using various machine learning approaches mentioned in Sect. 6 to identifying sarcasm in Twitter data. The attained results are compared with the state-of-the-art work.

7.1 *Experimental Results*

Experimental results show the performance of proposed algorithms towards feature extraction to train the classifiers. Table 3 shows sample learned phrases by PBLGA in four categories, namely Positive sentiment phrase from the positive sentiment file (PSF), Negative sentiment phrase from the negative sentiment file (NSF), Positive situation phase from positive situation file (psf), Negative situation phase from negative situation file (nsf).

Once the features are extracted, we train various machine learning approaches mentioned in Sect. 6 using the extracted features to perform classification task. To test how effectively these classifiers perform, four statistical parameters, namely *accuracy*, *precision*, *recall*, and *F-score* were considered. *Precision* is a statistical

Table 7 Testing results of PBLGA for each classifier in terms of *accuracy*, *precision*, *recall*, and *F-score*

Classifiers	Accuracy	Recall	Precision	F-score
Support vector machine	0.57	0.67	0.68	0.67
Naive Bayes	0.55	0.43	0.41	0.42
Maximum entropy	0.60	0.47	0.66	0.55
Decision tree	0.63	0.76	0.52	0.57

Table 8 Testing results of LDC (one user account's only) for each classifier in terms of *accuracy*, *precision*, *recall*, and *F-score*

Classifiers	Accuracy	Recall	Precision	F-score
Support vector machine	0.45	0.41	0.57	0.47
Naive Bayes	0.55	0.29	0.43	0.34
Maximum entropy	0.44	0.42	0.44	0.43
Decision tree	0.48	0.45	0.47	0.46

parameter that shows how many relevant pieces of information are identified. Similarly, *recall* shows how many pieces of extracted information are relevant. *F-score* is the harmonic mean of *precision* and *recall*. *Accuracy* determines how well these classifiers are working. *Accuracy*, *precision*, *recall*, and *F-score* are calculated using Eqs. (4), (5), (6), and (7).

$$Accuracy = \frac{T_p + T_n}{\text{Total tweets}} \tag{4}$$

$$Precision = \frac{T_p}{T_p + F_p} \tag{5}$$

$$Recall = \frac{T_p}{T_p + F_n} \tag{6}$$

$$F\text{-}Score = \frac{2 * Precision * Recall}{Precision + Recall} \tag{7}$$

where T_p = True positive, F_p = False positive, F_n = False negative.

Using the sample learned phrases as shown in Table 3, PBLGA was tested for each classifier in terms of *accuracy*, *precision*, *recall*, and *F-score* and the results are given in Table 7. Similarly, Tables 4, 5, and 6 show the sample learned phrases to test the results of LDC, TCUF, and TCTDF, respectively. The testing results for each classifier in terms of *accuracy*, *precision*, *recall*, and *F-score* are given in Tables 8, 9, and 10, respectively.

Table 9 Testing results of TCUF for each classifier in terms of *accuracy*, *precision*, *recall*, and *F-score*

Classifiers	Accuracy	Recall	Precision	F-score
Support vector machine	0.26	0.23	0.33	0.27
Naive Bayes	0.33	0.27	0.41	0.32
Maximum entropy	0.38	0.33	0.42	0.37
Decision tree	0.41	0.37	0.49	0.42

Table 10 Testing results of TCTDF for each classifier in terms of *accuracy*, *precision*, *recall*, and *F-score*

Classifiers	Accuracy	Recall	Precision	F-score
Support vector machine	0.28	0.24	0.38	0.29
Naive Bayes	0.39	0.33	0.43	0.37
Maximum entropy	0.43	0.37	0.46	0.41
Decision tree	0.48	0.41	0.52	0.46

A comparison of the proposed scheme with the state of the art is shown in Table 11. One can observe that PBLGA attains better result with SVM classifier as indicated in bold. However, for ME and DT, *recall* and *precision* value, respectively, attains better. Algorithms 3, 4, and 5 are proposed first time for sarcasm detection with respective features.

8 Conclusion

Sarcasm recognition is an extremely challenging work using machine learning approach. In this chapter, we make use of the supervised machine learning approach to identify sarcasm in tweets and analyzed the performances of various classifiers, namely Naive Bayes, SVM, decision tree, and maximum entropy. Earlier authors used SVM to identify sarcasm. Here, few more classifiers were used to analyze the performance of various machine learning approaches. The proposed algorithms attain a much better result set as compared to the previously existing work in this domain. The PBLGA approach gives us highest accuracy and recall using the decision tree and lowest for the Naive Bayes classifier. The Naive Bayes classifier gives consistently worse performance for all the three out of four approaches we have used and the decision tree gives us the best results for the remaining three. Thus we see that the approaches examined by us are much better suited to resolving the sentiment from the tweets we acquire and detecting the sarcasm element if present. In future, we will target to detect sarcasm in speech data and image data. Sarcasm detection is still open for research in several domains other than Twitter data.

Table 11 Comparison of proposed methods with state-of-art

Approach	*Precision*	*Recall*	*F-score*
Tungthamthiti et al. with Contradiction in sentiment score	0.46	0.52	0.49
with proposed features in SVM	0.53	0.54	0.53
with Uni-gram features in SVM	0.63	0.62	0.63
Riloff et al. system with positive verb	0.23	0.42	0.30
with negative situation	0.28	0.37	0.33
with SVM	0.37	0.67	0.47
PBLGA with SVM	**0.67**	**0.68**	**0.67**
with Naive Bayes	0.43	0.41	0.42
with Maximum Entropy	0.47	**0.66**	0.55
with Decision Tree	**0.76**	0.52	0.57
LDC with SVM	0.41	0.57	0.47
with Naive Bayes	0.29	0.43	0.34
with Maximum Entropy	0.42	0.44	0.43
with Decision Tree	0.45	0.47	0.46
TCUF with SVM	0.23	0.33	0.27
with Naive Bayes	0.27	0.41	0.32
with Maximum Entropy	0.33	0.42	0.37
with Decision Tree	0.37	0.49	0.42
TCTDF with SVM	0.24	0.38	0.29
with Naive Bayes	0.33	0.43	0.37
with Maximum Entropy	0.37	0.46	0.41
with Decision Tree	0.41	0.52	0.46

References

1. Barbieri, F., Saggion, H., Ronzano, F.: Modelling sarcasm in twitter, a novel approach. In: Proceedings of the 5th Workshop on Computational Approaches to Subjectivity, Sentiment and Social Media Analysis, pp. 50–58. ACL, Baltimore (2014)
2. Berger, A.L., Pietra, V.J.D., Pietra, S.A.D.: A maximum entropy approach to natural language processing. Computational Linguistics. **22**, 39–71 (1996)
3. Bharti, S.K., Babu, K.S., Jena, S.K.: Parsing-based sarcasm sentiment recognition in twitter data. In: International Conference on Advances in Social Networks Analysis and Mining (ASONAM), France, pp. 1373–1380. IEEE/ACM, New York (2015)
4. Bharti, S.K., Vachha, B., Pradhan, R.K., Babu, K.S., Jena, S.K.: Sarcastic sentiment detection in tweets streamed in real time: a big data approach. Digit. Commun. Netw. **2**(3), 108–121 (2016); Elsevier
5. Bhattacharyya, P., Verma, N.: Automatic lexicon generation through wordnet. In: International WordNet Conference (GWC), Brno, pp. 226–233 (2004)
6. Carvalho, P., Sarmento, L., Silva, M. J., De Oliveira, E.: Clues for detecting irony in user-generated contents: oh...!! it's so easy;-. In: Proceedings of the 1st International CIKM Workshop on Topic-Sentiment Analysis for Mass Opinion, pp. 53–56. ACM, New York (2009)
7. Chaumartin, F.R.: UPAR7: a knowledge-based system for headline sentiment tagging. In Proceedings of the 4th International Workshop on Semantic Evaluations, pp. 422–425, ACL, Stroudsburg (2007)

8. Davidov, D., Tsur, O., Rappoport, A.: Semi-supervised recognition of sarcastic sentences in twitter and amazon. In Proceedings of the Fourteenth Conference on Computational Natural Language Learning, pp. 107–116, ACL, Stroudsburg (2010)
9. Esuli, A., Sebastiani, F.: Sentiwordnet: A publicly available lexical resource for opinion mining. In: Proceedings of LREC, vol. 6, pp. 417–422 (2006)
10. Filatova, E.: Irony and sarcasm: Corpus generation and analysis using crowdsourcing. In: LREC, pp. 392–398 (2012)
11. Gautam, G., Yadav, D.: Sentiment analysis of twitter data using machine learning approaches and semantic analysis. In: Seventh International Conference on Contemporary Computing (IC3), India, pp. 437–442. IEEE, New York (2014)
12. González-Ibánez, R., Muresan, S., Wacholder, N.: Identifying sarcasm in twitter: a closer look. In: Proceedings of the 49th Annual Meeting of the Association for Computational Linguistics: Human Language Technologies: short papers, vol. 2, pp. 581–586 (2011)
13. Ingle, A., Maheshwari, N., Sutrave, N., Akumarthi, S., Bhitre, T.: Sentiment analysis: sarcasm detection of tweets. B.Tech thesis, pp. 1–35. VNIT, Nagpur, India (2014)
14. Justo, R., Corcoran, T., Lukin, S.M., Walker, M., Torres, M.I.: Extracting relevant knowledge for the detection of sarcasm and nastiness in the social web. Knowl.-Based Syst. **69**, 124–133 (2014)
15. Kreuz, R.J., Caucci, G.M.: Lexical influences on the perception of sarcasm. In: Proceedings of the Workshop on Computational Approaches to Figurative Language, pp. 1–4. ACL, Stroudsburg (2007)
16. Kreuz, R.J., Roberts, R.M.: Two cues for verbal irony: hyperbole and the ironic tone of voice. Int. J. Metaphor Symb. **10**, 21–31 (1995)
17. Kunneman, F., Liebrecht, C., Van Mulken, M., Van den Bosch, A.: Signaling sarcasm: from hyperbole to hashtag. Inf. Process. Manag. **51**(4), 500–509 (2015)
18. Liebrecht, C.C., Kunneman, F.A., van den Bosch, A.P.J.: The perfect solution for detecting sarcasm in tweets# not. In: 4th Workshop on Computational Approaches to Subjectivity, Sentiment and Social Media Analysis, Georgia, pp. 29–37. ACL, New Brunswick (2013)
19. Liu, P., Chen, W., Ou, G., Wang, T., Yang, D., Lei, K.: Sarcasm detection in social media based on imbalanced classification. In: 15th International Conference on Web-Age Information Management, China, pp. 459–471. Springer, Cham (2014)
20. Lukin, S., Walker, M.: Really? Well apparently bootstrapping improves the performance of sarcasm and nastiness classifiers for online dialogue. In: Proceedings of the Workshop on Language Analysis in Social Media, Georgia, pp. 30–40. ACL, Atlanta (2013)
21. Lunando, E., Purwarianti, A.: Indonesian social media sentiment analysis with sarcasm detection. In: International Conference on IEEE on Advanced Computer Science and Information Systems (ICACSIS), Bali, pp. 195–198, IEEE, New York (2013)
22. Marcus, M.P., Marcinkiewicz, M.A., Santorini, B.: Building a large annotated corpus of English: the Penn Treebank. Computational Linguistics. **19**, 313–330 (1993)
23. McCallum, A., Nigam, K.: A comparison of event models for naive Bayes text classification. In: AAAI-98 Workshop on Learning for Text Categorization, vol. 752, pp. 41–48 (1998)
24. Pang, B., Lee, L., Vaithyanathan, S.: Thumbs up?: sentiment classifcation using machine learning techniques. In: Proceedings of the Association for Computational Linguistics Conference on Empirical Methods in Natural Language Processing, vol. 10, pp. 79–86 (2002)
25. Pedersen, T.: A decision tree of bigrams is an accurate predictor of word sense. In: Proceedings of the Second Meeting of the North American Chapter of the Association for Computational Linguistics on Language Technologies, pp. 1–8. ACL, Stroudsburg (2001)
26. Pennebaker, J.W., Francis, M.E., Booth, R.J.: Linguistic inquiry and word count: LIWC 2001, Mahway: Lawrence Erlbaum Associates, **71**, 1–22 (2001)
27. Pielage, A.J.: Detecting sarcasm in english text. Artificial Intelligence MSc thesis, pp. 1–48 (2014)
28. Rajadesingan, A., Zafarani, R., Liu, H.: Sarcasm detection on twitter: a behavioral modeling approach. In: Proceedings of the Eighth ACM International Conference on Web Search and Data Mining, pp. 97–106, ACM, New York (2015)

29. Riloff, E., Qadir, A., Surve, P., De Silva, L., Gilbert, N., Huang, R.: Sarcasm as contrast between a positive sentiment and negative situation. In: Proceedings of the Empirical Methods in Natural Language Processing, vol. 13, pp. 704–714 (2013)
30. Rusu, D., Dali, L., Fortuna, B., Grobelnik, M., Mladenic, D.: Triplet extraction from sentences. In: Proceedings of the 10th International Multiconference Information Society-IS, pp. 8–12 (2007)
31. Strapparava, C., Valitutti, A.: Wordnet affect: an affective extension of wordnet. In: LREC, vol. 4, pp. 1083–1086 (2004)
32. Tayal, D.K., Yadav, S., Gupta, K., Rajput, B., Kumari, K.: Polarity detection of sarcastic political tweets. In: Computing for Sustainable Global Development (INDIACom), Delhi, pp. 625–628, IEEE, New York (2014)
33. Tripathy, A., Agrawal, A., Rath, S.K.: Classification of sentimental reviews using machine learning techniques. Proc. Comput. Sci. **57**, 821–829 (2015)
34. Tsur, O., Davidov, D., Rappoport, A.: ICWSM-a great catchy name: semi-supervised recognition of sarcastic sentences in online product reviews. In: Proceedings of the Fourth International AAAI Conference on Weblogs and Social Media, pp. 162–169 (2010)
35. Tungthamthiti, P., Kiyoaki, S., Mohd, M.: Recognition of sarcasm in tweets based on concept level sentiment analysis and supervised learning approaches. In: Proceedings of Pacific Asia Conference on Language, Information and Computing, Thailand, pp. 404–413. ACL, Stroudsburg (2014)
36. Utsumi, A.: Verbal irony as implicit display of ironic environment: distinguishing ironic utterances from non irony. J. Pragmat. **32**, 1777–1806 (2000)
37. Verma, N., Bhattacharyya, P.: Automatic lexicon generation through wordnet. In: International WordNet Conference (GWC), Brno, pp. 226–233 (2004)

The DEvOTION Algorithm for Delurking in Social Networks

Roberto Interdonato, Chiara Pulice, and Andrea Tagarelli

1 Introduction

All large-scale on-line social networks are characterized by a "participation inequality" principle [5, 17, 18], i.e., a disequilibrium between the niche of super contributors and the crowd of silent users, which just observe ongoing discussions, read posts, watch videos, and so on. In other words, the real audience of an SN does not actively contribute; rather, it *lurks*. A *lurker* is hence a member of an SN who gains benefit from others' information and services without significantly giving back to the community [5, 18]. In this respect, a major goal is to *delurk* such users, i.e., to develop a mix of strategies aimed at encouraging lurkers to return their acquired social capital, through a more active participation to the community life. This might contribute to help sustain the SN over time with fresh ideas and perspectives.

The Challenge of Delurking Social science and human–computer interaction research communities have widely investigated the main causes that explain lurking behaviors—cf., e.g., survey works proposed in [5, 20] for general overviews on the topic. Lurking is often due to a subjective reticence (rather than malicious motivations) to contribute to the community wisdom; a lurker often simply feels that gathering information by browsing is enough without the need of being further involved in the community [18]. Lurking can be expected or even encouraged because it allows users (especially newcomers) to learn or improve their understanding of the etiquette of an online community [5]. Sun et al. [20] have identified four types of factors related to lurking, namely: environmental influence determined by the online community, personal preference related to an individual's personality, relationships between the individual and the community, and privacy/security

R. Interdonato • C. Pulice • A. Tagarelli (✉)
DIMES, University of Calabria, 87036 Arcavacata di Rende (CS), Italy
e-mail: rinterdonato@dimes.unical.it; cpulice@dimes.unical.it; andrea.tagarelli@unical.it

R. Missaoui et al. (eds.), *Trends in Social Network Analysis*, Lecture Notes in Social Networks, DOI 10.1007/978-3-319-53420-6_4

considerations. By contrast, few suggestions have been given about *how to turn lurkers into participants/contributors*. Delurking actions can be broadly categorized into four types [13, 20]:

- reward-based external stimuli (e.g., tangible or intangible rewards),
- providing encouragement information (e.g., welcome statements, introduction to the netiquette rules),
- improvement of the usability and learnability of the system (to make it easier for users to participate),
- guidance from elders/master users to help lurkers become familiar with the system as quickly as possible.

Within this view, an important finding of some studies in social sciences and human–computer interaction (e.g., [2]) is that *the trustworthiness or credibility that lurkers perceive with regard to some members of the community* can indeed represent a key psychological factor to persuade lurkers to change their silent status. In addition, a further incentive for delurking would be represented by a habit of the community of nurturing newcomers and novices: in fact, often lurkers hold a belief that they will be ignored if they contribute, and hence they should be more willing to be persuaded to become active if they find that newcomers/novices will receive a reply to their posts in a constructive way [2].

The above considerations prompted us to investigate an effective computational approach to define a delurking strategy, in which lurkers are nurtured and persuaded to be more actively engaged in the SN community by other users. Our key idea in this work is *to conceptualize a delurking approach under a graph-based information diffusion model*. Research on information diffusion in SNs (e.g., [1, 7, 19, 27]) is well established due to a plethora of methods that have been developed in the last years, mostly upon the two seminal models, namely Independent Cascade (IC) and Linear Threshold (LT) [11, 26]. These assume that an information diffusion process would unfold in a static, directed graph, where each node can be "activated" or not (under a progressive assumption) and each edge is associated with a weight expressing the diffusion probability or influence degree from one node to another. Moreover, in an *influence maximization* (IM) framework (e.g., [6, 11, 14, 25]) the general objective is to find a set of initially activated users (also called early adopters, or seed users) which can maximize the spread of information through the network.

Contributions In this work,[1] we address a special case of IM, namely *targeted IM*; in particular, we develop a specific instance of targeted IM in which lurkers are regarded as the target of the diffusion process. Existing *lurker ranking* algorithms [21–23] can profitably be exploited to mine lurking behaviors in the network, and hence to associate users with a value (*lurking score*) indicating her/his lurking status. Given a budget k, our goal is to find a set of k nodes that are

[1] An abridged version of this paper appeared in [10].

capable of maximizing the likelihood of "activating" the target lurkers. Intuitively, the activation of a node means that a user is influenced by other users so to "become aware of" or "adopt" a piece of information. We observe that such an outcome of a diffusion process has a nice analogy in the lurking analysis setting under consideration: the information to spread corresponds to a delurking trigger action, activating nodes regarded as lurkers can be seen as delurking those users, while the lurking score of a user would represent the gain deriving from delurking that user. Our main contributions are summarized as follows:

- We propose the first computational framework that addresses the problem of delurking in social networks.
- We define the *delurking-oriented targeted influence maximization* problem. A key novelty in the formulated optimization problem is the objective function, which is defined in terms of the cumulative amount of the lurking scores associated with the nodes in the final active set, or *delurking capital*.
- Our delurking-oriented targeted IM problem shares the computational intractability with classic IM problems. However, since the proposed objective function is shown to be monotone and submodular under the LT model, we provide a greedy algorithm (with typical $1 - 1/e - \epsilon$ approximation ratio), named DEvOTION, which computes a k-node set that maximizes the delurking capital in the network, for a given minimum lurking threshold. We also point out that, to the best of our knowledge, our approach is the first to address a targeted IM problem under the LT diffusion model.
- We evaluate DEvOTION over SN datasets of different characteristics and sizes, assessing its performance in terms of estimation of the delurking capital, execution time, and seed characteristics. We also compare DEvOTION with baseline methods, state-of-the-art IM algorithms, namely SimPath [6] and TIM+ [25], and a recently proposed query-based targeted IM algorithm, namely KB-TIM [15].

Empirical evidence has shed light on the significance and effectiveness of DEvOTION. Our algorithm has shown to be robust w.r.t. the pruning of paths to be explored in the graph. A significant fraction of delurking capital can be achieved already with a small seed set, even for large network datasets. Compared with baseline methods, state-of-the-art IM algorithms (i.e., SimPath, TIM+) and targeted IM algorithm (i.e., KB-TIM), DEvOTION always obtains better quality seed sets, confirming its efficacy and uniqueness in delurking-oriented targeted IM contexts.

The rest of the paper is organized as follows. Section 2 first provides background on lurker ranking methods, then discusses related work focusing on targeted influence maximization algorithms. Section 3 presents our delurking-oriented targeted influence maximization method. Experimental evaluation methodology and results are reported in Sects. 4 and 5, respectively. Section 6 concludes the paper.

2 Targeted Influence Maximization

There is a relatively small body of research on IM that involves some notion of *target* of the diffusion process. A few studies have assumed that the target is unique and a-priori specified. In [9], Guo et al. address the problem of finding the top-k most influential nodes for a specific target user, using the IC model. Guler et al. [8] investigate optimal propagation policies to influence a target user positively, i.e., in favor of a given idea or message. Given a signed SN, where positive and negative signs correspond to friends and foes, and a selected pair of source and target nodes, the goal is to find a path between the two nodes which is optimal in terms of end-to-end propagation delay. The problem is solved via backward-induction dynamic programming. Yang et al. [28] address the problem of acceptance probability maximization, whereby a selected user (called initiator) wants to send a friendship invitation to a selected target which is not socially close to the initiator (i.e., the two nodes have no common friends). The goal is to find a set of nodes through which the initiator can best approach the target. Two main issues are how to model the invitation acceptance of users and how to compute the acceptance probabilities; the former is addressed through a topology-based social influence factor, while the latter is addressed using an extended IC model.

It should be noted that the above single-target IM problems stem from perspectives different from ours. Our proposed approach aims at maximizing the probability of activating a target set which can be arbitrarily large, by discovering a seed set which is neither fixed and singleton (unlike [8, 9, 28]) nor has constraints related to the topological closeness to a fixed initiator (unlike [28]).

In this respect, a work addressing an issue similar to ours has been recently proposed by Li et al. [15]. More specifically, the authors introduce the Keyword-Based Targeted Influence Maximization (KB-TIM) problem, which is to find a seed set that maximizes the expected influence over users who are relevant to a given advertisement. Each node is associated with a weighted term vector to capture its preference in different keywords and the advertisement is modeled as a keyword set. Thus, a user with keywords in common with the advertisement will belong to the target set. This problem has been solved by exploiting a state-of-the-art solution for the classic IM problem, named Reverse Influence Sampling (RIS) [3, 25], which provides theoretical guarantee on the quality of the results. More in detail, RIS consists of two steps: (1) computing, for a fixed number θ of nodes selected uniformly at random, the *reverse reachable* sets, i.e., the sets of nodes that can reach them; (2) using a standard greedy algorithm to select k nodes that cover the maximum number of reverse reachable sets previously generated. The rationale behind RIS is that by generating θ random reverse reachable sets and by selecting k nodes that maximize the coverage of these reverse reachable sets, the resulting set of size k should be a good solution to IM. Tang et al. [25] also show that, when θ is large enough, this set has high probability of being a near-optimal solution to IM. More in detail, they propose an algorithm, TIM+, that derives the parameter θ as function of a lower bound of the maximum expected spread.

TIM+ overcomes limitations of other solutions for IM, which suffer the drawback of getting soon impractical as the size of the graph increases. Indeed, TIM+ outperforms SimPath [6], which has been the state-of-the-art heuristic method for the LT model. SimPath is based on the computation of the possible paths on which the influence can spread starting from the seed nodes. The rationale behind SimPath is that, since the activation probability associated with a given path diminishes rapidly as the path gets longer, we can consider only the shortest paths from the seed set to the rest of the graph and, through a pruning threshold η, add up to the influence spread only the contributions of paths with probability not lower that η. This parameter represents the right trade-off between accuracy of spread estimation and running time.

The steps of KB-TIM are similar to TIM+ instead. However, as the former takes into account only users relevant to an advertisement, it defines a different lower bound for θ. Moreover, while in [3, 25] the random reverse reachable sets are sampled online, KB-TIM allows the sampling procedure to be performed offline by building a disk-based reverse reachable index for each keyword. When the seed set for a new advertisement is searched, the reverse reachable sets associated with the relevant keywords are loaded and merged. As a result, KB-TIM guarantees good execution times also on very large graphs.

Other approaches can be considered as related to targeted IM since they introduce in the diffusion process dimensions concerning the users' profiles. Lagnier et al. [12] propose a family of probabilistic diffusion models that involve vectors of features representing the content of information to be diffused and the profile of users. The problem tackled is to predict the diffusion of an information at a certain time given the diffusion status of the network at the previous time. Zhou et al. [29] adapt an extended IC to accommodate user preferences, which are learned from a set of users' documents labeled with topics. In [24], Tang et al. introduce node diversity into the IM task. Given a set of categories and the category distribution of each node, a set of diversity measures is proposed that satisfy monotonicity and submodularity. Li et al. [14] propose the conformity-aware cascade model. This is an IC model in which the influence probability from node u to v is computed proportionally to the product of u's influence and v's conformity, where the latter refers to the inclination of a node to be influenced by others. Influence and conformity values are calculated using a sentiment-analysis based algorithm, which assumes that edges in the network represent positive/negative attitudes of individuals toward opinions of others. The authors also describe a greedy algorithm, called CINEMA, which relies on a partitioning of the network to compute the node with maximum marginal gain in each subnetwork.

Compared to the latter group of works, our approach does not make any assumption on the network structure, nor depends on user characterization based on topic-biased or categorical distributions. Moreover, while some analogy could in principle be found between the notion of conformity in [14] and the proposed one of lurker activation, our method does not rely on sentiment analysis like CINEMA does.

Finally, it is worth emphasizing that, except KB-TIM, TIM+ and SIMPATH, *all the above works focus on the IC diffusion model*, while we propose a targeted IM problem under the LT model: as we shall discuss more in detail in Sect. 3.3, we argue that LT has a natural motivation to model the propagation of influence in a user delurking scenario.

3 Delurking-Oriented Targeted Influence Maximization

3.1 Problem Statement

Let $\mathscr{G}_0 = \langle \mathscr{V}, \mathscr{E} \rangle$ be the directed graph representing a social network, with set of nodes $\mathscr{V}$ and set of edges $\mathscr{E}$. We denote with $\mathscr{G} = \mathscr{G}_0(b, \ell) = \langle \mathscr{V}, \mathscr{E}, b, \ell \rangle$ a directed weighted graph representing the information diffusion graph associated with the social network $\mathscr{G}_0$, where $b : \mathscr{E} \rightarrow \mathbb{R}^*$ is an edge weighting function, and $\ell : \mathscr{V} \rightarrow \mathbb{R}^*$ is a node weighting function.

We assume that the node and edge weighting functions in the diffusion graph $\mathscr{G}$ are completely specified based upon the availability of a *lurker ranking* solution. This is produced by a method that is designed to identify and characterize lurkers in an SN, i.e., a method enabled to assign every user with a score that expresses the status of lurking behavior taken by that user. The intuition is that for any edge (u, v) in $\mathscr{G}$, the weight $b(u, v)$ will indicate how much node u has contributed to the v's lurking score calculated by the lurker ranking method, which resembles a measure of "influence" produced by u to v. Also, the node weight $\ell(v)$ will indicate the status of v as lurker, such as the higher the lurker ranking score of v the higher should be $\ell(v)$. We shall provide formal details about the lurker ranking method, namely LurkerRank, in Sect. 3.2, and about node and edge weighting functions in Sect. 3.5.

We denote with $\text{LS} \in [0, 1]$ a threshold value that indicates the *minimum lurking score* that a node in the network is required to have in order to be regarded as a target node. Moreover, given any seed set $S \subseteq \mathscr{V}$, we denote with $\mu(S)$ the *final active set*, i.e., the set of nodes that are active at the end of the diffusion process starting from S.

Upon the above defined quantities, we introduce a measure that will be essential to the definition of our targeted IM problem. This measure, we call *delurking capital* cumulated via S, is proportional to the amount of the lurking scores over all target nodes that are activated by the seed set S.

Definition 1 Given a set $S \subseteq \mathscr{V}$, the ***delurking capital*** $\text{DC}(\mu(S))$ associated with the final active set $\mu(S) \subseteq \mathscr{V}$ is defined as:

$$\text{DC}(\mu(S)) = \sum_{\substack{v \in \mu(S) \setminus S \, \wedge \\ \ell(v) \geq \text{LS}}} \ell(v) \tag{1}$$

Note that in Eq. (1) we do not consider nodes that belong to the seed set S. The ratio behind this choice is that the selection of the seed set should not be biased by nodes with highest lurking scores; this will be made clear next in the definition of our objective function, which in fact relies on $\mathrm{DC}(\mu(S))$.

We now formally define our proposed problem of delurking-oriented targeted influence maximization. Afterwards, we discuss the diffusion model and properties of the proposed objective function, and we end this section by presenting our developed algorithm.

Definition 2 (Delurking-Oriented Targeted Influence Maximization Problem) Given $\mathscr{G} = \langle \mathscr{V}, \mathscr{E}, b, \ell \rangle$, a diffusion model on $\mathscr{G}$, a budget k, and a lurking threshold LS, find a seed set $S \subseteq \mathscr{V}$ with $|S| \leq k$ of nodes (users) such that, by activating them, the final active set $\mu(S) \subseteq \mathscr{V}$ will have the maximum delurking capital:

$$S = \underset{S' \subseteq \mathscr{V} \ s.t. \ |S'| \leq k}{\mathrm{argmax}} \ \mathrm{DC}(\mu(S')) \tag{2}$$

3.2 Identifying and Characterizing Lurkers

In this section we describe how lurkers in an SN can be identified and their status quantified by means of a ranking method. To this aim, we will refer to our early work on ranking lurkers in social networks [21, 22]. The core of lurker ranking relies on a *topology-driven definition of lurking*, which is based solely on the network structure. Since lurking behaviors build on the *amount of information a node receives*, our key intuition is that the strength of a user's lurking status can be determined based on three basic principles, which are informally reported as follows:

Overconsumption: The excess of information-consumption over information-production. The strength of a node's lurking status is proportional to its in/out-degree ratio.
Authoritativeness of the information received: The valuable amount of information received from its in-neighbors. The strength of a node's lurking status is proportional to the influential (non-lurking) status of its in-neighbors.
Non-authoritativeness of the information produced: The non-valuable amount of information sent to its out-neighbors. The strength of a node's lurking status is proportional to the lurking status of its out-neighbors.

The above principles form the basis for three ranking models that differently account for the contributions of a node's in-neighborhood and out-neighborhood. A complete specification of the lurker ranking models is provided in terms of PageRank and AlphaCentrality based formulations. For the sake of brevity here, and throughout this paper, we will refer to only one of the formulations described in [21, 22], which is that based on the full *in-out-neighbors-driven lurker ranking*, hereinafter dubbed simply as *LurkerRank* (LR).

Given the social network graph $\mathcal{G}_0 = \langle \mathcal{V}, \mathcal{E} \rangle$, with set of nodes $\mathcal{V}$ and set of edges $\mathcal{E}$, the semantics of edge (u, v) is that v is "consuming" or "receiving" information from u. Given a node $v \in \mathcal{V}$, let us denote with $N^{\text{in}}(v)$ and $N^{\text{out}}(v)$ the set of in-neighbors and the set of out-neighbors of v, and with $\text{in}(v)$, $\text{out}(v)$ their respective sizes. The LurkerRank $\text{LR}(v)$ score of node v is defined as:

$$\text{LR}(v) = d[\mathcal{L}_{\text{in}}(v)\ (1 + \mathcal{L}_{\text{out}}(v))] + (1 - d)/(|\mathcal{V}|) \tag{3}$$

where $\mathcal{L}_{\text{in}}(v)$ is the in-neighbors-driven lurking function:

$$\mathcal{L}_{\text{in}}(v) = \frac{1}{\text{out}(v)} \sum_{u \in N^{\text{in}}(v)} \frac{\text{out}(u)}{\text{in}(u)} \text{LR}(u) \tag{4}$$

and $\mathcal{L}_{\text{out}}(v)$ is the out-neighbors-driven lurking function:

$$\mathcal{L}_{\text{out}}(v) = \frac{\text{in}(v)}{\sum_{u \in N^{\text{out}}(v)} \text{in}(u)} \sum_{u \in N^{\text{out}}(v)} \frac{\text{in}(u)}{\text{out}(u)} \text{LR}(u) \tag{5}$$

d is a damping factor ranging within [0,1], usually set to 0.85. To prevent zero or infinite ratios, the values of $\text{in}(\cdot)$ and $\text{out}(\cdot)$ are Laplace add-one smoothed.

As previously discussed, LurkerRank does not require any information other than the network topology, in which node (user) relationships are asymmetric and indicate that one node receives information from another one. The actual meaning of "received information" can depend on the specific context of network evaluation; in general, it refers to a social graph (i.e., $(u, v) \in \mathcal{E}$ means that v *is the follower of* u) or an interaction graph (e.g., v *likes* or *comments* u's posts); our LurkerRank has been in fact evaluated on both scenarios, as described in [22, 23]. For purposes of (targeted) IM, both social and interaction relations can be seen as indicator of user influence. However, we note that influence is normally produced regardless of actual, visible interaction between two users. Yet, information on interaction data might be significantly sparse in real SNs, causing a flawed setting for an IM task. In this work, we hereinafter assume that the graph $\mathcal{G}_0$ (on which LurkerRank is applied) is a followship graph, while evaluation on interaction graphs is left as further work.

3.3 Choosing the Information Diffusion Model

IC and LT models focus on different aspects of information diffusion. IC can be seen as a prototype for *contagion propagation* models and it is said to be *sender-centric*, since each active node independently infects its active neighbors with given diffusion probabilities.

By contrast, the LT model deals with situations in which exposure to multiple sources is needed for a user before taking a decision. More formally, we recall that a node v can be influenced by each neighbor u according to a weight $b(u, v)$ such that

$\sum_{u \in N^{\text{in}}(v)} b(u, v) \leq 1$, where $N^{\text{in}}(v)$ is the in-neighbor set of node v. At the very beginning of the diffusion process, each node v is assigned a threshold uniformly at random from [0, 1]. This represents the weighted fraction of v's neighbors that must become active in order for v to become active itself. Intuitively, the higher is this threshold, the harder will be the task of enrolling v in a new trend, since the total influence weight must exceed its threshold.

Within this view, we have chosen LT to model the process of influence propagation to pursue the goal of delurking (activating) users that have been identified and scored as lurkers (cf. Sect. 3.1). We believe that a natural motivation to use LT (rather than IC or IC-based models) stands in the ability of this model to reflect the cumulative effect of exposure to multiple sources of influence. This can be profitably exploited to maximize the likelihood of changing the status of a user into a more active role in the SN. Of course, as previously discussed in the Sect. 1, this can actually lead to delurking provided that the information to be diffused towards the target lurkers is of any type that concerns a well-established delurking action. In this regard, note that our approach is general as it can involve any type of delurking action in the form of piece of information to flow through the network. Note also that identifying the best delurking approach from a marketing or psychological perspective is beyond the scope of our work.

3.4 Properties of the Proposed Objective Function

The objective function of the problem in Eq. (2) differs from the ones in classic IM as it is defined in terms of the cumulative amount of the scores associated with the activated (target) nodes, i.e., $\text{DC}(\mu(S))$, instead of the size of the final active set (i.e., $|\mu(S)|$, commonly known as spread).

Example in Fig. 1 helps us motivate the different outcome obtained via LT for the classic IM and the proposed targeted IM based on LS. We assume for the sake of simplicity of the example that the node weights correspond to both the

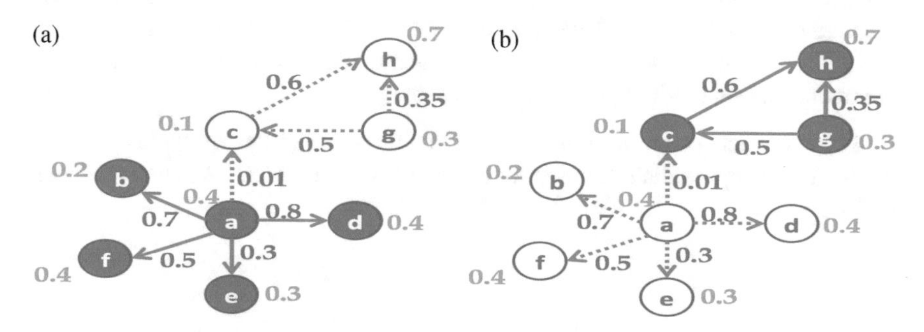

Fig. 1 Final active sets (*filled circles*) for different seeds. Edge weights (values in *blue*) and node weights (values in *green*) are computed by functions b and ℓ. To avoid cluttering of the figure, the node activation thresholds used by LT model here coincide with the node weights (best viewed in color). (**a**) $\mu(S)$ with $S = \{a\}$, (**b**) $\mu(S)$ with $S = \{g\}$

node activation thresholds (used in the LT model) and the lurking scores (ℓ). As shown on the left of the figure, the seed set $S = \{a\}$ is the best one to accomplish the influence maximization task as it causes four nodes to be activated during the process. However, the optimal seed set can be different in our setting of targeted IM: for example, if we set LS $= 0.6$, then the best seed set is $S = \{g\}$. Indeed, node h is the only one with lurking score ($\ell(h)$) greater than 0.6. This node is eventually activated (at the second step of the diffusion process) by the seed g due to the influence exerted jointly with c (which is in turn activated by g at the first step of the process).

The problem in Definition 2 preserves the complexity of the IM problem and, as a result, it is computationally intractable, i.e., it is still NP-hard. However, as for the classic IM problem, a greedy solution can be designed provided that the natural diminishing property holds for the considered problem. It is well known that for many diffusion models, including LT, the function $\sigma(A)$ mapping any subset $A \subseteq \mathcal{V}$ of nodes to the size of the final active set $\mu(A)$ satisfies monotonicity and submodularity by exploiting the *equivalent live-edge graph model* [11]. In the following, we provide a theoretical result proving that the function $\mathrm{DC}(\mu(A))$ mapping each active set $\mu(A) \subseteq \mathcal{V}$ to its overall delurking capital is monotone and submodular, for any LS $\in [0, 1]$.

Proposition 1 *The delurking capital function defined in Eq. (1) is monotone and submodular under the LT model.*

Proof Sketch By exploiting the equivalence between LT and the live-edge model shown in [11], for any set $A \subseteq \mathcal{V}$ we can express the expected delurking capital of the final active set $\mu(A)$ in terms of reachability under the live-edge graph:

$$\mathrm{DC}(\mu(A)) = \sum_{\forall X} \Pr(X)\mathrm{DC}(R^X(A)) \tag{6}$$

where $\Pr(X)$ is the probability that a hypothetical live-edge graph X is selected from all possible live-edge graphs, and $R^X(A)$ is the set of nodes that are reachable in X from A. Since for all $v \in \mathcal{V}$, $\ell(v)$ is a non-negative value, $\mathrm{DC}(R^X(A))$ is clearly monotone and submodular. Thus, the expected delurking capital under LT is a non-negative linear combination of monotone submodular functions, and hence it is monotone and submodular, which concludes the proof. □

As a consequence of the above result, a greedy approach can be applied in order to provide an approximate solution for our targeted IM problem. We shall present it in Sect. 3.6.

3.5 Modeling the Information Diffusion Graph

As anticipated in Sect. 3.1, the node weight $\ell(v)$ should reflect the status of v as lurker, such as the higher the lurker ranking score of v the higher should be $\ell(v)$. We define the node weighting function $\ell(\cdot)$ upon scaling and normalizing the stationary distribution produced by the LurkerRank algorithm over $\mathscr{G}_0$ (cf. Sect. 2). The scaling compensates for the fact that the lurking scores produced by LurkerRank, although distributed over a significantly wide range (as reported in [22]), might be numerically very low (e.g., order of 1.0e−3 or below). Moreover, we introduce a small smoothing constant in order to avoid that the highest lurking scores are mapped exactly to 1. Formally, for each node $v \in \mathscr{V}$, we define the *node lurking value* $\ell(v) \in [0, 1)$ as follows:

$$\ell(v) = \frac{\widetilde{\pi}_v - \min_r}{(\max_r - \min_r) + \epsilon_r} \tag{7}$$

where $\widetilde{\pi}$ denotes the stationary distribution of the lurker ranking scores (π) divided by the base-10 power of the order of magnitude of the minimum value in π, $\widetilde{\pi}_v$ is the value of $\widetilde{\pi}$ corresponding to node v, $\max_r = \max_{u \in \mathscr{V}} \widetilde{\pi}_u$, $\min_r = \min_{u \in \mathscr{V}} \widetilde{\pi}_u$, and ϵ_r is a smoothing constant proportional to the order of magnitude of the $\max_r$ value.

In order to define the edge weights so that they express a notion of strength of influence from a node to another (as normally required in an information diffusion model), we again exploit information derived from the ranking solution obtained by LurkerRank as well as from the structural properties of the social graph. Our key idea is to calculate the weight on edge $(u, v) \in \mathscr{E}$ proportionally to the fraction of the original lurking score of v given by its in-neighbor u:

$$b_0(u, v) = \left[\sum_{w \in N^{in}(v)} \frac{\mathrm{out}(w)}{\mathrm{in}(w)} \pi_w \right]^{-1} \frac{\mathrm{out}(u)}{\mathrm{in}(u)} \pi_u \tag{8}$$

The above edge weight definition has however two limitations: (1) it is independent of the v's lurking score, therefore the contribution of node u will be the same value for each of its out-neighbors with identical set of in-neighbors; (2) the constraint on the sum of incoming edge weights will be strictly equal to 1, which is not necessarily required. Therefore, we define the actual edge weighting function $b(\cdot)$ upon a modification of Eq. (8) according to the above remarks:

$$b(u, v) = b_0(u, v) \times e^{\ell(v)-1} \tag{9}$$

The above formula fulfills the requirement $\sum_{u \in N^{in}(v)} b(u, v) \leq 1$, moreover it takes into account $\ell(v)$ such that the resulting weight on (u, v) will be decreased (with exponential smoothing) for higher $\ell(v)$. This can be explained since the more a node acts as a lurker, the more active in-neighbors are needed to activate that node.

3.6 The DEvOTION Algorithm

Algorithm 1 shows our proposed greedy method, named DEvOTION (stands for **DE**lurking **O**riented **T**argeted **I**nfluence maximizati**ON**). Following the lead of the study in [6], DEvOTION exploits as well the search for shortest paths in the diffusion graph, however in a backward fashion. Along with the information diffusion

---

Algorithm 1: DElurking Oriented Targeted Influence maximizatiON—DEvOTION

Input: A graph $\mathcal{G} = \langle \mathcal{V}, \mathcal{E}, b, \ell \rangle$, a budget (seed set size) k, a lurking threshold $LS \in [0, 1]$, a path pruning threshold $\eta \in [0, 1]$.
Output: Seed set S.
1: $S \leftarrow \emptyset$
2: $T \leftarrow \mathcal{V}$ *{nodes that can reach target nodes}*
3: $TargetSet \leftarrow \emptyset$ *{stores the target nodes at current iteration}*
4: **for** $u \in \mathcal{V}$ **do**
5:   **if** $\ell(u) \geq LS$ **then**
6:     $TargetSet \leftarrow TargetSet \cup \{u\}$
7:   **end if**
8: **end for**
9: **while** $|S| < k$ **do**
10:   $bestSeed, bestSeed.DC \leftarrow -1$ *{keeps track of the node with the highest spread}*
11:   **for** $u \in T \setminus S$ **do**
12:     $u.DC \leftarrow 0$ *{initializes each node's spread to zero}*
13:   **end for**
14:   $T \leftarrow \emptyset$;
15:   **for** $u \in TargetSet \setminus S$ **do**
16:     backward($\langle u \rangle, 1, \ell(u)$)
17:   **end for**
18:   **if** $bestSeed \neq -1$ **then**
19:     $S \leftarrow S \cup \{bestSeed\}$
20:   **else**
21:     **break**
22:   **end if**
23: **end while**
24: **return** S

25: **procedure** backward($\mathcal{P}, pp, score$)
26: $u \leftarrow \mathcal{P}.last()$
27: $T \leftarrow T \cup \{u\}$
28: **while** $v \in N^{in}(u) \wedge v \notin S \cup \mathcal{P}.nodeSet()$ **do**
29:   $pp \leftarrow pp \times b(v, u)$ *{updates the path probability}*
30:   **if** $pp \geq \eta$ **then**
31:     $v.DC \leftarrow v.DC + pp \times score$
32:     **if** $v.DC > bestSeed.DC$ **then**
33:       $bestSeed \leftarrow v$ *{sets the best seed node as v}*
34:     **end if**
35:     backward($\mathcal{P}.append(v), pp, score$)
36:   **end if**
37: **end while**

---

graph $\mathscr{G}$, the budget integer k, and the minimum lurking score LS, DEvOTION takes in input a real-valued threshold η. This parameter is used to control the size of the neighborhood within which paths are enumerated: in fact, the majority of influence can be captured by exploring the paths within a relatively small neighborhood; note that for higher η values, less paths are explored (i.e., paths are pruned earlier) leading to smaller runtime but with decreased accuracy in spread estimation.

The key idea of DEvOTION is to perform a backward visit of the graph starting from the nodes identified as target (i.e., the nodes u with $\ell(u) \geq$ LS). To this end, all nodes are initially examined to compute *TargetSet* (Lines 4–8). In order to yield a seed set S of size at most k, DEvOTION works as follows. During each iteration of the main loop (Lines 9–19), DEvOTION computes the set T of nodes that reach the target ones and keeps track, into the variable *bestSeed*, of the node with the highest marginal gain (i.e., delurking capital DC). The former allows an efficient reset of the spread of each non-seed node contained in T (Lines 11–13). The latter is found at the end of each iteration upon calling the subroutine backward over all nodes in *TargetSet* that do not belong to the current seed set S (Lines 15–17). This subroutine takes a path $\mathscr{P}$, its probability *pp*, and the lurking score of the end node in the path (i.e., a target node), and extends $\mathscr{P}$ as much as possible (i.e., as long as *pp* is not lower than η). Initially (Line 16), a path is formed by one target node, with probability 1. Then (Lines 28–37), the path is extended by exploring the graph backward, adding to it one, unexplored in-neighbor v at time, in a depth-first fashion. The path probability is updated (Line 29) according to the LT-equivalent "live-edge" model [6, 11], and so the delurking capital (Line 31). The process is continued until no more nodes can be added to the path.

Consider the example in Fig. 1, where the target set is $\{h\}$. By applying DEvOTION, assuming for simplicity to set $k = 1$ and $\eta = 0$, the target node can be reached via a (with a.DC $= [0.01 \times 0.6] \times 0.7$), c (with c.DC $= 0.6 \times 0.7$), and g (with g.DC $= [0.35 + 0.5 \times 0.6] \times 0.7$). Node g has the highest DC, since it has the largest chance of success ($0.35 + 0.5 \times 0.6$), and hence it is chosen as seed node.

Note that moving backward from the target nodes has two positive side effects: indeed, as all the paths starting from nodes that cannot reach any target are ignored, we get a further path pruning that reduces DEvOTION running time without affecting its accuracy. It is also worth noting that (1) if LS is set to zero and the node lurking values are uniformly distributed, our problem reduces to classic IM problem, and (2) if the target set includes a single node, i.e. $|\mathit{TargetSet}| = 1$, our problem is equivalent to a single-target IM problem.

Estimating Delurking Capital The delurking capital of a seed set is estimated by simulating the LT diffusion process several times. Given a graph $\mathscr{G}$, a lurking threshold LS and a seed set S, Algorithm 2 estimates the final delurking capital DC by running a certain number R of Monte Carlo simulations. The estimation procedure works by propagating the active status from the seed set S to its neighborhood, and iteratively from the newly activated nodes to the rest of the graph, according to the LT model steps. When a node is reached for the first time (e.g., one of its neighbors has been activated), the procedure generates a random threshold

Algorithm 2: Monte Carlo estimation of delurking capital

Input: A graph $\mathcal{G} = (\mathcal{V}, \mathcal{E}, b, \ell)$, a lurking threshold LS $\in [0, 1]$, seed set S, number of Monte Carlo iterations R
Output: Delurking capital $DC(\mu(S))$

1: $curr\_DC \leftarrow 0$
2: **for** $u \in S$ **do**
3: $u.isActive \leftarrow$ **true**
4: **end for**
5: **for** $j = 1$ *to* R **do**
6: **for** $v \in \mathcal{V} \setminus S$ **do**
7: $v.isActive \leftarrow$ **false**
8: $v.receivedInf \leftarrow 0$
9: $\vartheta_v \leftarrow -1$
10: **end for**
11: $temp \leftarrow S$
12: **while** $temp \neq \emptyset$ **do**
13: $u \leftarrow temp.remove(0)$
14: **for** $v \in N^{out}(u) \ \wedge \ v.isActive =$ **false do**
15: $v.receivedInf \leftarrow v.receivedInf + b(u, v)$
16: **if** $\vartheta_v = -1$ **then** {node v has been reached for the first time during the current simulation}
17: choose $\vartheta_v \sim U[0, 1]$
18: **if** $v.receivedInf \geq \vartheta_v$ **then**
19: $v.isActive \leftarrow$ **true**
20: $temp \leftarrow temp \cup \{v\}$
21: **if** $\ell(u) \geq LS$ **then**
22: $curr\_DC \leftarrow curr\_DC + \ell(v)$
23: **end for**
24: **end while**
25: **end for**
26: **return** $curr\_DC/R$;

which determines the activation of the node based on its received influence (i.e., sum of the influence weights on the incoming edges corresponding to its already activated in-neighbors). Algorithm 2 sums up the lurking scores of the target nodes activated after the diffusion ends, and then takes the average of these sums over the R simulations to obtain the final delurking capital DC.

4 Experimental Evaluation

4.1 Evaluation Methodology

We conducted four main evaluation stages in order to assess: the significance of our DEvOTION and its sensitivity to the various parameters (stage 1); its relevance and effectiveness through a comparative analysis with respect to baseline methods (stage 2) and other targeted and non-targeted IM methods (stages 3 and 4). A summary of each of the evaluation stages is reported in the following.

In the first evaluation stage, we studied the impact of the various parameters on the performance of DEvOTION, i.e., the size of seed set (k), the lurking threshold (LS) for the selection of the target set, and the path pruning threshold (η). Note that, to simplify the interpretation of LS, we will instead use symbol LS-perc to denote a percentage value that determines the setting of LS such that the selected target set corresponds to the top-LS-perc of the distribution of node lurking scores (ℓ values). We will show results that correspond to LS-perc $\in \{5\%, 10\%, 25\%\}$.

In the second evaluation stage, we compared DEvOTION with three baseline methods, dubbed Random, LargestDegree, and bottom-LR. Method Random calculates the delurking capital obtained for 100 randomly selected seed sets for each value of k, then the final delurking capital is averaged over the number of random extractions. Method LargestDegree selects, for each value of k, the k nodes in the graph with the largest out-degree. Finally, the bottom-LR algorithm selects, for each value of k, the k nodes with the lowest lurking value, i.e., the k most active users in the network (according to the definition of lurking developed in the LurkerRank algorithms [22]); the ratio behind this baseline is that the highly active users might be seen as potential candidates for the seed set.

In the third evaluation stage, we compared DEvOTION with SimPath [6] and TIM+ [15], two well-known state-of-the-art algorithms for IM problems, whereas in the fourth evaluation stage, we compared DEvOTION with KB-TIM [25], a recently proposed query-based targeted IM algorithm. The three mentioned algorithms have been previously discussed in Sect. 2. Note that KB-TIM and TIM+ are able to work under both the IC and LT diffusion models, but were used in their LT version for compatibility with DEvOTION. By contrast, other (targeted) IM algorithms discussed in Sect. 2 could not have been taken into account in our evaluation since they are based only on the IC model (thus, any comparison with our LT-based model would be like comparing "apples and oranges").

All experiments were carried out on an Intel Core i7-3960X CPU @3.30 GHz, 64 GB RAM machine. All algorithms were written in C++. All competing algorithms refer to the original source code provided by their authors.

4.2 *Experimental Setting*

Comparing DEvOTION with (non-targeted) IM algorithms like SimPath and TIM+ required to evaluate the quality of seed sets produced by the two competing algorithms under a *targeted* scenario. Therefore, instead of reporting the spread over the entire graph, we computed the delurking capital over different target sets in accord with the setting of DEvOTION. SimPath was carried out with maximum path length equal to four (as suggested in the original paper) and $\eta \in \{1.0\text{e}{-}03, 1.0\text{e}{-}04\}$, according to the values used for DEvOTION as well. Concerning TIM+, we considered two configurations corresponding to minimum and maximum approximation, i.e., $\epsilon = 0.1$ (minimum approximation) and $\epsilon = 1.0$ (maximum approximation); we used default settings of the other parameters in TIM+.

As concerns the comparison with KB-TIM, a pre-processing phase was required to set up a fair comparison with DEvOTION, given the keyword-based target selection used by this algorithm. In KB-TIM, each node is assigned a list of (*keyword*, *occurrences*) pairs, which drives the query-based target selection. In order to make it equivalent to our threshold-based target selection, we associated fixed keywords to those users having highest lurking scores (i.e., keywords corresponding to the LS-perc values used to identify a certain top-ranking). We then used the lurking score of each node to replace the occurrence values (conveniently normalized to assume integer values). For each dataset, we ran three queries corresponding to the LS-perc values used as keywords (i.e., $\{5\%, 10\%, 25\%\}$), in order to obtain for each query a target set which exactly matches the threshold-based one used for DEvOTION. Moreover, we used the incremental reverse-reachable index (IRR) in KB-TIM. Regarding the Monte Carlo based procedure of DC estimation, we set the number of runs R to 10,000.

4.3 Data

We used *FriendFeed*, *GooglePlus*, and *Instagram* network datasets to conduct our analysis. As major motivations underlying our data selection, we maintain continuity with our previous studies [22, 23] and we use publicly available datasets. Table 1 summarizes main structural characteristics of the evaluation network datasets.

FriendFeed dataset refers to the latest (2010) version of the dataset studied in [4]; note that due to the recognized presence of spambots in this dataset, we filtered out users with an excessive number of posts (above 20 posts per day) as suggested in [4].

GooglePlus dataset is the one studied in [16]. It was originally collected from users who had manually shared their circles using the *share circle* feature, and the topology was built by combining the edges from each node's ego network.

Table 1 Main structural characteristics of the evaluation network datasets

Data	# Nodes	# Links	Avg in-degree	Avg path length	Clustering coefficient	Assortativity	# Sources # sinks
FriendFeed	493,019	19,153,367	38.85	3.82	0.029	−0.128	41,953 292,003
GooglePlus	107,612	13,673,251	127.06	3.32	0.154	−0.074	35,341 22
Instagram	17,521	617,560	35.25	4.24	0.089	−0.012	0 0

Instagram dataset refers to our latest 2014 dump,[2] Note that, by construction, this dataset is the one of the three that contains the largest number of loosely connected components. Therefore, for the sake of significance of the information diffusion process (which necessarily relies on strong connectivity of the components) we decided in this case to use the induced subgraph corresponding to the maximal strongly connected component of the original network graph.

5 Results

5.1 *Impact of Parameters in* DEvOTION

We evaluated the performance of DEvOTION in terms of delurking capital obtained by varying all three parameters involved, i.e., *k*, LS-perc, η. We initially focused on the understanding of the impact of the η parameter, by varying it from 1.0e−03 to 0; note that η = 1.0e−03 is the default value used in other IM algorithms (e.g.,[6]), while $\eta = 0$ indicates no path pruning. Figure 2 shows results obtained on Instagram, for three settings of η. At a first glance over all three plots in the figure, it can be noted that no significant effects on DC are yielded by varying η and keeping fixed the other two parameters. This indicates that no significant gain in spread (DC) is obtained for lower values of η. This fact achieves particular importance when it is coupled with the time performance results shown in Fig. 3a: several orders of magnitude in the runtime are saved when $\eta > 0$. Within this view, η = 1.0e−03 might be be preferred to η = 1.0e−04 when trade-off between DC performance and runtime has to be ensured. Analogous observations were equally raised from the DC and time performance over the other two datasets for the various η; for the sake of simplicity of presentation, we do not report the $\eta = 0$ cases for FriendFeed (Fig. 3b) and GooglePlus (Fig. 3c).

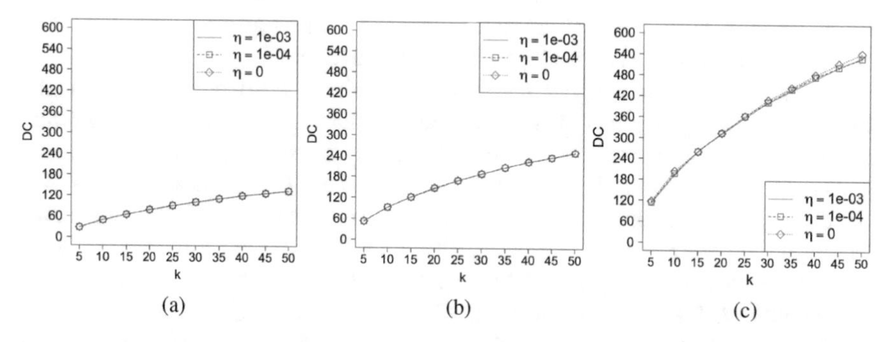

Fig. 2 Delurking capital in function of *k* and η, with LS-perc set to (**a**) 5%, (**b**) 10%, and (**c**) 25%, on Instagram

[2]Available at http://uweb.dimes.unical.it/tagarelli/data/.

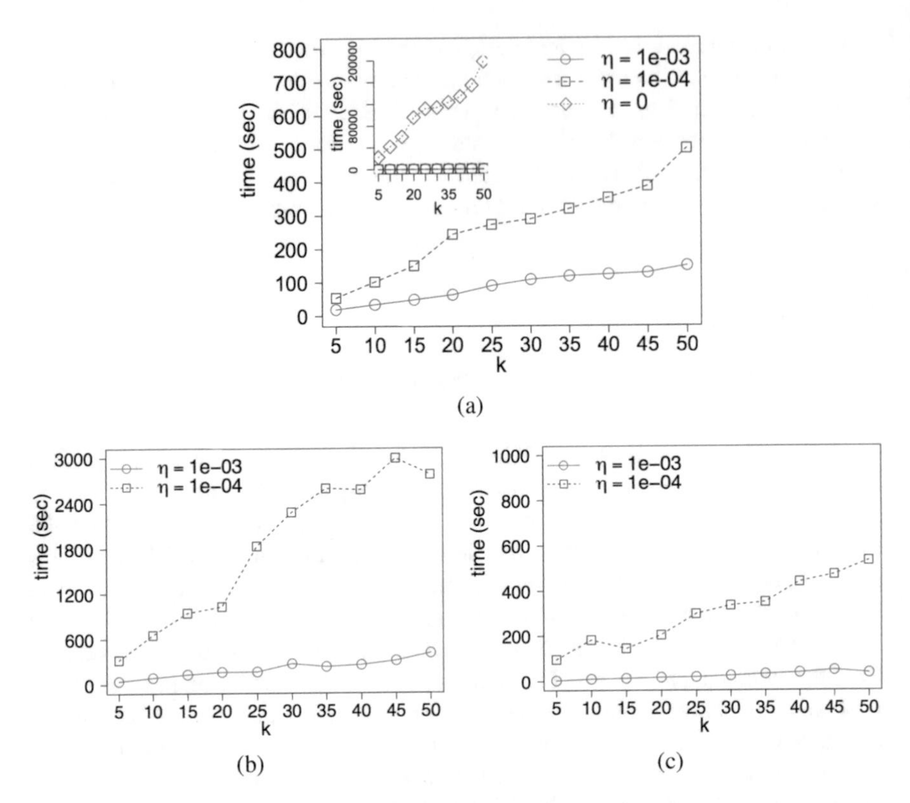

Fig. 3 Execution time of DEvOTION in function of k, with LS-perc $= 5\%$ and for varying η, on (**a**) Instagram, (**b**) FriendFeed, and (**c**) GooglePlus. The *inset* in the Instagram plot shows the overall results that include $\eta = 0$, while the main plot zooms in to focus on $\eta > 0$

Concerning the other two parameters, as expected, the delurking capital increases with both the size of the target set (LS-perc) and the size of the seed set (k). Detailed values are shown in Tables 2, 3, and 4, wherein the first group of three columns correspond to DEvOTION results on the various datasets.

Besides the obviously higher values of DC obtained on FriendFeed and GooglePlus than on Instagram, it is interesting to also understand the effect of the network scale combined with the LS-perc. Specifically, by considering the pair-wise gain percentage computed on the DC value averaged over k (i.e., values in the last row of the tables), from lower to higher LS-perc, we obtained the following: 0.878 from LS-perc $= 5\%$ to LS-perc $= 10\%$, and 1.124 from LS-perc $= 10\%$ to LS-perc $= 25\%$, on Instagram; 0.924 and 1.074, on GooglePlus; 0.706 and 0.578, on FriendFeed. It hence turns out that, when considering higher size of target set, the difference in the DC average gains tends to decrease or even become negative with larger network scale.

Another important remark is that on all datasets a significant fraction of delurking capital, for a given choice of LS-perc, can be achieved just using low k. This appears

Table 2 Delurking capital performances on Instagram, for various k and LS-perc configurations, with $\eta = 1.0e{-}04$ for both DEvOTION and SimPath

	DEvOTION			SimPath			TIM+			KB-TIM		
k	5%	10%	25%	5%	10%	25%	5%	10%	25%	5%	10%	25%
5	2.78e+01	5.23e+01	1.14e+02	2.55e+01	4.91e+01	1.09e+02	2.58e+01	4.96e+01	1.10e+02	2.79e+01	5.27e+01	1.12e+02
10	4.90e+01	9.18e+01	1.97e+02	4.44e+01	8.51e+01	1.88e+02	4.70e+01	8.95e+01	1.94e+02	4.35e+01	7.91e+01	1.66e+02
15	6.48e+01	1.21e+02	2.60e+02	5.86e+01	1.12e+02	2.49e+02	6.12e+01	1.17e+02	2.56e+02	5.92e+01	1.10e+02	2.29e+02
20	7.86e+01	1.48e+02	3.14e+02	7.21e+01	1.38e+02	3.05e+02	7.34e+01	1.40e+02	3.07e+02	7.31e+01	1.36e+02	2.75e+02
25	9.06e+01	1.70e+02	3.62e+02	8.51e+01	1.62e+02	3.56e+02	8.57e+01	1.63e+02	3.56e+02	8.66e+01	1.61e+02	3.26e+02
30	1.01e+02	1.89e+02	4.03e+02	9.55e+01	1.82e+02	3.97e+02	9.61e+01	1.82e+02	3.97e+02	9.77e+01	1.81e+02	3.77e+02
35	1.10e+02	2.07e+02	4.40e+02	1.05e+02	2.00e+02	4.35e+02	1.06e+02	2.01e+02	4.35e+02	1.07e+02	1.98e+02	4.09e+02
40	1.19e+02	2.23e+02	4.75e+02	1.15e+02	2.17e+02	4.71e+02	1.15e+02	2.18e+02	4.71e+02	1.15e+02	2.11e+02	4.36e+02
45	1.26e+02	2.36e+02	5.03e+02	1.22e+02	2.31e+02	5.01e+02	1.23e+02	2.31e+02	5.00e+02	1.19e+02	2.20e+02	4.58e+02
50	1.33e+02	2.49e+02	5.28e+02	1.28e+02	2.43e+02	5.26e+02	1.29e+02	2.44e+02	5.28e+02	1.25e+02	2.30e+02	4.73e+02
Avg	9.00e+01	1.69e+02	3.59e+02	8.52e+01	1.62e+02	3.54e+02	8.62e+01	1.64e+02	3.55e+02	8.54e+01	1.58e+02	3.26e+02

$\epsilon = 0.1$ and index IRR are used for TIM+ and KB-TIM, respectively

Table 3 Delurking capital performances on FriendFeed, for various k and LS-perc configurations, with $\eta = 1.0e{-}04$ for both DEvOTION and SimPath

	DEvOTION			SimPath			TIM+			KB-TIM		
k	5%	10%	25%	5%	10%	25%	5%	10%	25%	5%	10%	25%
5	3.88e+03	6.11e+03	7.92e+03	3.04e+03	4.69e+03	7.24e+03	3.08e+03	4.71e+03	7.26e+03	5.89e+01	3.05e+03	4.12e+03
10	4.93e+03	8.07e+03	1.11e+04	4.10e+03	6.76e+03	1.05e+04	4.14e+03	6.79e+03	1.05e+04	2.25e+03	3.37e+03	5.62e+03
15	5.34e+03	9.02e+03	1.35e+04	4.58e+03	7.81e+03	1.26e+04	4.62e+03	7.84e+03	1.26e+04	2.50e+03	4.56e+03	6.72e+03
20	5.61e+03	9.59e+03	1.48e+04	5.47e+03	9.40e+03	1.48e+04	5.51e+03	9.43e+03	1.49e+04	2.79e+03	6.73e+03	9.51e+03
25	5.84e+03	1.00e+04	1.59e+04	5.52e+03	9.56e+03	1.55e+04	5.56e+03	9.60e+03	1.56e+04	2.95e+03	8.29e+03	1.10e+04
30	6.03e+03	1.04e+04	1.68e+04	5.61e+03	9.79e+03	1.62e+04	5.65e+03	9.83e+03	1.63e+04	4.46e+03	8.90e+03	1.38e+04
35	6.19e+03	1.07e+04	1.75e+04	5.70e+03	1.00e+04	1.69e+04	5.74e+03	1.01e+04	1.70e+04	5.02e+03	9.27e+03	1.53e+04
40	6.34e+03	1.10e+04	1.82e+04	5.81e+03	1.03e+04	1.76e+04	5.86e+03	1.03e+04	1.77e+04	5.29e+03	9.30e+03	1.55e+04
45	6.47e+03	1.12e+04	1.88e+04	5.87e+03	1.04e+04	1.80e+04	5.92e+03	1.05e+04	1.81e+04	5.34e+03	9.50e+03	1.60e+04
50	6.59e+03	1.15e+04	1.94e+04	5.95e+03	1.06e+04	1.86e+04	6.00e+03	1.07e+04	1.87e+04	5.36e+03	9.63e+03	1.62e+04
Avg	5.72e+03	9.76e+03	1.54e+04	5.16e+03	8.93e+03	1.48e+04	5.21e+03	8.97e+03	1.49e+04	3.60e+03	7.26e+03	1.14e+04

$\epsilon = 0.1$ and index IRR are used for TIM+ and KB-TIM, respectively

Table 4 Delurking capital performances on GooglePlus, for various k and LS-perc configurations, with $\eta = 1.0e{-}04$ for both DEvOTION and SimPath

	DEvOTION			SimPath			TIM+			KB-TIM		
k	5%	10%	25%	5%	10%	25%	5%	10%	25%	5%	10%	25%
5	4.29e+02	8.68e+02	1.79e+03	4.13e+02	8.42e+02	1.78e+03	4.13e+02	8.42e+02	1.78e+03	3.18e+02	7.68e+02	1.58e+03
10	5.94e+02	1.18e+03	2.54e+03	5.82e+02	1.18e+03	2.54e+03	5.83e+02	1.18e+03	2.54e+03	5.25e+02	9.15e+02	1.93e+03
15	7.07e+02	1.40e+03	2.94e+03	6.67e+02	1.33e+03	2.88e+03	6.67e+02	1.33e+03	2.88e+03	5.28e+02	1.07e+03	1.94e+03
20	7.93e+02	1.54e+03	3.20e+03	7.51e+02	1.49e+03	3.18e+03	7.51e+02	1.49e+03	3.18e+03	5.31e+02	1.08e+03	2.39e+03
25	8.67e+02	1.67e+03	3.44e+03	8.12e+02	1.59e+03	3.38e+03	8.14e+02	1.60e+03	3.39e+03	5.58e+02	1.11e+03	2.50e+03
30	9.27e+02	1.78e+03	3.65e+03	8.84e+02	1.72e+03	3.61e+03	8.78e+02	1.71e+03	3.60e+03	6.25e+02	1.22e+03	2.58e+03
35	9.79e+02	1.87e+03	3.84e+03	9.41e+02	1.82e+03	3.80e+03	9.50e+02	1.83e+03	3.82e+03	6.55e+02	1.37e+03	2.81e+03
40	1.02e+03	1.94e+03	4.00e+03	9.82e+02	1.89e+03	3.96e+03	9.82e+02	1.89e+03	3.96e+03	7.26e+02	1.41e+03	2.98e+03
45	1.06e+03	2.01e+03	4.13e+03	1.03e+03	1.97e+03	4.09e+03	1.03e+03	1.97e+03	4.09e+03	7.70e+02	1.49e+03	3.18e+03
50	1.09e+03	2.07e+03	4.26e+03	1.06e+03	2.03e+03	4.21e+03	1.06e+03	2.03e+03	4.21e+03	7.85e+02	1.57e+03	3.36e+03
Avg	8.47e+02	1.63e+03	3.38e+03	8.12e+02	1.59e+03	3.34e+03	8.12e+02	1.59e+03	3.35e+03	6.02e+02	1.20e+03	2.52e+03

$\epsilon = 0.1$ and index IRR are used for TIM+ and KB-TIM, respectively

evident, especially for lower LS-perc, in Fig. 2 by observing the increasing slope of the line fitting the DC curves as LS-perc increases. Concerning the other two datasets, we observe from Tables 3 and 4 that low k is enough to obtain a DC value close to or with the same order of magnitude of DC values achieved by using highest k.

5.2 Comparison with Baselines

Figure 4 shows the delurking capital obtained by the three baselines in function of k and LS-perc, on Instagram; here we refer to this dataset only, since analogous findings were derived from the other two datasets.

At a first glance we observe, as expected, that the delurking capital trends are increasing with both the size of seed set and target set; although, in the Random case, the delurking capital obtained is substantially independent of the size of the target set. Consider however that both Random and bottom-LR methods achieve very low DC (below 50 for the Random, and 180 for bottom-LR, for the largest target set, i.e., LS-perc = 25%). In particular, results confirm our intuition that choosing globally low-ranked lurkers (i.e., users regarded as highly active in the SN) does not ensure a good influence spread towards a selected target set.

The best performing baseline is LargestDegree, whose DC values are of similar order of magnitude as DEvOTION although always lower than our algorithm. Moreover, even though peaking seeds with high out-degree might increase the probability of spreading the influence towards the target set, the reachability of target nodes is not guaranteed, like for the other baselines.

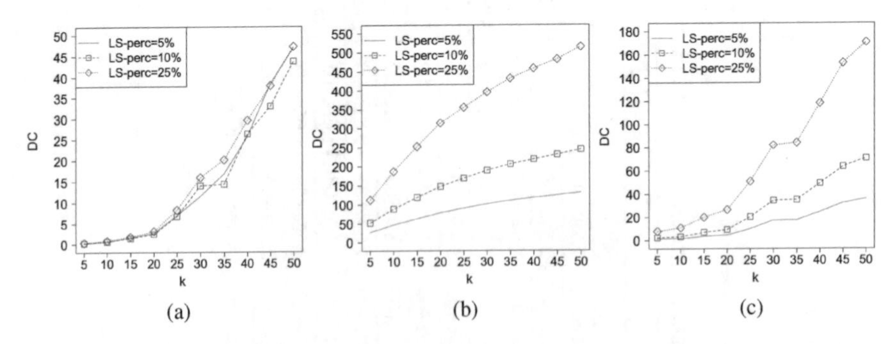

Fig. 4 Delurking capital in function of k and LS-perc: (**a**) Random, (**b**) LargestDegree, and (**c**) bottom-LR baselines, on Instagram

5.3 *Comparison with Influence Maximization Algorithms*

Comparison with SimPath SimPath shares with DEvOTION the kind of algorithmic approach to LT-based influence maximization, which includes the involvement of parameter η for controlling the enumeration of the paths through the diffusion graph. We compared the two algorithms with identical settings of η, specifically focusing on $\eta = 1.0\text{e}{-}03$ and $\eta = 1.0\text{e}{-}04$. Results of the comparative evaluation for $\eta = 1.0\text{e}{-}04$, in terms of DC performances, are reported in Tables 2, 3, and 4 (second group of three columns) and summarized in Table 5.

A first important remark that stands out is that the delurking capital obtained by SimPath is always lower than the one obtained by DEvOTION, for all datasets and configurations. The largest increment in DC is obtained for FriendFeed, with percentage of increment ranging from 3.97% (for LS-perc = 25%) to 10.8% (for LS-perc = 5%). Significant increment is observed also w.r.t. the other datasets, with percentages of increment ranging from 1.06% (for LS-perc = 25%) to 4.27% (for LS-perc = 5%) on GooglePlus, and from 1.6% (for LS-perc = 25%) to 5.69% (for LS-perc = 5%) on Instagram.

Impact of parameter η in the evaluation turns out to be marginal. Nevertheless, it should be noted that even though performance corresponding to $\eta = 1.0\text{e}{-}3$ (results not shown) are slightly lower in terms of DC, the increment w.r.t. SimPath is generally higher, indicating that DEvOTION is more robust than SimPath to path pruning.

As the size of target set increases, a general decreasing trend can be observed in the gap between DEvOTION and SimPath DC values. This is actually an expected outcome, since a larger target set means a larger overlap with the entire node set, thus letting the seed nodes picked up by a non-targeted algorithm reach a larger part of the target set.

Table 5 Summary of delurking capital performances of competitors against DEvOTION

	FriendFeed			GooglePlus			Instagram		
	5%	10%	25%	5%	10%	25%	5%	10%	25%
Gain vs. SimPath	557.69	832.02	587.30	34.67	47.90	35.44	4.85	6.64	5.67
Incr. vs. SimPath	10.80%	9.32%	3.97%	4.27%	3.02%	1.06%	5.69%	4.10%	1.60%
Gain vs. TIM+	513.14	791.25	518.70	34.32	47.01	33.72	3.82	5.08	4.00
Incr. vs. TIM+	9.85%	8.82%	3.49%	4.22%	2.96%	1.01%	4.43%	3.11%	1.12%
Gain vs. KB-TIM	2120.24	2503.13	4007.25	244.75	432.92	854.93	4.65	10.73	33.27
Incr. vs. KB-TIM	58.87%	34.48%	35.23%	40.66%	36.07%	33.86%	5.44%	6.79%	10.20%

Aggregate values are derived from Tables 2 and 3

Regarding comparison in terms of running times, DEvOTION turns out to be two orders of magnitude faster than SimPath, for the same choice of η. Considering the largest dataset, FriendFeed, DEvOTION shows average running time of $47,278$ s smaller than SimPath in its faster configuration (η = 1.0e–03, LS-perc = 5%), and of $182,131$ s in its slowest configuration (η = 1.0e–04, LS-perc = 25%).

Comparison with TIM+ TIM+ follows an approach to influence maximization that was designed to improve efficiency in previously existing algorithms, including SimPath. Therefore, it was not surprising to find out that TIM+ is indeed faster than DEvOTION. Taking FriendFeed as case in point, there is an average saving of 204 s for the fastest configurations (η = 1.0e–03 and LS-perc = 5% for DEvOTION, $\epsilon = 1.0$ for TIM+) , and of 6100 s for the slowest configurations (η = 1.0e–04 and LS-perc = 25% for DEvOTION, $\epsilon = 0.1$ for TIM+).

Nevertheless, DEvOTION always outperforms TIM+ in terms of DC on all datasets and for all configurations. The performance gain achieved by DEvOTION is more significant on FriendFeed, with average percentage of increment of 9.85% (for LS-perc = 5%), 8.82% (LS-perc = 10%) and 3.49% (LS-perc = 25%). Analogously to what we observed for the evaluation with SimPath and for the same reason, the gap in performance between DEvOTION and TIM+ decreases for larger target sets. Smaller but significant increment is also obtained on GooglePlus (average of 4.22% for LS-perc = 5%) and Instagram (average of 4.43% for LS-perc = 5%).

5.4 Comparison with KB-TIM

DEvOTION always performs significantly better than KB-TIM in terms of DC, on all datasets and with all configurations. Surprisingly, the gain in performance between DEvOTION and KB-TIM is larger than the one observed for other non-targeted competitors (i.e., SimPath, TIM+). Again, as previously seen for the comparative evaluation with IM algorithms, the highest increment corresponds to FriendFeed. As shown in Table 5, the percentage of average increment in DC ranges from 35.23% (for LS-perc = 25%) to 58.87% (for LS-perc = 5%). The increment is from 33.86% (for LS-perc = 25%) to 40.66% (for LS-perc = 5%) for GooglePlus, and from 5.44% (for LS-perc = 5%) to 10.20% (for LS-perc = 25%) for Instagram.

It should be noted that the comparison between DEvOTION and KB-TIM on Instagram is the only one following an opposite trend, w.r.t. other datasets and competitors, with the size of the target set: larger increments are achieved for larger target sets. This is probably due to the structural characteristics of Instagram and it is confirmed by a greater overlap between the seed sets of DEvOTION and KB-TIM than the seed overlap observed for the other datasets, as we shall discuss in the next section. As regards the execution times, KB-TIM performs comparably to TIM+, thus running always faster than DEvOTION and SimPath.

5.5 Seed Characteristics

Besides evaluation of the algorithms' performance in terms of spread (i.e., delurking capital), we also investigated possible commonalities and differences among DEvOTION and the competing methods in terms of characteristics of the identified seed sets. Specifically, we conducted two analyses: (1) a pair-wise evaluation of the overlaps between seed sets produced by the various algorithms, and (2) an inspection of topological characteristics of the seeds. For the latter analysis, we considered basic measures of node centrality, namely (out)degree, closeness, and betweenness. We will focus our study again on DEvOTION and the influence maximization algorithms.

Overlap of Seed Sets The normalized overlap of seed set between DEvOTION and the other algorithms has a generally increasing trend with the seed set size (k). (This is of course expected since the likelihood of overlap naturally increases with larger sets of nodes). Remarkably, there are two main observations that stand out, which in part can also be raised from Fig. 5 for some selected cases (i.e., $k = \{5, 50\}$ and LS-perc $= 5\%$).

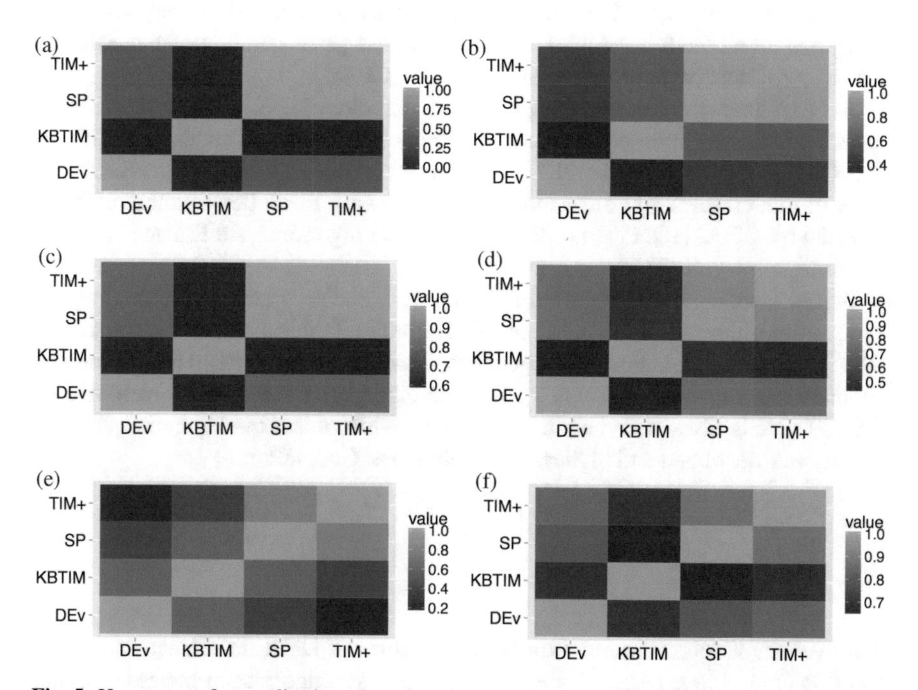

Fig. 5 Heat-maps of normalized overlap of seed sets produced by DEvOTION and competing IM algorithms. Results correspond to LS-perc $= 5\%$. (**a**) FriendFeed, with $k = 5$. (**b**) FriendFeed, with $k = 50$. (**c**) GooglePlus, with $k = 5$. (**d**) GooglePlus, with $k = 50$. (**e**) Instagram, with $k = 5$. (**f**) Instagram, with $k = 50$ (Abbreviations stand for: SimPath (SP), DEvOTION (DEv))

First, the minimum overlap of seed sets produced by DEvOTION is reached w.r.t. KB-TIM in all cases on all datasets (with the exception of LS-perc $= 5\%, k = 5$ on Instagram, where the minimum overlap is obtained with SimPath and TIM+). For larger k, the normalized overlap is within medium regimes, while it can be close or equal to zero on FriendFeed, but is medium to high on GooglePlus and Instagram for $k = 5$. Generally, KB-TIM behaves the most different from any other algorithm, regardless of the size of both seed sets and target sets.

Second, on all network datasets, SimPath and TIM+ produce seed sets that are very close to each other. This holds in general for every seed set size. In particular, considering the smallest and largest size of seed set, the two algorithms yield identical seed sets for $k = 5$ on FriendFeed and GooglePlus, while the overlap is still high for $k = 50$: 49, 45, and 42 out of 50, respectively on FriendFeed, GooglePlus, and Instagram. DEvOTION can have relatively high overlap with SimPath and TIM+, with the former as the major competitor, for higher LS-perc, on all datasets; specifically, the normalized overlap is up to about 0.75 on FriendFeed, and 0.9 on GooglePlus and Instagram. However, for lower LS-perc, the overlap is low (for smaller k) to medium (for higher k).

Topological Characteristics of Seeds Closeness is not a discriminating feature to distinguish between seeds identified by the various algorithms. It has very negligible variations for all seeds and for all algorithms in Instagram, and it is even constant in the other datasets (with order of magnitude of 1.0e−05).

Betweenness distribution is always much left-skewed for all algorithms in all datasets. One exception corresponds to GooglePlus where all seeds identified by DEvOTION (and TIM+ as well) have zero betweenness, which means that seeds were source nodes in the diffusion graph. In general, mean betweenness of seeds detected by DEvOTION is lower than in the other algorithms (in FriendFeed, even one order of magnitude lower).

Figure 6 shows the histogram of outdegree of the seed nodes identified by the various algorithms. Note that, while SimPath and TIM+ have very close trends, DEvOTION shows a less heterogeneous distribution w.r.t. KB-TIM. Moreover, compared to the other algorithms, minimum outdegree of seed nodes identified by DEvOTION is always higher in all datasets, while mean outdegree is generally smaller in FriendFeed and higher in Instagram and GooglePlus.

5.6 Discussion

The evaluation stages centered on the comparison of DEvOTION with other LT-based IM algorithms demonstrated how the proposed algorithm represents a better solution for a delurking task in online social networks, since it always shows higher performance in terms of delurking capital, on all datasets and setting of seed set and target size parameters.

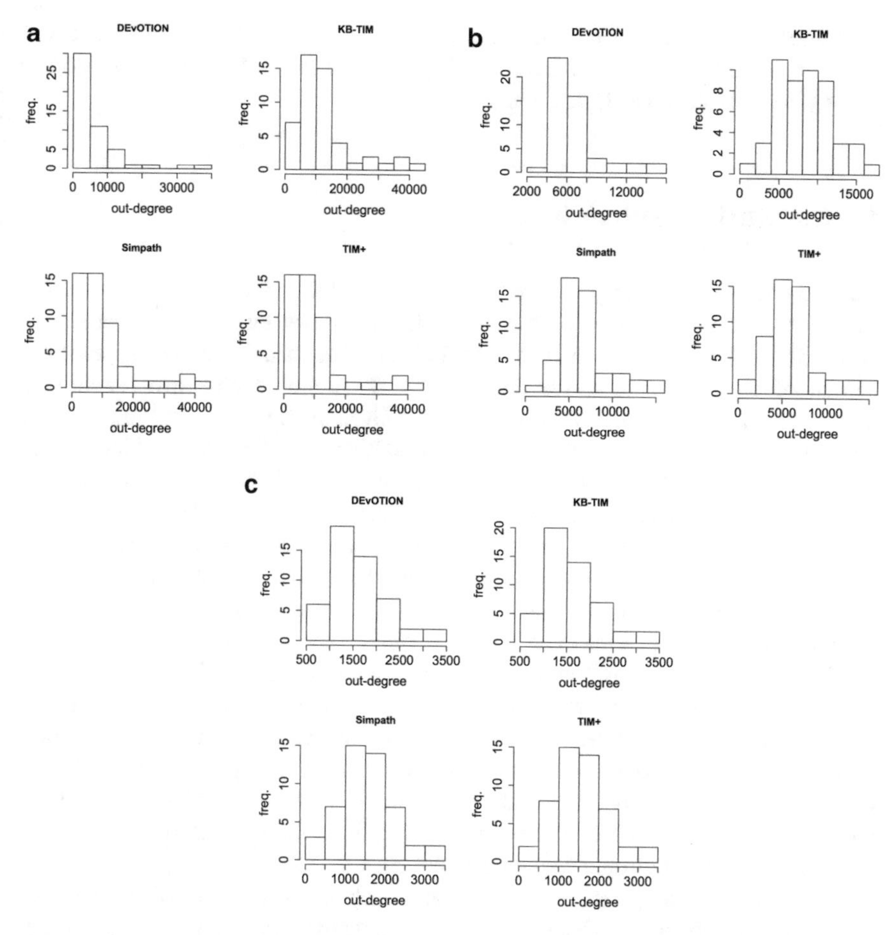

Fig. 6 Histograms of outdegree of the seed nodes identified by the various algorithms on (**a**) FriendFeed, (**b**) GooglePlus, and (**c**) Instagram

It is also worth noting that non-targeted IM algorithms (i.e., SimPath and TIM+) obtain a better DC score than a targeted algorithm like KB-TIM. The latter also tends to detect seed sets that differ the most from DEvOTION, especially for larger datasets. These results suggest that delurking users in online social networks is a complex task, for which ad-hoc solutions like DEvOTION might preferably be designed, and they cannot be equated to a standard keyword/query-based task like the one addressed by KB-TIM. Moreover, it turns out that adapting a targeted algorithm like KB-TIM to a slightly different task than the one for which it was devised leads to worse performance than using general purpose IM algorithms like SimPath and TIM+.

Also, concerning the time performance, DEvOTION runs significantly faster than SimPath, which is the closest competitor in terms of graph exploration and path enumeration strategy. DEvOTION is, as expected, slower than the other two algorithms using different heuristics, TIM+ and KB-TIM. The opportunity of developing similar heuristics for a delurking task, than obtaining a faster version of DEvOTION, is leaved as future work.

6 Conclusions and Future Work

We proposed the first computational approach to delurking, i.e., to engaging users that take on a lurking role in SNs. We defined a novel targeted IM problem in which the objective function to be maximized is defined in terms of delurking capital of the target users. We proved that the proposed objective function is monotone and submodular, by using the LT-equivalent live-edge graph model, and developed an approximate algorithm, DEvOTION, to solve the problem under consideration. Significance and effectiveness of DEvOTION have been assessed, also in comparison with baselines and state-of-the-art non-targeted and targeted LT-based influence maximization methods. Experimental evaluation was conducted on network graphs that were originally built upon data crawled/fetched from real SN web sites. Remarkably, our evaluation datasets are all publicly available, which is an important aspect in terms of repeatability of the experiments.

There are several opportunities of research that could be realized in the future. For instance, we plan to extend our objective function by taking into account structural features of the seed nodes to be selected, which might impact on their diversity. The latter could also benefit of the integration in our approach of user characterization based on topic modeling and/or sentiment analysis. Further, it would be interesting to investigate on reducing the running time of DEvOTION, by exploiting efficient implementation based on, e.g., RIS (cf. Sect. 2).

We would like to point out that our approach is independent of the particular strategy of delurking (being it based on rewards, welcome messages, etc.). Demonstrating DEvOTION on a real case study is beyond the scope of this paper, nevertheless we believe that the development of a web-based system embedding our approach can really aid to support engagement of users. The proposed approach is also versatile as it can easily be generalized to deal with other targeted IM scenarios. Therefore, we envisage further developments from various perspectives, including human–computer interaction, marketing, and psychology.

References

1. Bakshy, E., Rosenn, I., Marlow, C., Adamic, L.A.: The role of social networks in information diffusion. In: Proceedings of World Wide Web Conference (WWW), pp. 519–528 (2012)
2. Bishop, J.: Increasing participation in online communities: a framework for human-computer interaction. Comput. Hum. Behav. **23**, 1881–1893 (2007)
3. Borgs, C., Brautbar, M., Chayes, J.T., Lucier, B.: Influence maximization in social networks: towards an optimal algorithmic solution. CoRR abs/1212.0884 (2012)

4. Celli, F., Lascio, F.M.L.D., Magnani, M., Pacelli, B., Rossi, L.: Social network data and practices: the case of friendfeed. In: Proceedings of International Conference on Social Computing, Behavioral Modeling, and Prediction (SBP), pp. 346–353 (2010)
5. Edelmann, N.: Reviewing the definitions of "lurkers" and some implications for online research. Cyberpsychol. Behav. Soc. Netw. **16**(9), 645–649 (2013)
6. Goyal, A., Lu, W., Lakshmanan, L.V.S.: SIMPATH: an efficient algorithm for influence maximization under the linear threshold model. In: Proc. IEEE International Conference on Data Mining (ICDM), pp. 211–220 (2011)
7. Guille, A., Hacid, H., Favre, C., Zighed, D.A.: Information diffusion in online social networks: a survey. SIGMOD Rec. **42**(2), 17–28 (2013)
8. Guler, B., Varan, B., Tutuncuoglu, K., Nafea, M.S., Zewail, A.A., Yener, A., Octeau, D.: Optimal strategies for targeted influence in signed networks. In: Proceedings of International Conference on Advances in Social Networks Analysis and Mining (ASONAM), pp. 906–911 (2014)
9. Guo, J., Zhang, P., Zhou, C., Cao, Y., Guo, L.: Personalized influence maximization on social networks. In: Proceedings of ACM Conference on Information and Knowledge Management (CIKM), pp. 199–208 (2013)
10. Interdonato, R., Pulice, C., Tagarelli., A.: "Got to have faith!": the DEvOTION algorithm for delurking in social networks. In: Proceedings of International Conference on Advances in Social Networks Analysis and Mining (ASONAM), pp. 314–319 (2015)
11. Kempe, D., Kleinberg, J.M., Tardos, E.: Maximizing the spread of influence through a social network. In: Proceedings of ACM SIGKDD International Conference on Knowledge Discovery and Data Mining (KDD), pp. 137–146 (2003)
12. Lagnier, C., Denoyer, L., Gaussier, E., Gallinari, P.: Predicting information diffusion in social networks using content and user's profiles. In: Proceedings of European Conference on Information Retrieval (ECIR), pp. 74–85 (2013)
13. Lai, H., Chen, T.T.: Knowledge sharing in interest online communities: a comparison of posters and lurkers. Comput. Hum. Behav. **35**, 295–306 (2014)
14. Li, H., Bhowmick, S.S., Sun, A., Cui, J.: Conformity-aware influence maximization in online social networks. VLDB J. **24**, 117–141 (2015)
15. Li, Y., Zhang, D., Tan, K.: Real-time targeted influence maximization for online advertisements. PVLDB **8**(10), 1070–1081 (2015)
16. McAuley, J.J., Leskovec, J.: Learning to discover social circles in ego networks. In: Neural Information Processing Systems (NIPS), pp. 548–556 (2012)
17. Nonnecke, B., Preece, J.J.: Lurker demographics: counting the silent. In: Proceedings of ACM Conference on Human Factors in Computing Systems (CHI), pp. 73–80 (2000)
18. Preece, J.J., Nonnecke, B., Andrews, D.: The top five reasons for lurking: improving community experiences for everyone. Comput. Hum. Behav. **20**(2), 201–223 (2004)
19. Saito, K., Ohara, K., Yamagishi, Y., Kimura, M., Motoda, H.: Learning diffusion probability based on node attributes in social networks. In: Proceedings of International Symposium on Methodologies for Intelligent Systems (ISMIS), pp. 153–162 (2011)
20. Sun, N., Rau, P.P.L., Ma, L.: Understanding lurkers in online communities: a literature review. Comput. Hum. Behav. **38**, 110–117 (2014)
21. Tagarelli, A., Interdonato, R.: "Who's out there?": identifying and ranking lurkers in social networks. In: Proceedings of International Conference on Advances in Social Networks Analysis and Mining (ASONAM), pp. 215–222 (2013)
22. Tagarelli, A., Interdonato, R.: Lurking in social networks: topology-based analysis and ranking methods. Soc. Netw. Anal. Min. **4**(230), 27 (2014)
23. Tagarelli, A., Interdonato, R.: Time-aware analysis and ranking of lurkers in social networks. Soc. Netw. Anal. Min. **5**(1), 23 (2015)
24. Tang, F., Liu, Q., Zhu, H., Chen, E., Zhu, F.: Diversified social influence maximization. In: Proceedings of International Conference on Advances in Social Networks Analysis and Mining (ASONAM), pp. 455–459 (2014)

25. Tang, Y., Xiao, X., Shi, Y.: Influence maximization: near-optimal time complexity meets practical efficiency. In: Proceedings of ACM SIGMOD International Conference on Management of Data (SIGMOD), pp. 75–86 (2014)
26. Watts, D.J.: A simple model of global cascades on random networks. PNAS **99**, 5766–5771 (2002)
27. Weng, L., Ratkiewicz, J., Perra, N., Gonçalves, B., Castillo, C., Bonchi, F., Schifanella, R., Menczer, F., Flammini, A.: The role of information diffusion in the evolution of social networks. In: Proceedings of ACM SIGKDD International Conference on Knowledge Discovery and Data Mining (KDD), pp. 356–364 (2013)
28. Yang, D., Hung, H., Lee, W., Chen, W.: Maximizing acceptance probability for active friending in online social networks. In: Proceedings of ACM SIGKDD International Conference on Knowledge Discovery and Data Mining (KDD), pp. 713–721 (2013)
29. Zhou, J., Zhang, Y., Cheng, J.: Preference-based mining of top-*k* influential nodes in social networks. Futur. Gener. Comput. Syst. **31**, 40–47 (2014)

Social Engineering Threat Assessment Using a Multi-Layered Graph-Based Model

Omar Jaafor and Babiga Birregah

1 Introduction

Social engineering could be understood as the science of getting people to comply with one's wishes [1]. It relies on human biases bypassing technological security mechanisms [2, 3] which renders it particularly difficult to detect. Hence, a social engineering attack might occur without security professionals ever knowing about it until the organization starts suffering the damages caused by it. There exists a wide array of successful social engineering attacks, many of which are intuitive like dumpster diving or shoulder surfing [4]. Other attacks are more elaborate requiring the nurturing of relationships with victims. The renown security expert Kevin Mitnick described a variety of attacks that use authority and empathy to deceive people [5].

Social engineers usually follow a pattern consisting of four phases: information gathering, relationship development, execution, and exploitation [3, 5]. The three first phases are likely to occur outside an organization's border, allowing the social engineer to take advantage of multiple web based tools like social network search mechanisms or forum and blog aggregators. The availability of web automation and web scraping tools has further reduced the skills and efforts required to launch automated social engineering attacks on web platforms like on-line social networks [6, 7]. These platforms contain valuable information about the lifestyle, schedule, and relationships of victims who often disclose it without being aware [8, 9]. This information helps launching context aware attacks that would make the victim naturally believe the authenticity of the messages he receives [10]. Some context aware attacks are as simple as name dropping and others involve impersonating

O. Jaafor (✉) • B. Birregah
Charles Delaunay Institute, UMR CNRS 6281, University of Technology of Troyes, Troyes, France
e-mail: omar.jaafor@utt.fr; babiga.birregah@utt.fr

R. Missaoui et al. (eds.), *Trends in Social Network Analysis*, Lecture Notes in Social Networks, DOI 10.1007/978-3-319-53420-6_5

people a victim might know [11]. In the case of phishing attacks for example, an experiment conducted by Jagatic et al. [12] suggests that they are four times more likely to succeed if the victim is solicited by someone who appears to be an acquaintance.

As part of a social engineering prevention policy, it is recommended to perform accurate vulnerability assessment of a system to social engineering attacks [13, 14]. Despite the harmful consequences of social engineering, modeling threats from these attacks has received little attention which leaves security officers with models that were developed for other purposes like network vulnerability assessment. Monitoring different blogs, forums, and social networks would allow the disclosure of valuable information in regard to the channels favored by social engineers, the users most vulnerable to social engineering attacks and trends and most common approaches used by social engineers.

Network vulnerability assessment models usually link an action to a set of observable consequences of this action. This is typically achieved by considering nodes to represent states in an information system. Hence, a security expert would only monitor state variables to deduce a set of possible attacks that could be performed. The logic of assessing social engineering threats is different as many of these attacks do not leave any observable trace in an information system (or the monitored segment of it). Representing these actions is useful as it would allow assessing the efforts and skills required to perform attacks in addition to estimating their likelihood and effects. Other reasons are to tailor security awareness programs to account for these attacks, and to make an informed decision on the sources of information that security professionals should monitor (social networks, blogs, specific websites).

Multi-layered graphs allow modeling interconnections between actions in different contexts where each context is represented on a separate layer. This paper describes a model for social engineering vulnerability assessment based on these graphs. This model allows both the representation of different contexts of an attack and the representation of steps and substeps of attacks which do not change the state of the monitored information system. It extends the work presented in [15] by allowing to model identity edges within the same context. Many public figures and companies take the decision to have several accounts on the same social network (context), and these accounts are often linked to different users and face different threats. It is hence worthwhile linking these accounts by identity edges, as they constitute for a social engineer different routes to reach an entity. This paper also presents more detailed case studies and a method that uses the proposed model for the detection of malicious content.

This paper is organized as follows: Sect. 2 presents some related works about vulnerability assessment. To our knowledge, there exists no model that is constructed specifically for social engineering vulnerability assessment. We will nevertheless present some models related to social network vulnerability assessment, network vulnerability assessment, and phishing assessment that could also be adapted for social engineering vulnerability assessment. We then describe the proposed multi-layered graph based model in Sect. 3. Finally, Sect. 4 presents three case studies where different activities related to vulnerability assessment are performed using the proposed graph model.

In the first case study, a method for detecting malicious users that relies on the proposed model is presented and used for the detection of stolen credit card resellers. It highlights the advantages of detecting malicious users using the different contexts where they interact. These users, often part of a malicious organization, are driven by requirements to be profitable and effective. This makes them as explicit and visible as a social network provider or blog allows them to be. It is hence effective to detect them in contexts where they are less concerned about being detected and perform entity mappings to find their accounts in other social networks and blogs where they have a user base and would be more cautious about their activities. The second case study concerns the assessment of threats from a social engineering attack (cross-site cloning) and a reverse social engineering attack (recommendation based attack). The third case study is focused on the detection of vulnerable users to multiple cross-site attacks.

2 Related Work

This section first starts by presenting related works in regard to vulnerability assessment models. It then presents some multi-layer graph models that could be adapted for vulnerability assessment.

2.1 Vulnerability Assessment

There has been very little research that concerns vulnerability assessment from social engineering. We present in this section models aimed at vulnerability assessment from various types of threats and that could be adapted to social engineering.

Jakobsson [10] proposed a graph based model for phishing attacks. His phishing digraph is composed of two different sets of nodes. The first representing access to some information and the second access to some resource. Edges represent actions that an adversary can perform. Hence a directed edge (n_1, n_2) would signify that access to the resource or information n_2 could be achieved by an attacker given he accessed n_1. Although this model could be used to assess complex phishing attacks, some of its drawbacks is that it doesn't account for potential victims (or groups of victims), multiple attackers for coordinated phishing attacks [16] or different contexts.

Vida et al. [17] used a multi-layered approach to assess malware infection on a multilayer network composed of a social network of statistical physicists. They showed that the interconnection between layers increases the infectivity of viruses in the studied network. Their model assesses the evolution of malware infection without offering the means to analyze and visualize specific types of infections, like ones resulting from a particular attack. Also, their multilayer network assumes that a node should be in the same state in every layer, which would not allow to realistically model social engineering attacks as nodes on different layers could be in completely different states. For example, a user could have a compromised social network account and a safe email account.

Jaafor et al. [18] provided a privacy threat assessment framework for on-line social network aggregators. This framework only allows the assessment of privacy threats coming from aggregators without evaluating the threats emanating from the use of their underlying social networks. It is thus to be considered as a complementary framework that focuses on a specific type of vulnerabilities.

Perez et al. [19] proposed a framework for the detection of potentially harmful profiles from the ego-network of smartphone users. This framework is based on the analysis of the presence of user contacts on different layers of interaction (phone book, social networks) in order to assess their legitimacy as contacts. Hence, it provides a method of performing one type of vulnerability assessment and could be complementary to a comprehensive framework.

There are several models aimed at assessing network vulnerabilities that could be adapted to the assessment of the vulnerabilities from social engineering [20–23].

Phillips and Swiler [20] proposes a graph based model which represents each state of a system as a distinct node, while edges represent actions that change the state of the system. These edges have conditions attached to them like the occurrence of an event or the level of expertise of the attacker for example. They categorize attacker profiles into different types. Hence, vulnerabilities could be assessed based on the expertise and resources that would be required of an attacker allowing the ranking of vulnerabilities based on the ease by which they could be performed. One of the drawbacks of this model if used to assess social engineering threats is that it renders difficult deducing the relationships and similarities between resources and users one wishes to protect, given that an entire state of a system is modeled by the attributes of a single node.

A graph based model proposed by Ammann et al. [21] considers exploits as atomic transformations that generate a set of postconditions when a set of preconditions are verified. This graph based model considers nodes to represent the preconditions and postconditions of an attack. An exploit is associated with (1) a set of nodes (preconditions and postconditions), (2) an action, and (3) a set of users performing the action. The set of labeled edges going from the precondition nodes to the postcondition nodes model an exploit. An attack needing several exploits is hence represented by nodes on several layers where initial precondition nodes of the attack would be in the first layer and the final postcondition node of the attack would be in the last layer. Nodes at each intermediary layer would be the preconditions of exploits having their postconditions on the subsequent layer. Layers in this model are only used to split an attack into several steps rather than account for the different context of an attack. Although this model would allow to show relationships between resources and users, it would not be able to model attacks that do not change the state of the supervised system and would not account for the different contexts that could be used for an attack.

2.2 *Multilayer Graphs for Social Networks*

Several multi-layer models for social networks could be adapted for vulnerability assessment. As most of these models are aimed at modeling human interactions, many would not represent valuable aspects of interactions for vulnerability assessment like disjunction (only one interaction has to occur) or conjunction (all interactions must occur), etc.

Multi-layered models that represent interactions on social networks [24–26] generally spur from the sociological conception of these networks: a set of users having different types of relationships. The only inter-layer connections in these models are the ones mapping an entity to its occurrences in different layers, making these models not suited for representing the diversity of inter-layer connections in a vulnerability assessment model. Boccaletti et al. [27] provides a literature review on different models that could be adapted to represent interactions on multi-layered networks (networks of networks for example).

Kivelä et al. [28] proposed a model based on a multi-dimensional layer structure. They consider each dimension (which they call an aspect) to hold a set of "elementary layers." A layer would be composed of an elementary layer on each dimension. An example is a multi-layered network with two dimensions: (1) type of relationship (2) time. Considering that type of relationship is composed of the following elementary layers: {friend, colleague, kin} and that the time layer contains the following elementary layers: {time 1, time 2}, a layer would be identified by one elementary layer in each dimension: {colleague, time 2}. One of the limitations of this model when applied to vulnerability assessment is that it considers nodes to be of the same type unless they are on different layers. This would lead to increasing the number of layers making the model complex to analyze. It also is unable to represent conjunction and disjunction.

3 A Multi-Layered Model for Social Engineering Vulnerability Assessment

The objective of the proposed multi-layer model $\mathscr{M}$ is to bridge the gap between multi-layered graph models for social networks that allow to represent the variety of user interactions and the multiplicity of their identities and models for vulnerability assessment that were tailored for assessing a variety of vulnerabilities but are not adapted to social engineering vulnerability assessment. The proposed model is inspired from [28] and from models for network vulnerability assessment [20, 21].

A way to represent different types of nodes in the model proposed by Kivelä et al. [28] would consist of defining a different dimension for each type of node. As one of the design criteria for a vulnerability assessment model is simplicity, the proposed graph-based model is based on a reduced number of layers and different types of nodes in every layer. The proposed model also allows for the representation of disjunction and conjunction edges that are very useful for vulnerability assessment.

This model comprises the following three layer dimensions:

- The contexts that an attacker could use $C = \{c_1, c_2, \ldots, c_N\}$. Examples of contexts are social networking platforms (Facebook, Twitter, LinkedIn, etc.), phone, email, blogs, human interaction, etc. Whether an attacker leverages a network of friends and friends of friends (Facebook, Orkut, etc.), a status update network (Twitter) or email is important in determining his ability to gain the trust of victims. Also, when assessing vulnerabilities of a set of users, the contexts where they could be reached helps determining the threats they are exposed to.
- The different states allowing a successful attack $S = \{s_1, s_2, \ldots, s_M\}$, where a state is a snapshot of a system at a particular step of an attack. Some models store all the information about a state of a system in a node, which is not suited for social engineering given the wide range of information that every single node would have to store. For this reason, state layers are used. These layers also represent actions that do not change the state of a system, which are of great value to assessing social engineering attacks that heavily rely on these actions.
- The modeled attack scenarios $A = \{a_1, a_2, \ldots, a_H\}$. Some attacks resemble sufficiently each other to group them in one attack scenario, allowing the reuse of components that model an attack in modeling other ones. This has several advantages like reducing the time and effort needed to model attacks and highlighting the similarities between attacks.

The set of layers in $\mathscr{M}$ is defined by $L = \{C \times S \times A\}$.
We consider two types of nodes V_{actor} and V_{resource} in $\mathscr{M}$:

- V_{actor}: the set of vertexes representing specific or generic victims or attackers.
- V_{resource}: the set of vertexes representing a resource potentially used by an attacker. Examples of resources are email servers, web pages, search engines, or phone directories.

Let $\mathrm{V} = V_{\text{actor}} \cup V_{\text{resource}}$ be the set of distinct entities (actors and resources) in $\mathscr{M}$ that could be present on one or more layers in L. The set of nodes $V_{\mathscr{M}}$ in the multi-layered network $\mathscr{M}$ is a subset of $V \times Ł$. A node in $V_{\mathscr{M}}$ could hence be described by the pair (v_i, l_α) where v_i indicates the actor or resource in $\mathrm{V}{=}V_{\text{actor}} \cup V_{\text{resource}}$ and l_α indicates in which layer the node appears.

Nodes are described by a set of attributes that hold information about their state. An example would be a node representing a user on an on-line social network that is described by attributes indicating the users that could download his profile, the groups he is member of and the email address he used to register in the social network.

We define two different sets of edges $E_{\mathscr{M}} = E_{\text{action}} \cup E_{\text{identity}}$. Both sets could either contain inter-layer or intra-layer edges.

- E_{action}: contain edges that represent the actions required by an attacker to perform a step in an attack. These edges have two labels attached to them describing the actions that should be performed (action label) and the attack the actions are part of (attack label). Edges also have attributes attached to them that could indicate a variety of elements like the probability of success of an action or skills and effort required by an attacker to perform an action. An action edge could go from an actor node or a resource node to a different actor or resource node. These edges could only cross state layers forward or remain in the same state layer. This means that an action could go from state s_α to state $s_{\alpha+1}$ but not the reverse. This spurs from the idea that no successful action requires an attacker to backtrack [21].

Also, action edges do not cross layers belonging to different attack scenarios.

Two action edges in the same state layer with the same action label and the same attack label indicate that either one of the actions needs to be performed in order for an attack to move to the subsequent state (disjunction). Actions in the same state layer with different action labels and the same attack label indicate that all of these actions need to be performed in order for the attack to move to the next state layer (conjunction).

- E_{identity}: contain edges that map nodes representing the same entity together. An example would be a user that has accounts on different social networks in addition to an email account. These edges would link the different social network accounts and the email account together. Identity edges could either link an actor node to another actor node or a resource node to another resource nodes.

Edges in $E_{\mathscr{M}}$ can be described by $(v_{i_1}^{l_{\alpha_1}}, v_{i_2}^{l_{\alpha_2}})$, where $v_i \in V$ indicates the entity the edge is incident to and $l_\alpha \in L$ indicates the layer in which that entity is present.$(v_{i_1}^{l_{\alpha_1}}) \in V_{\mathscr{M}}$ represent the node an edge is incident to. An intra-layer edge would imply that $l_{\alpha_1} = l_{\alpha_2}$ and an inter-layer edge would imply that $l_{\alpha_1} \neq l_{\alpha_2}$. If $v_{i_1} \neq v_{i_2}$, the edge would belong to E_{action} and if $v_{i_1} = v_{i_2}$, it would belong to E_{identity}.

The multilayer network $\mathscr{M}$ is defined by its set of nodes $V_{\mathscr{M}}$, its set of edges $E_{\mathscr{M}} \subset V_{\mathscr{M}} \times V_{\mathscr{M}}$ and its layers $L = C \times S \times A$.

$$\mathscr{M} \subset V_{\mathscr{M}} \times E_{\mathscr{M}} \times L \tag{1}$$

Figure 1 illustrates two different layers of $\mathscr{M}$ in the same attack scenario and state but in different contexts. The circles represent attackers and the triangles represent

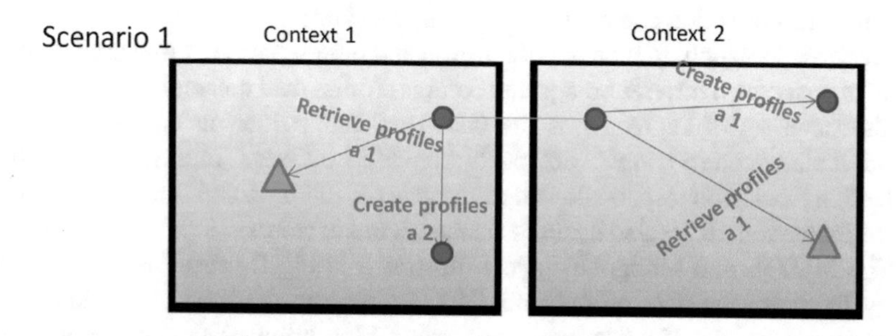

Fig. 1 An example of two layer in the vulnerability assessment model

resources. Two different attacks are modeled in this attack scenario. This is indicated by edge labels “a 1” and “a 2” which signify that the edges are part of attack 1 or attack 2. In order for attack 1 to move to a subsequent state layer, the attacker has to create new profiles indicated by the topmost right edge with the label “Create profiles.” He also must either retrieve user profiles from the context represented by the rightmost layer or the context represented by the leftmost layer. This is shown by two edges having the same action label “Retrieve profile” and having the same attack label “a 1”. These edges represent two different possibilities for an actor to conduct an action where only one of them should succeed for the attack to move to the subsequent state. The two “create profile” edges have different attack labels and are thus considered as two unrelated actions despite having the same action label.

4 Case Studies

We will provide three case studies where the proposed model is used to (1) detect malicious stolen credit card resellers, (2) to evaluate threats from a social engineering attack and a reverse social engineering attack, and (3) to detect vulnerable users to a specific social engineering attack.

4.1 Data Collection

The data collection process is illustrated in Fig. 2. The first step was to provide a set of keywords to the data collector that sends requests to the Twitter Streaming API and to the Webhose.io API. The former was queried twice, on the 29th of June 2015 at 12:30 opening a stream of data until the 1st of July at the same time and on 27th of December at 4:15 pm to the 7th of January at 5 am Greenwich time. During the first data collection, a total of 1,919,806 tweets were retrieved with 625,367 distinct profiles. The second 12 day data collection period resulted in the retrieval of 3,103,907 tweets. The keywords that were used in the data collection were related to bank cards and on-line payment: mastercard, visa, amex, ccv, ccv2, paypal and the first data collection contained also the keyword dump.

Webhose.io which offers a popular forum and blog search API was used twice to retrieve posts from different blogs and forums. A first data collection was launched on the same set of keywords as the ones used for Twitter on the 1st of July at 14:35. It allowed retrieving 73,860 posts from 6557 different forums. A second data collection from Webhose.io allowed to retrieve 62,205 blog and forum posts on the 4th of January at 10:30 also using the same set of keywords.

The MySQL and MongoDB layers illustrated in Fig. 2 were used to store and query the collected data on demand. The Twitter and the intra-blog relationship mappers construct a Twitter graph and many blog graphs based on interactions between users of these platforms (retweets, replies, comments). The inter-blog

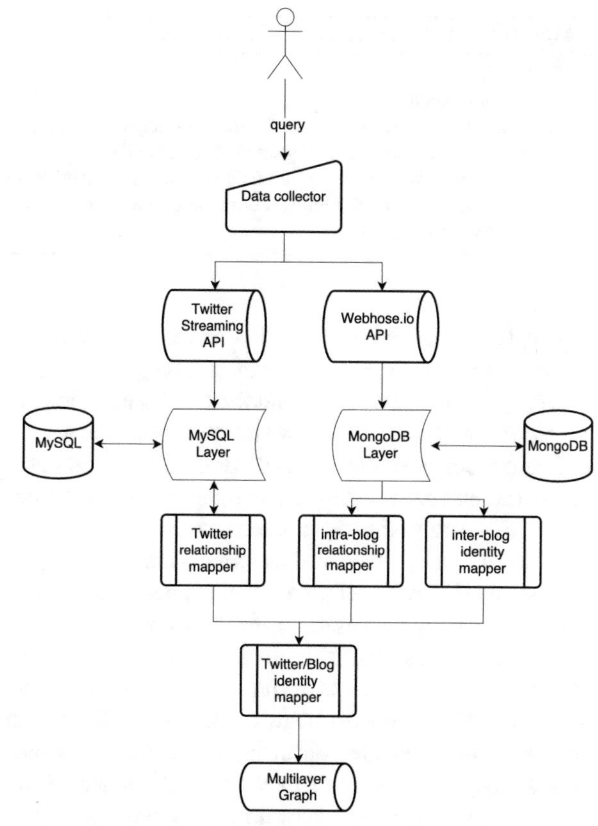

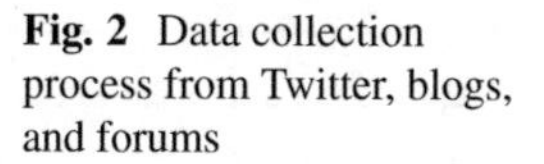
Fig. 2 Data collection process from Twitter, blogs, and forums

identity mappers then connect the many blog graphs by creating identity edges between users on blogs. The Twitter/blog identity mapper links the blog and forum multilayer graph to the Twitter graph by creating identity edges between Twitter nodes and blog nodes, which results in the generation of multilayer graphs $\mathcal{M}_x$.

4.2 Detection of Stolen Credit Card Resellers

Blogs, forums, and on-line social networks have become popular for the exchange of different kinds of stolen goods. Among these goods, the exchange of stolen credit card information (carding) is very common. These can be easily found on different blogs and social networks with prices from 1 to 40 dollars depending on their balances [29]. Stolen credit card resellers might pose a great threat to the security of a network as many have access to malware such as key loggers or might host phishing websites that allows them to steal credit card credentials. Hence, users that could interact with credit card resellers for malicious purposes, out of curiosity

Algorithm 1: Detection of malicious users

Repeat:
1: Train Classifier
2: Classify posts from social networks and blogs.
3: Construct a multi-layered graph model from the classified data :
4: Perform Entity mapping on user accounts (inter/intra layer).
5: Detect users that interact with malicious posts.
6: Update training set

or simply because they couldn't perceive that a user is a credit card reseller could become vehicles to spread malware and illegal activities. This case study focuses on the analysis of card resellers and the forums and blogs that relay their posts.

Algorithm 1 describes how the proposed model could be leveraged to detect different types of malicious users present in social networks and blogs (depending on the training set). It starts by training a classifier to detect malicious users, which are credit card resellers in this case study. The next step is to build a multi-layered graph model using the classified data. This requires performing entity mapping on user accounts. Each account that is mapped to an account that published malicious content is usually also used for malicious purposes. Then, interactions on the multi-layered graph are analyzed where accounts that relay malicious posts are often either malicious accounts trying to give credibility to the malicious content or are vulnerable users. An analysis on the constructed multi-layered graph allows obtaining many insights on malicious users, like which blogs or social networks they interact in or whether they are well connected in the graph which increases their danger. This would permit security professionals to evaluate the threats they pose and to respond accordingly. For example, they could block the blogs and social networks leveraged by malicious users, or block the URLs posted by malicious users. The final step in this proposed method is to update the training set, and when new data is available, update the dataset that will be used for the next classification.

The first step of this case study was performed by building a binary supervised classification model to predict the labels of a bag of words representation of posts from the second data collection on Webhose.io (launched on the 4th of January at 10:30). Each post was stripped from nonalphabetical characters and every word was transformed to its lowercase.

The training set was built from different posts that were collected on the 10th of December using Webhose.io and boardreader. It was composed of 486 posts that were not carding posts and 172 that were carding posts. Although the training dataset is not very big, it was enough to detect carding posts given the relative homogeneity of data that was collected. The results of many binary classifications using Weka 3.7 with tenfold validation are summarized in Table 1. The default parameters of Weka 3.7 were used for every classification.

We selected the model built by Random Forest with a 100 trees for this experimentation as it gave with logistic regression the best results in addition to having a lower false negative rate. The model constructed using the Random Forest

Table 1 Classification of carding posts

Algorithm	Well classified (%)	False pos. (%)	False neg. (%)
Naive Bayes	95.28	10.46	2.67
Logistic regression	98.32	2.9	1.23
AdaBoost	96.50	8.72	1.64
LibSVM	96.20	13.37	0.41
J48	96.96	5.81	2.05
RandomForest	98.17	5.23	0.2

Table 2 Summary of the training dataset and predictions by Random Forest

Indicator	Value
# of positives	600
# of negatives	61,605
# of users with a carding post	249
# of users	31,711
# of users with no carding post	31,710
# of forums	8709
# of forums with a carding post	82
# of forums with no carding post	8627

algorithm predicted that 600 posts were about selling stolen credit cards which is 0.96% of the total posts that were classified. Table 2 summarizes the information about the test dataset in addition to predictions obtained using the Random Forest model. As the number of forums and blogs where carding posts were found is relatively small in comparison with the total number of forums collected, it is easier for security professionals to monitor or block some of these sites.

The next steps of this experimentation involve using the proposed graph based multi-layer model to represent the collected posts. The objective of modeling this data is to gain more information on malicious users, forums, and blogs that host carding posts.

We consider a multi-layer graph model $\mathscr{M}_1$ composed of the three following elementary layers:

- $\mathscr{C} = \{c_1, c_2, \ldots, c_{8709}\}$ is the context elementary layer. We consider in this experimentation each context to be a specific blog or forum. Hence users who have posted in a thread of a blog or forum are part of the same context layer.
- It would be possible to consider multiple state layers (one per month, or one per week, etc.) in order to gain insights on the evolution of stolen card reseller communities, but for simplicity, we only used one state layer $S = \{s_1\}$.
- One attack scenario is considered for this case study $A = \{a_1\}$.

The built model consists of one type of nodes and two types of edges. Every node holds either a label “legitimate user” or a label “malicious user” for card resellers. An edge exists between two nodes if they posted a message on a thread of the same forum or blog.

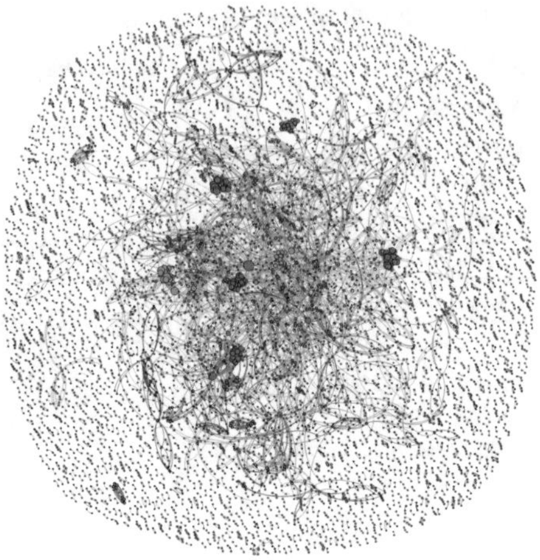

Fig. 3 Graph G_1 of interactions discussing credit card topics on blogs and forums

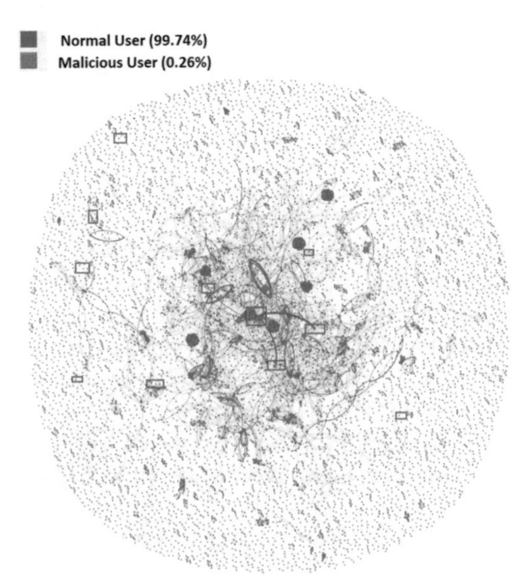

Fig. 4 Graph G_1 highlighting spammers that attempt to sell stolen credit cards

Figure 3 illustrates a graph G_{blogs} that was constructed from $\mathscr{M}_1$. Nodes that have posted on a thread and did not get any replies were omitted from G_{blogs} for the sake of visualization, but despite not interacting with any other node, they are considered as part of the context layer (which represents a blog or forum) in which they have posted their message in $\mathscr{M}_1$. Every color in this figure refers to a different blog or forum (context) and the size of nodes is proportional to their degree. Figure 4 highlights credit card resellers that have the highest degree. These nodes have either responded to posts or were replied to more often than other nodes and could constitute the highest threats for nonmalicious nodes. Nonmalicious nodes

Table 3 Description of G_{blogs}

Indicator	Value
Number of nodes	8876
Number of edges	15,742
Average degree	3.08
Average weighted degree	4.02
Number of connected components	2847

with a high degree are also central as if compromised, they would allow to rapidly infect the network.

Table 3 presents some indicators that describe graph G_{blogs}. We could see from this table that the weighted average degree is not very high in comparison with the average degree, indicating that a very small subset of nodes interact regularly with each other. Given the high number of nodes with a small degree. Monitoring these nodes with regular interactions would allow to rapidly react in case the network is infected.

Entity mapping on blogs, forums, and social network allows to link user accounts in different platforms to one entity. While there are multiple methods that could be considered for this study [30–32], we performed a basic entity mapping method for the sake of this experimentation that considers two accounts to be linked to the same user if they have posted a message containing the same email address or the same ICQ (I seek You) identifier. ICQ is an instant messaging computer program that is very popular among stolen credit card resellers.

After the use of entity mapping on ICQ, it was found that 91 posts that were initially classified by the random forest algorithm as not related to credit card resellers had a common ICQ identifier with an account that has posted a message classified by the Random Forest algorithm as malicious. We manually verified each one of the 91 post and found that they were all attempting to sell stolen credit card information. Hence, a basic entity mapping based on ICQ identifiers allowed to increase the number of carding posts found by more than 15%.

We also performed an entity mapping on email addresses, but obtained much lesser matches between posts that were classified as related to carding and posts classified as unrelated to carding. Only ten posts that were not classified as carding posts were identified to be carding posts using entity mapping on emails. These ten posts were also found using entity mapping on ICQ identifiers.

We then mapped malicious users found on forums to the users in the second Twitter dataset comprised of more than 3.1 million tweets and collected on the 27th of December at 4:15 pm to the 7th of January at 5 am. We did not perform a classification of Tweets to detect malicious content, but rather only focused on finding accounts belonging to the already detected malicious users of the dataset retrieved from Webhose. After performing entity mapping on ICQ, we found 31 malicious users containing accounts on both forums and blogs from Webhose and Twitter.

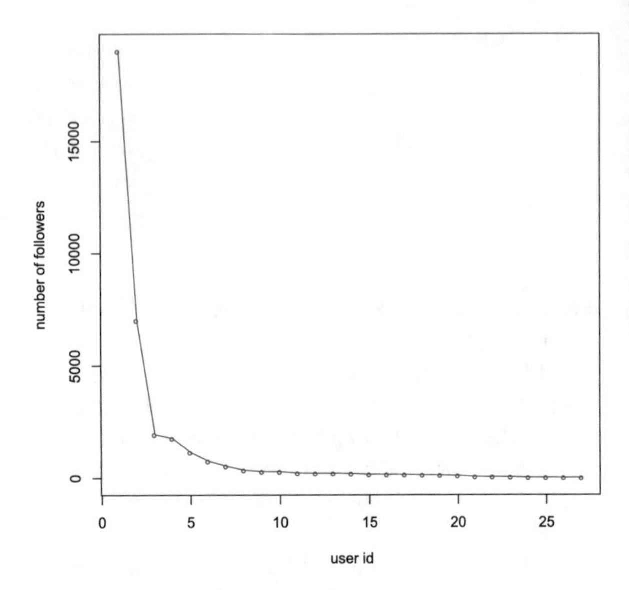

Fig. 5 Distribution of malicious Twitter users by their number of followers

A manual inspection of every account confirmed that these Twitter accounts were selling stolen credit cards except one that was deleted or suspended. 16 of the 30 users hide the date of creation of their account. For the 14 remaining users, one opened his account in 2008, 5 created their accounts in 2009, and the remaining created their accounts after this date and before 2013. This suggests that these malicious users managed to bypass the security measures of Twitter for many years without being detected. We noticed that a majority of these users are much more cautious to hide their illegal activities on Twitter in comparison to forums. They often publish many legitimate posts before posting a SPAM post. One reason might be to avoid being detected as they would consider that Twitter has improved security measures in comparison with other forums. Another reason would be linked to the consequences of being detected on Twitter in which they have a follower base. On most forums, there is no follower relationships between users, and hence the consequences of being detected is simply to create a new account on the forum. Figure 5 illustrates the distribution of malicious Twitter users by their number of followers. We can see that one user has more than 1.9K followers, another had more than 7000 and 11 had above a hundred followers. As most followers of these users are not malicious (confirmed by manually inspecting a sample), these followers would be vulnerable to malware threats, encouraging illegal activities without noticing and even to damages to their reputation in some cases.

The next step of this case study was to investigate 78 blog and forum nodes that were not classified as malicious, were not mapped to malicious nodes (ICQ or email) but that posted a message in the same thread as malicious nodes. These nodes are colored in blue in Fig. 4 and have an edge with a malicious node colored in pink. Figure 6 illustrates the distribution of nodes that were not classified as malicious and that posted in the same thread as malicious nodes. We can see that there are

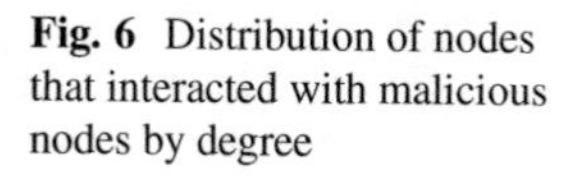

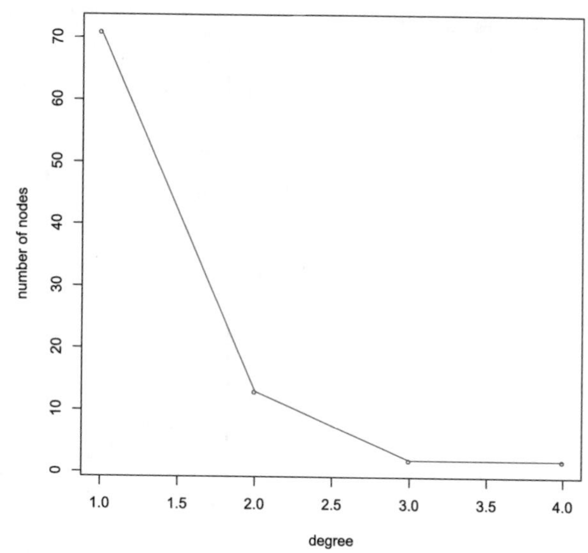

Fig. 6 Distribution of nodes that interacted with malicious nodes by degree

Table 4 Malicious nodes by method of detection

Indicator	Value
Random Forest	600
Entity mapping	111
Interactions	78

very few of these nodes with a degree not exceeding 4 which would encourage their monitoring and their study. After manually labeling posts published by these nodes, we found that every single one of these users posted messages concerning stolen credit cards and other fraudulent activities.

Table 4 summarizes the number of malicious nodes that were found by the classifier, by entity mapping and by studying the threads containing malicious posts. We can see that studying interactions among carding communities enables to highly increase the number of users classified as malicious.

Figure 7 illustrates the distribution of the number of malicious posts by forum. We could see that there are very few forums with a high number of malicious posts (more than 50) and the majority of forums have a very small number of malicious posts (one or two). This suggests that some forums have poor security measure or a high number of posts compared to the number of moderators. Closely monitoring or blocking some of these forums would reduce the exposure of users to malicious posts linked to carding activity. The forums with most malicious posts is chromeplugins.com, bmx-forum.com which is a sports forum and www.physforum.com which is a physics forum. We also found 69 forums that contained legitimate posts about credit cards, like advertising for example and malicious posts. The forums that regroup both malicious and legitimate posts about the same topic might be more effective in tricking a legitimate user as their posts often have a shared vocabulary.

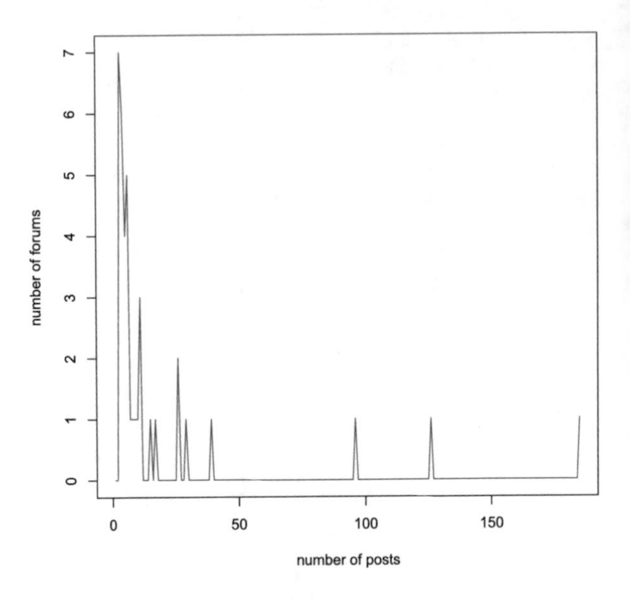

Fig. 7 Distribution of malicious posts by forum

Malicious users have many constraints like time and the need to be profitable leading them to diversify the channels from which they could reach victims. Driven by these constraints, they would make their posting habits explicit and repetitive if not facing the threat of being blocked, or if being blocked would not make them lose their audience. It is also cumbersome for these malicious nodes to have a separate email account, skype account, ICQ account, etc. for every forum or social network. They would most likely use a small set of accounts they reuse on different platforms. This could be leveraged with other entity mapping methods to link their multiple accounts together. Hence, modeling forum posts in a multi-layer graph model could be very useful as it is possible to easily detect malicious nodes on contexts where the profitability drive would push them to be aggressive and explicit and link these accounts to other contexts where these users are more cautious and difficult to detect.

The method we presented is suited for detecting different types of malicious users. It is sufficient to supply a training set to the classifier containing a sample of posts from the malicious users on wishes to detect, and perform the steps described in Algorithm 1. Examples of malicious users that could be detected are those who attempt phishing attacks, spammers, users that propagate malware, etc.

Detecting malicious posts is critical for the protection of an information system. It allows blocking the content they share at the gates of an information system reducing the possibility of infection by viruses. A second use would be to find the blogs, forums, and social networks with a weak security policy and possibly block them. This practice is very common in large companies where many websites are usually blocked for security reasons. Another use would be to detect vulnerable users that interact on the same blogs and social networks as malicious ones and tailor a security awareness program for them.

4.3 Threat Assessment from a Social Engineering Attack and a Reverse Social Engineering Attack

In this section, we will analyze the threats linked to a social engineering attack and a reverse social engineering attack using the proposed model.

4.3.1 Cross-Site Profile Cloning

This automated social engineering attack was used by Bilge et al. [33] to clone profiles between XING and LinkedIn. It is a simple attack that consists of cloning profiles across social networks and convincing users of a social network that the attacker is one of their contacts in the other social network. It is composed of the following steps:

1. Select two social networks that would be used for the attack.
2. Identify users that are registered in one social network but not the other.
3. Impersonate their identities in the social network where they are not registered.
4. Add contacts of the impersonated users that are registered on both social networks.

4.3.2 Recommendation System Based Attack

This reverse social engineering attack described in [34] relies on influencing the recommendation system of a social network in order for it to suggest attacker profiles as legitimate contacts to victims. Social network services encourage users to increase their number of contacts (friends, followees, subscribers, etc.) by recommending new contacts to them. These recommender systems use multiple criteria in order to assess whether a user might be interested in connecting with another user. An attacker could guess some of the criteria used by a recommendation system and perform some actions that would lead to being recommended to the victim. Irani et al. [34] used the search by email mechanism of Facebook that considered that if a user knows the email of another user, they should probably have some relationship and should hence be recommended to each other. This attack is composed of the two following steps:

1. Retrieving emails of users the attackers wishes to interact with.
2. Performing a search by email using the Facebook web based search.

4.3.3 Evaluation of These Attacks

We will model in this section a profile cloning attack between Twitter and a forums we named $forum_x$ in addition to a recommendation system based attack on $forum_x$.

We constructed a multi-layered graph $\mathscr{M}_2$ with three elementary context layers C={web, $forum_x$, Twitter}, four elementary state layers S={state 1, state 2, state 3, state 4} and one elementary attack scenario layer T={test 1}. We consider five generic actors in the network:

- Victims 1: victims that have an account on both $forum_x$ and Twitter.
- Victims 2: victims that have an account on $forum_x$ and for which the emails they used to create their accounts could be retrieved.
- Attacker 1: malicious users performing cross-site profile cloning.
- Attacker 2: the cloned victim's accounts that could be found on Twitter but not on $forum_x$.
- Attacker 3: a malicious user performing the recommendation system based attack.

This attack also makes use of three resources:

- Email directory: will be used by attacker 3 to look up email addresses.
- Search engines: also used by attacker 3 to retrieve email addresses.
- $forum_x$ search: the web based search by email mechanism used by attacker 1.
- $forum_x$ recommendation: the recommendation system used by attacker 3.
- Twitter search: refers to both the web based Twitter search interface and the Twitter search API. It is used by attacker 1 to retrieve the Twitter profiles of Victims 1.

Figure 8 offers an illustration of the two attacks. Attributes describing the state of nodes are not shown in this figure for simplicity. The colors of nodes (red) indicate that both attacks could be successful in state 4. These colors represent the node labels in the model indicating whether a node has been compromised by an attack. The node shapes indicate the type of the node, where the triangle shape represents resource nodes and the circle shape represents actor nodes. The actions required for the attacks to unfold are indicated by edge labels. We remind that the edges having the same label: "retreive mail" are bound together by an OR operator meaning that either of the actions needs to be performed to move to the next state (disjunction). The actions that have different labels are bound together by AND operators meaning that the actions they represent should all be performed to move to the next state (conjunction). The sequencing of actions is indicated by state layers, as actions in a layer at a specific state are only meaningful for an attack if actions on all layers at the prior state were successfully performed.

The cross-site profile cloning starts by a malicious user (in purple) retrieving a set of profiles and their followers from Twitter. He searches for these retrieved profiles in state 2 using the $forum_x$ search represented by a triangle. He then creates fake profiles (Attacker 3) of some users he didn't find on $forum_x$ but found on Twitter in state 3. The fake profiles he creates are then used to add the $forum_x$ accounts belonging to the users that are present on Twitter and $forum_x$ in state 4 which then leads the nodes he befriends to become compromised.

Concerning the recommendation system based attack, the attacker represented by the purple node on context 1 should search for email addresses of the users he wishes

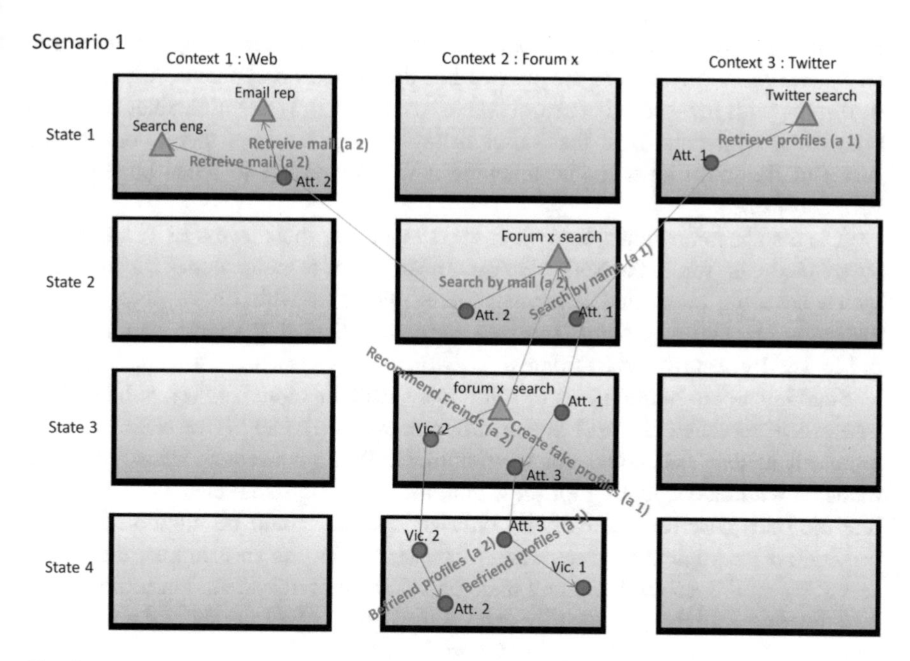

Fig. 8 An attack scenario from two social engineering attacks

to befriend. The attacker could look up email addresses of employees of a company or search for documents containing their mail like CVs for example. Attacker 2 then searches for the users he wants to befriend (Victims 2) using the $forum_x$ search mechanism. The $forum_x$ recommendation system would then recommend him in state 3 to Victims 2. Finally, the users that he was recommended to are labeled in red in state 4 meaning that if they befriend him, the attack would be successful.

We can see that a variety of criteria could be used to assess the threats of an attack like the number of actions performed by the attacker, the cost or skills required for every action (stored in edge attributes), the number of resources accessed in an attack, the users impacted, etc. If the skills required for an attack are taken into account, the cross-site profile cloning attack would be considered more likely. The reason is that action edges in the recommendation based attack that require retrieving emails that were used to create forum accounts would necessitate higher skills than the actions required in the cross-site profile cloning. Hence, the cross-site profile cloning attack should be focused on in a security awareness program.

The critical nodes that would allow the unfolding of different attacks should also be identified using this model. For resource nodes, the $forum_x$ search node is mandatory for both the attacks presented above. It is hence advisable to secure this node by different means like encouraging users to avoid emails containing their first and last names in forums and using pseudonyms for example.

Concerning the identification of critical actor nodes, the analysis of interactions on the multi-layer graph offers a variety of criteria that could be used to this aim.

Examples would be to consider nodes that have accounts on multiple forums and social networks, nodes that are part of big connected components, who interact in forums with low security measures, etc. (see Sect. 4.2) as vulnerable. It is then possible to aggregate users that suffer from different levels of threats in separate nodes of the multi-layer model allowing a security expert to focus on the most vulnerable ones.

It is also recommended to rank context layers by their levels of vulnerability where contexts (forums or SNS) that offer a variety of resource nodes for attackers, which have big connected component like Twitter or which have poor security measures like chromeplugins.com (see case study Sect. 4.2) should be monitored or blocked by security professionals.

State layers are homologous to state nodes used in some network vulnerability assessment models [21], with major differences like the ability of state layers to represent actions that could not be monitored. While a network attack could be modeled without accounting for these actions, modeling social engineering attacks like the ones described above with only actions that could be monitored would provide an incomplete representation of these attacks. As an example, the action search by mail does not leave any trace on the monitored system, but representing it in the model might give security experts the insight of encouraging the use of an anonymous email address for accessing social networks.

4.4 Detection of Vulnerable Users to a Social Engineering Attack

After assessing threats from a type of malicious users (Sect. 4.2) and evaluating specific social engineering attacks (Sect. 4.3), we will now perform an assessment to find critical nodes from a sample of users vulnerable to a specific social engineering attack (cross-site cloning). This will highlight how the proposed model bridges between multi-layer social network models [24–28] and models for vulnerability assessment [20–23].

Security professionals rarely perform vulnerability assessment without a set of specific attacks in mind. These attacks could be the most likely, the most dangerous or a mix of both. Hence it is by knowing how a set of specific attacks unfold that a security professional would detect vulnerable and dangerous nodes. The reason is that knowledge of an attack would allow a security professional to know what to look for in order to detect vulnerable users or to protect a system against malicious ones.

The dataset used in this case study is composed of 1,919,806 tweets of the first data collection on Twitter launched on the 29th of June 2015 and the 73,860 forum posts retrieved in the first data collection from Webhose launched on the 1st of July 2015.

A multi-layer graph $\mathscr{M}_2$ is extended with 1759 context layers, one for the Twitter context layer and the rest for each blog and forum. It also contains one state layer and one scenario layer:
$\mathscr{M}_1 \subset \{Twitter, Forum_1, ..Forum_{1758}\} \times \{state1\} \times \{scenario1\}$.

The nodes in $V_{\mathscr{M}_2}$ refer to users that have published a post in Twitter or in a blog or forum. The two types of edges in $E_{\mathscr{M}_3}$ are:

- Action edges: On the blog and forum layers, an action edge exists between two nodes if they have published a post, commented, or replied in the same thread of a forum or blog. On the Twitter layer, an edge exists between two nodes if one of them has retweeted or replied to the other (using an @ followed by the screen name).
- Identity edges: These edges exist between two nodes of different context layers (Twitter, blogs, and forums) if it is likely that they belong to the same user. For this case study, we considered an identity edge between two nodes of different layers if they have the same user name and used the same keywords in their posts, although more advanced identity mapping methods could be used to link nodes from different social platforms together.

We now perform a vulnerability assessment on a set of users by both analyzing their interactions and by mapping the user attributes necessary for the unfolding of the cross-site profile cloning attack presented in Sect. 4.3. Vulnerability assessment in this case study is based on the two following questions:

1. Which nodes in the assessed dataset could be represented by node Vic. 1 in $\mathscr{M}_2$ (Sect. 4.3)?
2. Which nodes are the most critical in infecting a network?

From the assessment performed in the previous section, we know that users vulnerable to cross-site profile cloning are the ones who have accounts on multiple social networks or blogs. Hence, the condition that a user could be mapped to node Vic. 1 is that he has more than one account on different social networks.

We hence compute a simple indicator which is the number of adjacent layers of a user. A user is adjacent to a layer if the former has an interaction with an another user on that layer. This indicator would be able to detect users vulnerable to cross-site profile cloning as they are adjacent to more than one layer. Also, a wide variety of other social engineering attacks leverage vulnerabilities emanating from the presence of a user on different layers, which emphasizes the importance of monitoring these users.

Figure 9 illustrates the distribution of nodes by their adjacent layer. The nodes having only one adjacent layer were omitted from the figure. Their number is 208,155 which represents around 99.5% of nodes. Hence, the number of adjacent layers per node puts the light on a small subset of users that could introduce a wide array of inter-layer vulnerabilities into the network.

After finding the users that could be represented by node Vic 1 which are the ones vulnerable to cross-site cloning, we will evaluate the levels of danger of different nodes by studying their interactions in the network.

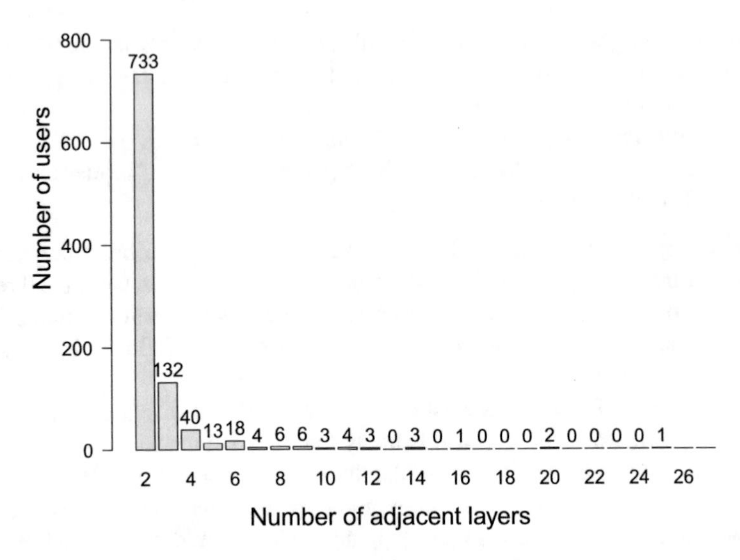

Fig. 9 The distribution of nodes by their adjacent layers

The users that posted in Twitter, blogs, or forums and that have no interaction (retweet, comment, reply) with other users were omitted from $\mathscr{M}_2$ as they are less likely to be of value in assessing vulnerabilities. They are in fact very unlikely to propagate malicious content given that they form disconnected components. Hence, we remove nodes from $\mathscr{M}_2$ of degree 0.

Figure 10 illustrates the number of nodes by connected component. We can see from this figure that there are a few connected components that have more than ten nodes associated with them. Nodes in the greatest connected components are the ones that could lead to the infection of multiple nodes if compromised.

Figure 11 illustrates the biggest connected component of a graph G_2 built from $\mathscr{M}_2$. Both the Twitter layer and the forum and blog layers are illustrated on the same window. We could see that this component is largely composed of Twitter nodes that are connected with a small number of blog and forum nodes.

Table 5 provides a description of G_2. We can see from this table that G_2 has a high number of connected components relative to its number of nodes. Its greatest connected component contains more than 50k nodes and the rest of the connected components contain less than 6k nodes. Hence, nodes in the greatest connected component have a much higher potential to infect a large number of users in the network either by launching attacks on other nodes or by being compromised by malicious nodes.

Given the high standard deviation of the number of nodes per layer compared to the mean, there is a very big concentration of nodes on a couple of layers leaving the rest of the layers of the network with very few nodes and interactions. As supervising each layer requires time and effort, this suggests that a security officer could only monitor a couple of layers (Twitter and a couple of forums) where most

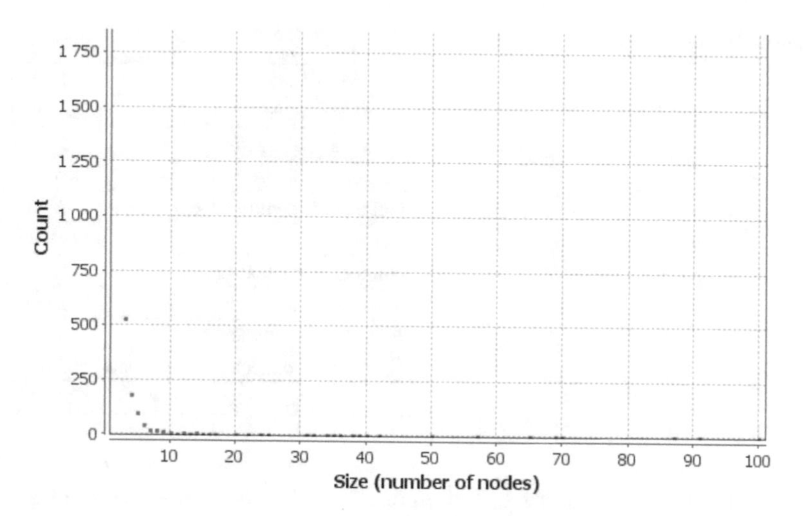

Fig. 10 The distribution of the number of nodes by the number of connected components

Fig. 11 Greatest connected component of G_2

interactions occur. They could also focus on only the nodes in the layers where most interactions occur that are also part of the greatest connected components which highly reduces the efforts required to monitor the network but would still allow security officers to limit the damages that would be caused by an attack.

Table 5 Description of G_2

Indicator	Value
Number of nodes	211,252
Number of edges	204,027
Percentage of Twitter edges	85.75%
Number of blog/forum edges	5.42%
Number of identity edges	9.83%
Density	1.01410^{-5}
Number of context layers	1759
Mean of nodes per layer	119.09
S.D. of nodes per layer	4858.33
Number of connected components	62,673
Max nodes per layer (Twitter)	203,763

From the analysis of user interactions on $\mathcal{M}_2$ and of the cross-site profile cloning attack, the number of users with more than one adjacent layer that are part of the largest connected component are the most critical ones as they are vulnerable to cross-sit profile cloning and could infect the highest number of nodes. Given the number of such nodes is very small, it would be advised to monitor them as they are also vulnerable to other attacks that leverage inter-layer connections.

5 Conclusion

It is possible to consider social engineering as one of the major security threats faced by information systems. Despite its high success rates and the damages it causes, it has been given very little attention when it comes to vulnerability assessment modeling. Most models only focus on network security ignoring the peculiarities related to social engineering attacks. One of the drawbacks of using network vulnerability assessment models for social engineering is that they usually focus on monitoring state variable of a system, while many steps in social engineering attacks do not leave traces in a monitored system. These actions are nevertheless useful to model as they allow assessing the likelihood and success rate of attacks and to tailor security awareness programs to counter them.

Social engineering attacks are characterized by the wide range of channels (email, social networks, phone, mail box) that could be leveraged. Although many channels would not be monitored by security professionals, it is useful to assess the threats emanating from these channels in order to perform an informed decision on which channels ought to be monitored. Also, modeling these attacks allows to better assess the threats an information system is subject to, as actions performed via different channels require different resources and levels of expertise from the attacker and have different success rates. Another reason would be to tailor security awareness programs from threats linked to specific channels.

Multi-layered graphs allow to easily capture the diversity of channels and contexts an attacker uses to approach a victim. They also have the advantage of easily depicting inter-connections between these different channels and contexts.

We present in this paper a graph-based multi-layered model for social engineering vulnerability assessment. This model is built from layers that depict different attack scenarios, different states in an attack and multiple contexts that could be used. It allows putting into lights the inter-connections between the different elements in an attack (actions performed, users involved, and resources used) and represents actions that don't necessarily leave traces in a monitored system.

Using this model to analyze the multiple identities and interactions of users allows to disclose attack vectors a malicious user could leverage. The case study that concerned the assessment of threats from credit card resellers allowed to increase the true negatives of a random forest classifier by more than 20% using basic entity mapping methods, to rank forums by the threats they pose and users by their levels of threat. Modeling interactions between Twitter users and forum and blog users has allowed the ranking of these users on the basis of their vulnerabilities to cross-site attacks.

Acknowledgements This work was supported by the French Investment for the future project REQUEST (REcursive QUEry and Scalable Technologies) and the region of Champagne-Ardenne.

References

1. Heikkinen, S.: Social engineering in the world of emerging communication technologies. In: Proceedings of Wireless World Research Forum, pp. 1–10 (2006) [Online]. Available: http://www.cs.tut.fi/~sheikki/docs/WWRF-Heikkinen-SocEng.pdf
2. Luo, X., Brody, R., Seazzu, A., Burd, S.: Social engineering. Inf. Resour. Manag. J. **24**(3), 1–8 (2011)
3. Kvedar, D., Nettis, M., Fulton, S.P.: The use of formal social engineering techniques to identify weaknesses during a computer vulnerability competition. J. Comput. Sci. Coll. **26**(2), 80–87 (2010)
4. Granger, S.: Social engineering fundamentals, part I: hacker tactics. In: SecurityFocus (2001)
5. Mitnick, K.D., Simon, W.L.: The art of deception: controlling the human element in security (2003) [Online]. Available: http://www.bmj.com/content/347/bmj.f5889
6. Kaul, P., Sharma, D.: Study of automated social engineering, its vulnerabilities, threats and suggested countermeasures. Int. J. Comput. Appl. **67**(7), 13–16 (2013) [Online]. Available: http://research.ijcaonline.org/volume67/number7/pxc3886726.pdf
7. Huber, M., Kowalski, S., Nohlberg, M., Tjoa, S.: Towards automating social engineering using social networking sites. In: 2009 International Conference on Computational Science and Engineering, vol. 3 (2009)
8. Malagi, K., Angadi, A., Gull, K.: A survey on security issues and concerns to social networks. Int. J. Sci. Res. **2**(5), 256–265 (2013)
9. Gross, R., Acquisti, A.: Information revelation and privacy in online social networks (Facebook case). In: Proceedings of the 2005 ACM Workshop on Privacy in the Electronic Society, pp. 71–80 (2005) [Online]. Available: http://portal.acm.org/citation.cfm?id=1102214
10. Jakobsson, M.: Modeling and preventing phishing attacks. Lecture Notes in Computer Science, vol. 3570, no. 578, pp. 1–19. Springer, Berlin (2005) [Online]. Available: http://citeseerx.ist.psu.edu/viewdoc/download?doi=10.1.1.64.1926&rep=rep1&type=pdf

11. Thornburgh, T.: Social engineering: the dark art. In: Proceedings of the 1st Annual Conference on Information Security Curriculum Development, pp. 133–135 (2004) [Online]. Available: http://dl.acm.org/citation.cfm?id=1059554
12. Jagatic, T.N., Johnson, N.A., Jakobsson, M., Menczer, F.: Social phishing. Commun. ACM **50**(10), 94–100 (2007)
13. Dolan, A.: Social engineering. SANS Institute InfoSec Reading Room, p. 18 (2004) [Online]. Available: http://www.google.fr/url?sa=t&rct=j&q=&esrc=s&source=web&cd=2&cad=rja&uact=8&ved=0CCgQFjAB&url=http://www.sans.org/reading-room/whitepapers/engineering/social-engineering_1365&ei=aaIqVbbvKtLoaJqOgYgE&usg=AFQjCNEVDB7ZW2F5BkYk2HVichZECspwhQ&sig2=C-MU7GL542m
14. Gragg, D.: A multi-level defense against social engineering. SANS Institute Infosec Reading Room (December 2002). https://www.sans.org/reading-room/whitepapers/engineering/multi-level-defense-social-engineering-920
15. Jaafor, O.: Multi-layered graph-based model for social engineering vulnerability assessment. In: ASONAM, Paris, France, pp. 1480–1488. ACM, New York (2015) [Online]. Available: http://link.springer.com/bookseries/8768
16. Shashidhar, N., Chen, L.: A phishing model and its applications to evaluating phishing attacks. In: Second International Cyber Resilience conference (2011), Perth, Australia, pp. 63–69 (2011)
17. Vida, R., Galeano, J., Cuenda, S.: Vulnerability of multi-layer networks under malware spreading. arXiv preprint arXiv:1310.0741, pp. 1–5 (2013) [Online]. Available: http://arxiv.org/abs/1310.0741
18. Jaafor, O., Birregah, B., Perez, C., Lemercier, M.: Privacy threats from social networking service aggregators. In: Cybercrime and Trustworthy Computing Conference (CTC). IEEE Computer Society, New York (2014)
19. Perez, C., Birregah, B., Lemercier, M.: The multi-layer imbrication for data leakage prevention from mobile devices. IEEE Computer Society, New York (2012)
20. Phillips, C., Swiler, L.P.: A graph-based system for network-vulnerability analysis. In: Proceedings of the 1998 Workshop on New Security Paradigms, pp. 71–79. ACM (1998)
21. Ammann, P., Wijesekera, D., Kaushik, S.: Scalable, graph-based network vulnerability analysis. In: Proceedings of the 9th ACM Conference on Computer and Communications Security - CCS'02, p. 217 (2002) [Online]. Available: http://dl.acm.org/citation.cfm?id=586110.586140
22. Tidwell, T., Larson, R., Fitch, K., Hale, J.: Modeling internet attacks. Network **1**, 5–6 (2001) [Online]. Available: http://citeseerx.ist.psu.edu/viewdoc/download?doi=10.1.1.108.9040&rep=rep1&type=pdf
23. Wang, L., Singhal, A., Jajodia, S.: Toward measuring network security using attack graphs. In: Proceedings of the 2007 ACM Workshop on Quality of Protection - QoP'07, p. 49 (2007) [Online]. Available: http://portal.acm.org/citation.cfm?doid=1314257.1314273
24. Magnani, M., Rossi, L.: Multi-stratum networks: toward a unified model of on-line identities. arXiv preprint arXiv:1211.0169, pp. 1–18, November 2012 [Online]. Available: http://arxiv.org/abs/1211.0169v1
25. Cardillo, A., Zanin, M., Gómez-Gardeñes, J., Romance, M., García del Amo, A.J., Boccaletti, S.: Modeling the multi-layer nature of the European Air Transport Network: resilience and passengers re-scheduling under random failures. Eur. Phys. J. Spec. Top. **215**, 23–33 (2013)
26. Perez, C., Birregah, B., Lemercier, M.: A smartphone-based online social network trust evaluation system. Soc. Netw. Anal. Min. **3**(4), 1293–1310 (2013) [Online]. Available: http://link.springer.com/10.1007/s13278-013-0138-4
27. Boccaletti, S., Bianconi, G., Criado, R., del Genio, C.I., Gómez-Gardeñes, J., Romance, M., Sendiña-Nadal, I., Wang, Z., Zanin, M.: The structure and dynamics of multilayer networks. Phys. Rep. **544**(1), 1–122 (2014) [Online]. Available: http://dx.doi.org/10.1016/j.physrep.2014.07.001
28. Kivelä, M., Arenas, A., Barthelemy, M., Gleeson, J.P., Moreno, Y., Porter, M.A.: Multilayer networks. p. 37 (2014) [Online]. Available: http://arxiv.org/abs/1309.7233

29. Stolen credit card details available for 1 each online, 2015. [Online]. Available: http://www.theguardian.com/technology/2015/oct/30/stolen-credit-card-details-available-1-pound-each-online
30. Golbeck, J., Rothstein, M.: Linking social networks on the web with FOAF: a semantic web case study. In Proceedings of the 23rd National Conference on Artificial Intelligence, vol. 2, pp. 1138–1143 (2008)
31. Raad, E., Chbeir, R., Dipanda, A.: User profile matching in social networks. Proceedings - 13th International Conference on Network-Based Information Systems (NBiS 2010), pp. 297–304 (2010)
32. Rowe, M., Ciravegna, F.: Disambiguating identity through social circles and social data. In: CEUR Workshop Proceedings, vol. 351, pp. 80–93 (2008)
33. Bilge, L., Strufe, T., Balzarotti, D., Kirda, E., Antipolis, S.: All your contacts are belong to us: automated identity theft attacks on social networks. Www 2009, pp. 551–560 (2009)
34. Irani, D., Balduzzi, M., Balzarotti, D., Kirda, E., Pu, C.: Reverse Social Engineering Attacks in Online Social Networks. Lecture Notes in Computer Science (Lecture Notes in Artificial Intelligence and Lecture Notes in Bioinformatics), vol. 6739, No. March, pp. 55–74. Springer, Berlin (2011)

Through the Grapevine: A Comparison of News in Microblogs and Traditional Media

Byungkyu Kang, Haleigh Wright, Tobias Höllerer, Ambuj K. Singh, and John O'Donovan

1 Introduction

Over the last decade, microblogs have evolved from an online communication channel for personal use to a central hub for information exchange between users. On microblogging platforms, users produce or share information with friends or strangers. Recent studies revealed that the greater part of today's internet users rely on information on microblogs [19] (e.g., Twitter and Reddit) as a primary source of a wide range of information, particularly news. Accordingly, this new paradigm highlights the importance of automated tools that detect reliable and newsworthy information on microblogs.

Going beyond typical information consumers, professional journalists also admit to relying heavily on social media streams for their news stories [18, 32]. During the last decade, microblogs have been studied by researchers in communication and journalism as an essential news gathering tool and several guidelines are proposed.[1] Many users favor to browse microblogs such as Reddit and Twitter on a daily basis since these platforms provide personalized news content based on their previous browsing patterns. Recent research also highlights that traditional news outlets still play an important role in the provision of reliable, well-curated news content [19].

However, news outlets are typically biased in some way or other, and do not always act as the best information filters in all cases. A recent study by Budak et al. [11] highlights the polarizing political bias that exists across most of the

[1] http://asne.org/Files/pdf/10_Best_Practices_for_Social_Media.pdf.

B. Kang (✉) • H. Wright • T. Höllerer • A.K. Singh • J. O'Donovan
Department of Computer Science, University of California, Santa Barbara, CA 93106-5110, USA
e-mail: bkang@cs.ucsb.edu; wright@cs.ucsb.edu; holl@cs.ucsb.edu; ambuj@cs.ucsb.edu; jod@cs.ucsb.edu

R. Missaoui et al. (eds.), *Trends in Social Network Analysis*, Lecture Notes in Social Networks, DOI 10.1007/978-3-319-53420-6_6

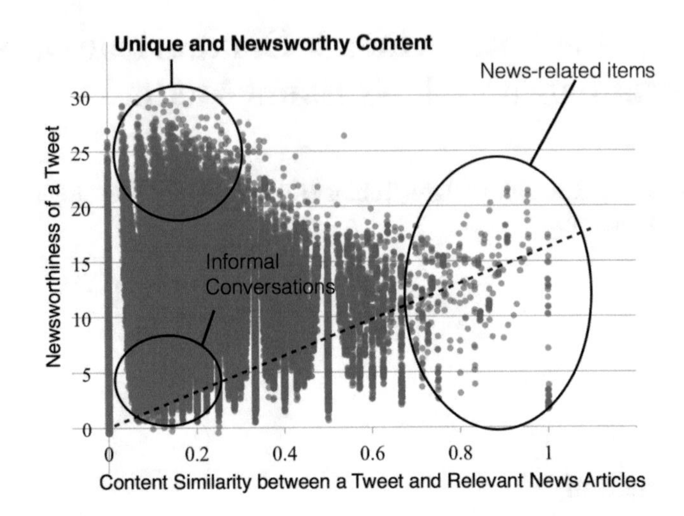

Fig. 1 Overview of approach to filtering unique and newsworthy content. *Y*-axis tweet newsworthiness is computed from NLTK and from human evaluation. *X*-axis is tweet similarity to mainstream news

top US traditional news outlets. Despite the possibility for bias, we believe that curated news from a variety of sources can be leveraged to help identify and classify newsworthy messages in social media streams. In particular, we propose a novel method for identifying *niche* user-provided topics from social media that is (a) not reported in traditional curated news, and (b) is newsworthy information. Figure 1 shows an overview of our first approach. Each data point represents a Twitter post, located on the *x*-axis by similarity to a target set of news articles, and on the *y*-axis by general newsworthiness of the message content. The distribution shows a linear trend indicating the correlation of newsworthiness and similarity to curated content, as we would expect to see. In this case however, we are interested in the highlighted "niche content" section in the top left of the graph, which contains those unique messages that are *not similar* to mainstream media, but do have newsworthy content based on other metrics. This content could be found through a series of text based search queries, but defining relevant keywords is difficult, and may potentially only uncover a given slice of the true overlap between the data sources. To explore this concept, we study a variety of topics from 37 million Twitter posts and 6112 New York Times articles and attempt to answer the research questions below. The authors would like to note that an earlier version of this study has been published in [22]. The novel contribution in this manuscript includes all of the research on network-based (LDA and BPR) analysis of news in social media networks (Fig. 2).

1. **RQ1** How can we best detect newsworthy information in social media that is not covered by traditional media?
2. **RQ2** How do information consumers perceive the detected information?

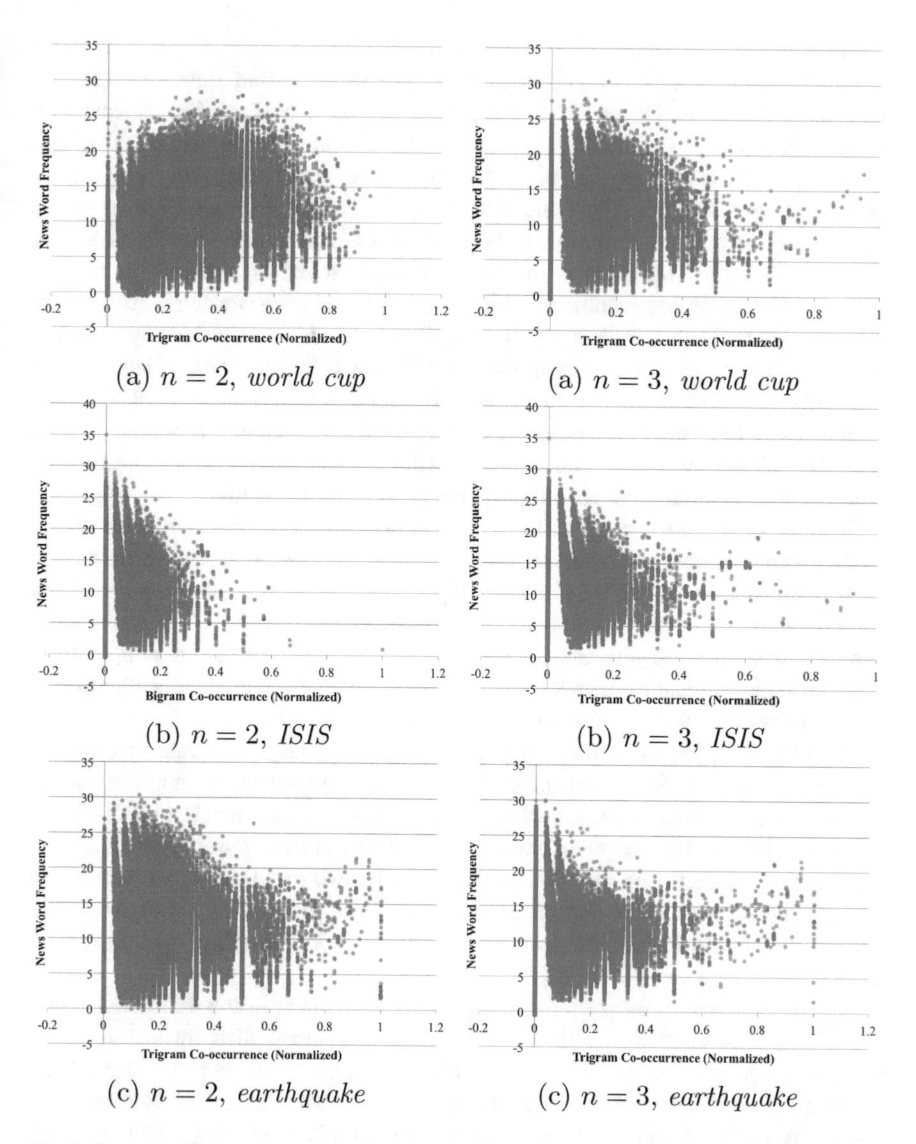

(a) $n = 2$, *world cup* (a) $n = 3$, *world cup*

(b) $n = 2$, *ISIS* (b) $n = 3$, *ISIS*

(c) $n = 2$, *earthquake* (c) $n = 3$, *earthquake*

Fig. 2 News word frequency on tweets and *n*-gram (*n* = 2, 3) co-occurrence with mainstream news articles (NYT) on different topics

3. **RQ3** How do the niche information get propagated differently from traditional news in the network?

In this paper, we propose two distinct approaches to capture unique news content on microblogs. First approach is based on a variety of content-similarity metrics. Simply put, we compute different content-based similarity metrics on microblog posts and a corpus of traditional news articles. Using these similarity metrics with our newsworthiness scores computed on individual tweets, we can locate the niche (unique and newsworthy) contents and analyze them to find important features that can be utilized for developing automated detection algorithm. Specifically we describe two experiments: first, an automated evaluation is performed to test a variety of mechanisms that predict overlap between a microblog post and a corpus of news articles. These include manipulations on *n*-grams, part-of-speech tags, stop words, and stemming techniques. A co-occurrence score is produced for each message, which is in turn compared to a set of manually annotated newsworthiness scores, combined with a content-based newsworthiness score. The different strategies are ranked by the resulting distance and the best approach is used for experiment 2. Manual annotations of newsworthiness were collected using a crowd-sourced study described in [30]. The second experiment samples data in various ways from the highlighted areas of Fig. 1 for a range of topics and presents an A/B style questionnaire about newsworthiness, similarity to traditional media content, and personal focus to 200 participants in an online study.

Results of experiment 1 show that a simple *n*-gram approach with word-stemming but without stop word removal produced the most accurate approximation of the manual annotations. Results from experiment 2 show that there is a significant difference in reported "similarity to mainstream news content" for messages sampled from the top left area of Fig. 1 compared with a random sample from the right side, indicating that the method is capable of automatically identifying newsworthy content that is not covered by mainstream media.

To address **RQ3**, the second approach—network analysis on microblog news contents—has been demonstrated in Sect. 4. In this approach, we apply a variety of commonly used network measures of structural and functional connectivity to microblog information to unveil unique characteristics that represent both niche and generic news contents on microblogs. Particularly, two experiments are performed on the collection of 2.4M Twitter dataset to find the differences between the two groups (niche and traditional groups) in network topology (`Exp 1`) and topical association across users (`Exp 2`).

Results of `Exp 1` show that the majority of subgraphs in the traditional group have long retweet chains with a giant component surrounded by a number of small components. On the other hand, unique contents typically propagate from a dominating node with only a few multi-hop retweet chains observed. Furthermore, results from `Exp 2` indicate that strong and dense topic associations between users are frequently observed in the graphs of the traditional group, but not in the unique group.

The differences between the unique and traditional news groups that we found in this study will benefit future studies for intelligent and scalable algorithms to automatically classify or predict unique or interesting news in microblogs. We will discuss our future work and possible applications for which our model can be applied in Sects. 5 and 6.

2 Related Work

With the increasing reliance on user-provided news content from microblogs, recent research has focused on the relationship between microblogging platforms and traditional news outlets [14, 24, 35]. As we briefly discussed in the previous section, news content, including opinions and conversations about news now comprise a significant portion of overall content on microblogs. Hermida et al. [18] conducted a large-scale online survey and unveiled behaviors of news consumers on social media including microblogs. According to many studies, including [18], microblogs such as Twitter have become a major source of news information for individual consumers and also for professional journalists who rely on the dynamic content for story-hunting and marketing.

2.1 *Microblogs and Traditional Media*

Over their short history, microblogs have been a communication channels upon which users share useful information that they discover elsewhere, such as online news media, blogs, or forums. Recent studies have focused on the relationship between microblogs and traditional news outlets since both end-users and journalists rely on microblogs for information. To understand the relation between these sources, researchers investigated association using topic modeling algorithms such as LDA [14, 35]. Furthermore, since microblog users not only reproduce and forward original information but sometimes re-shape content by adding additional value such as personal opinion or on-site images of an event, "produsage" (the hybridization of production and consumption) behavior and its byproducts have been studied [20] on different types of news contents: soft and hard news.

2.2 *Newsworthiness*

Shoemaker [29] argues that news and newsworthiness have different underlying concepts. However, they also admit that newsworthiness is one of the important components that makes news public. In this study, we assume that newsworthiness is a core information attribute that categorizes a piece of content in terms of usefulness to the general public.

Quality of information in microblogs has been widely studied in the information retrieval community, and remains releveant in this research. André et al. [4] studied microblog content through the first large corpus of follower ratings on Twitter updates collected from real users. They found that 64% of tweets are reported as not worth reading or middling, which implies that users tolerate a large amount of useless information on microblogs. In addition, factors that make microblog content "useful" and "not useful" were investigated through a qualitative study in search tasks [21]. We revisit their question about the content value in microblogs with particular focus on their unique role in news consumption. In other words, we examine microblog contents and pan for niche content which only exists on microblogging platforms, not others. Community feedback was also exploited to automatically identify high-quality content in the Yahoo! Answers [1] community question/answering platform.

Our research examines several low-level features of microblog posts to arrive at a good classifier. Castillo et al. [13] also explored features that can be exploited to automatically predict newsworthiness of information on microblogs. Participants of their crowd-sourced online study were asked to label a group of microblog messages with either a "news" or "non-news" category. The tweets labeled with news category were then annotated with newsworthiness score in 5 Likert scale in the subsequent annotation task. This study showed the possibility of automated identification of newsworthy information through machine learning. Moreover, the authors revealed important features which can be directly obtained or processed from microblog contents and metadata, without the need for human-labeled examples.

2.3 Content Similarity

Due to the scale and complexity of microblog and news data, it would require a huge effort for an end user to capture newsworthy content in a microblog that is not covered in traditional media using a series of traditional text-based search queries. Our automated approach to filtering for newsworthy information relies heavily on content matching techniques. A wide range of content similarity metrics have been studied and proposed for many years, ranging from simple string-based measures [2, 15] to semantic similarity [26], structural similarity such as stop word n-grams [31] and text expansion mechanisms [8, 23]. In particular, in the context of microblog content analysis,

Herdağdelen [17] proposed n-gram based approach to Twitter messages, which we build on in this research. Another well-known approach by Becker et al. [6] learns useful similarity metrics that can be used for event detection in social media data. This approach contrasts to our work in that we adopt a simpler per-tweet similarity. For future work we will examine automated event extraction algorithms and evaluate our uniqueness approaches at the event rather than message level. Other approaches such as Anderson et al. [3] and Guy et al. [16] take a user-based approach to similarity for social media content analysis. Our approach contrasts to this work by focusing only on similarity at the content level, but we believe that user-level analysis has significant potential in this area, in particular by supporting discovery of broader, more diverse content, and supporting serendipitous discovery of new, unique new content.

Our methods apply several content similarity metrics including normalized word n-grams to determine and measure how two information sources–*microblog* and *traditional news outlet*–are quantitatively associated. We carefully consider the limited nature of microblog contents: the limited number of characters and embedded items. Our choice of metrics for content were proposed in [5]. Bar et al. [5] evaluate different content similarity metrics and report effectiveness and efficiency of the composite of multiple metrics using supervised machine learning approach in their study.

2.4 Network Measures and Metrics

Recent studies have focused on either network structure or retweet behavior [27] by looking at the characteristic of the information diffusion. For example, [27] has shown that call for action type of retweets generated sparse graphs while tweets sharing information generated a denser network during their propagation. Also, [33] proposed a model that measures speed, scale, and range of information diffusion by analyzing survival of each message in the network.

Both content and network-based features considered, a recent work done by Canini et al. [12] shows a good example of how various features can be used to measure the quality of information on microblogging platforms.

2.5 Topical Similarity and Object Association

In the network analysis, several different approaches are considered to model the topic space of each group (traditional and niche) in the given network. We first apply Latent Dirichlet Allocation (LDA) topic-modeling algorithm [9, 28] in order to generate n words on each group of messages.

Secondly, we compare one from another in terms of structural similarity of the topics extracted from the content. Specifically, we tokenize each message into individual elements and compute stopwords and word n-grams as used in [25].

Afterward, Bipartite Projection via Random Walks (BPR) [34] is applied to construct topic-similarity network for each group. BPR is a method that produces associations between objects, defined in [34]. This method performs random walks on a two-mode bipartite network. In this network, edges exist only between nodes of different modes, and these edges represent an association between these two nodes. The result of this random walk is a one-mode unipartite network that captures the similarity between nodes of the chosen mode. This method takes into account the overall structure of the original bipartite network.

3 Content Similarity Based Approach

This section describes our approach to filtering unique and newsworthy content from microblog streams based on comparison with contents from mainstream media. According to the study in [29], Shoemaker claims that newsworthiness is not the only attribute which represents news. However, in this study, we adopt newsworthiness as the central indicator of news contents in general. Basically, we assume here that curated news articles are newsworthy. Our first approach exploits news articles as a reference to identify Twitter postings about a target topic that are newsworthy but are not the focus of curated mainstream news. We begin by exploring a set of mechanisms for computing similarity between a microblog post and a topic-specific corpus of news articles.

3.1 Data Collection

To examine real-world microblog messages and news contents, we choose "Twitter" and "New York Times" as representative examples for microblogging platforms and traditional media outlets. Both provide well-documented application program interfaces (APIs)[2] through which we can retrieve microblog messages or news articles as well as a rich set of metadata (e.g., keywords, embedded multimedia items, urls). With the two APIs we collected about 35 million (35,553,515) microblog messages from Twitter and 6112 news articles from New York Times and other sources such as Reuters and Associated Press (AP). An overview of this data collection is shown in Table 1. Before the crawling stage, we selected major news events such as natural disasters, world cup and various political issues over the course of 4 years (2012–2015) to examine how both media differ from each other and see if there is topic-specific bias across different events. We collected topic-specific data sets[3] using related keywords to retrieve microblog messages and

Table 1 Overview of the data sets collected from New York Times and Twitter

Topic	World cup	ISIS	Earthquake	Hurricane sandy
Tweets	22,299,767	8,480,388	921,481	3,851,879
Articles	4097	422	329	1264
From	6/24/14	1/20/15	1/20/15	10/29/2012
To	7/17/14	3/29/15	3/31/15	12/31/2012
Days	24	69	71	64

[2]New York Times Article Search API: http://developer.nytimes.com/docs Twitter API http://dev.twitter.com.

[3]*Dataset available upon email request.*

news articles from Twitter and New York Times databases. In particular, for Twitter data, we used the Streaming API to monitor transient bursts in the message stream while we collected regular data about the events.

3.2 Similarity Computation

A key challenge in this approach is to discover meaningful mappings between a short microblog post and a larger corpus of news articles. Since traditional text-matching mechanisms such as TF-IDF or topic modeling do not work well with short messages, a variety of simpler mechanisms were evaluated. Table 4 shows an overview of the mechanisms tested and their performance with respect to manually labeled "ground truth" assessments of newsworthiness. An initial pre-processing was applied to all messages to remove superfluous content such as slang and gibberish terms.

Word *n*-grams Next, a set of word *n*-grams as described in [5] were computed, varying n from 1 to 3. Part-of-Speech (POS) tagging was applied to identify potentially useful noun, verb, pronoun, and adjective terms. A standard stop-word list was identified and systematically removed as shown in Table 4. A Twitter-specific stop-word list was compiled from a manual analysis of posts. This list contained platform-specific terms such as "twitter," "rt," "retweet," "following," etc., based on a term frequency analysis. In total, 24 combinations of lightweight NLP techniques were applied to 4 topic-specific collections of twitter posts and NYT news articles. These are detailed in Table 4. Each method computed a content-based similarity score between a *single* microblog post and a larger collection of news articles.

For each topic studied, we obtained thousands of n-grams from the NYT article collection and use it as a corpus of news n-grams ($n = 1, 2, 3$). Next, we applied n-gram extraction on the entire tweet collection and computed the number of co-occurrences of n-grams from each post with those in the news n-gram corpus. To account for length deviation, this score (*Score*) was normalized by the total number of n-grams in each tweet.

Newsworthiness In this study, we apply a two-dimensional approach to newsworthiness: (1) news term frequency in each tweet ($News_{Term}$) and (2) newsworthiness score labeled by real-world microblog users ($News_{User}$) in [0–5] Likert scale.

For $News_{Term}$, we compute number of tokens that contain news terms using the Reuters news word corpus in NLTK[4] and divide this number by total number of tokens in each message.

$News_{User}$ is the human-annotated newsworthiness score, and is also normalized by the maximum score. Normalization is performed on both metrics in order to eliminate bias of different message sizes in tweets and take the average of the two metrics for Eq. (1). Table 2 shows the selected set of similarity metrics that we employ in this study.

[4]NLTK Reuters Corpus has 1.3M words, 10K news documents categorized. http://www.nltk.org.

Table 2 Metrics analyzed in the study

Metrics	Nomenclature	Description
n-gram similarity	*Score*	Number of n-grams that co-occur between news article corpus and a tweet
News word frequency	$News_{Term}$	News word frequency with NLTK Reuters corpus
Newsworthiness score	$News_{User}$	Human annotated newsworthiness score [0–5] on a tweet

3.3 Strategy Selection

We define a simple inverse distance metric in order to evaluate our content-based similarity measure (*Score*) and select the best performer among 24 candidates. This metric is then applied to the composite sets of multiple metrics to select the best feature based on the linear relationship between the similarity score and newsworthiness of a message. We discuss the procedure in detail in this section. Afterwards, we explain our evaluation method and procedure in Sect. 3.5.

Definition 1 Each event-specific data collection T contains N messages where $T = \{m_1, m_2 \ldots m_N\}$, and we represent individual message as m where $m \in T$. Inverse distance of a message between newsworthiness and content similarity to news corpus is represented as *InvDist*.

$$InvDist(m_i, c_N) = \frac{1}{|News(m_i, c_R) - Score(m_i, c_N)| + 1} \tag{1}$$

where $News(m_i, c_R)$ is:

$$News(m_i, c_R) = \frac{News_{Term}(m_i, c_R) + News_{User}(m_i)}{2} \tag{2}$$

Please note that c_N and c_R are a corpus of news articles on a topic and the Reuters news vocabulary corpus in NLTK, respectively.

Since we compare one strategy against others in the selection procedure, we use the average of inverse distance for a strategy over all messages, computed using Eq. (1).

We apply a fractional function to the inverse distance metric in Eq. (1). Intuitively, this approach maximizes gain in highly correlated messages and, likewise, penalize un-correlated messages between newsworthiness $News(m)$ and content similarity $Score(m)$. As briefly mentioned earlier in this section, we believe that both $News_{Term}$ and $News_{User}$ represent different aspects of newsworthiness. Unlike the n-gram co-occurrence (*Score*), which reflects the word-based association on a specific-event, $News_{Term}$, which is corpus-based news word frequency, represents

Table 3 Correlation coefficients between newsworthiness *News*(m) (arithmetic mean of news word frequency and user annotated newsworthiness score) and n-gram co-occurrence score *Score*(m) (all metrics normalized [0,1])

	Correlation coeff.	2-Tailed test significance
Pearson	0.47063	$< 1e{-}10$
Spearman	0.41414	$< 1e{-}10$

Algorithm 1: n-Gram strategy evaluation (best feature selection)

```
Result: Best performing strategy
initialization;
for all n-gram strategies do
    for all message m where m ∈ T do
        nGram ← computeNGramScore(m, strategy, corpusNYT);
        newsTerm ← computeNewsTerm(m, corpusReuters);
        news ← mean(newsUser, newsTerm);
        similarity ← computeSimilarity(news, nGram);
    end
    similarity‾ ← 1/n Σ_{i=1}^{N};
end
best ← argmax_{strategy} similarity‾;
return best
```

topic-independent association between a microblog message and the Reuters news word corpus. To validate our inverse distance metric, we performed Pearson and Spearman correlation tests with the best feature selected and they are shown in Table 3. The best feature selection is summarized in Algorithm 1.

As shown in Table 4, *unigram with stemmer only* feature has the highest correlation. Therefore, we select this feature for our user experiment and evaluation.

3.4 Experimental Setup

In this paper, we aim to identify unique newsworthy contents on microblogs that differs from those in mainstream news media like New York Times. So far we have explored different features based on content similarity metrics and text processing techniques. To validate our approach discussed in the previous section, we conduct an experiment including a crowd-sourced user study.

Table 4 [*n*-gram table] Comparison of different NLP mechanisms applied to computing co-occurrence between a microblog message and a news corpus (topic:*occupysandy*)

Avg # terms in news	Avg # terms in tweets	# Co-occurence	Stopword removal	Stemming	Noun only (POS-tag)	*n*-gram	Inverse distance
3863	17.952	10.509	N	N	N	1	0.774
12,085	16.965	1.713	N	N	N	2	0.814
15,246	16.011	0.162	N	N	N	3	0.689
1719	7.401	2.678	N	N	Y	1	0.777
4596	6.532	0.144	N	N	Y	2	0.75
5792	5.762	0.014	N	N	Y	3	0.714
1596	*17.952*	*3.868*	*N*	*Y*	*N*	*1*	***0.960***
4592	16.965	0.165	N	Y	N	2	0.758
5790	16.011	0.006	N	Y	N	3	0.740
1564	7.401	2.654	N	Y	Y	1	0.736
4509	6.532	0.145	N	Y	Y	2	0.8
5678	5.762	0.014	N	Y	Y	3	0.769
1557	11.161	1.744	Y	N	N	1	0.857
4495	10.171	0.068	Y	N	N	2	0.714
5664	9.251	0.006	Y	N	N	3	0.666
1557	6.217	1.473	Y	N	Y	1	0.857
4495	5.345	0.057	Y	N	Y	2	0.8
5664	4.611	0.007	Y	N	Y	3	0.666
1557	11.161	2.949	Y	Y	N	1	0.857
4495	10.171	0.163	Y	Y	N	2	0.833
5664	9.251	0.013	Y	Y	N	3	0.8
1557	6.217	1.99	Y	Y	Y	1	0.857
4495	5.345	0.136	Y	Y	Y	2	0.8
5664	4.611	0.015	Y	Y	Y	3	0.666

Each row in this table represents a different combination of text-matching mechanisms that were evaluated in our study

3.4.1 Random Sampling

For the experiment, we randomly sample 10,000 tweets from each collection. This sampling task allows us to avoid possible scalability issue from the high volume of our data sets and fit the experiments and user study. We sampled tweets that are primarily written while events were taking place or shortly thereafter. For the NYT articles, however, we aggregate them together first before we compute similarity features.

3.4.2 Niche Content Extraction

Our hypothesis is that, in general, newsworthy contents on microblogs do not completely overlap with mainstream news contents. In this study, the term "niche content" was coined for microblog exclusive (unique) newsworthy information. As the coined term implies, we assume that this type of information has a unique value and, thus, we believe that it is worth to investigate. The aim of this study is to find the unique characteristics of the niche content on microblogs and exploit our findings to provide a guideline to design more effective newsworthy information filtering algorithm in many applications.

We apply both statistical and heuristic approaches, including manual inspection on the contents with semantic relatedness in mind, to the experiment. Specifically, we manually inspect frequently used unigrams (see Table 6) after removing noisy information via stop word removal. Next, we classify these frequent terms into three different groups. Exploratory analysis such as frequency and burst analysis was also performed to scrutinize the data collections and compare contents from different categories with the features. We then sample microblog messages from two different groups: contents with high/low similarity with regard to mainstream news media contents. To perform this second-phase sampling task, we choose 20 and 80 percentile in n-gram feature distribution as the thresholds. We will provide some insights into the distinction that we interpreted from the experiment and discuss limitations later in Sect. 3.5.

3.4.3 User Study

Following our content extraction and comparative analysis, we conduct a crowd-sourced user study to validate our hypothesis. In the user study, the participants were shown two groups of ten tweet messages. Each group of tweets was randomly sampled from the messages with high similarity and low similarity to mainstream news media contents in $News_{n\text{-}gram}$ metric, respectively. The participants were then asked to answer six different questions regarding (1) similarity to traditional news articles, (2) newsworthiness, and (3) how personal the shown content is. They were also asked to answer to general questions such as demographic information (gender, age, education level, etc.) and their microblog usage.

3.5 *Evaluation*

We now discuss evaluation of the research questions posed earlier. Using the best performing co-occurrence method from the 24 mechanisms for computing similarity between a short Twitter message and a larger collection of news, showing in Table 4, we conducted a user experiment to assess perceived differences between messages sampled from the niche areas shown in Fig. 1 and a general sampling of messages in the topic. The experiment consisted of two conditions: (1) message sampling along the 20th and 80th percentiles of the x-axis from Fig. 1 (i.e.: the co-occurrence score between a tweet and the NYT article corpus), and (2) messages sampled from the top left corner of Fig. 1. I.e.: co-occurrence score combined with a content-based newsworthiness score for the message. This area represents messages that are inherently newsworthy but do not frequently occur in the mainstream corpus. In both conditions, the samples were shown alongside randomly sampled messages about the topic and user perception was evaluated. Information consumers can perceive newsworthiness differently over time, so we first examine a sample of temporal distributions of topics across the two domains (NYT and Twitter) (Table 5).

3.5.1 Frequency Analysis

Figure 3 shows a frequency analysis of Twitter postings and NYT articles related to the 2014 world cup. Multiple peaks on both line plots show sudden bursts of discussions (on microblogs) or reports (from news outlets) on the corresponding topic (*world cup*). In this representative example, both streams follow a similar trend, but the bursts are more pronounced on Twitter than in traditional news. This trend in bursts is representative of several analyzed topics, so, while Twitter appears to be more reactive to events in terms of bursts, both streams show peaks of interest for critical events (semi-final and final in this case), indicating that newsworthiness of events is similar on both sources.

3.5.2 Study Participants and Procedure

Participants for the user experiment were recruited though Amazon's Mechanical Turk (MTurk). A total of 200 participants took the study which lasted an average of 8 min. 48% of participants were male and 52% were female. All participants were active microblog users. Age ranged between 18 and 60, with the majority between 25 and 50 (78%). 69% of participants reported having a 4-year college degree or higher. Participants were all located within the United States and had completed a minimum of 50 previous successful tasks on the MTurk platform.

Participants were shown a Qualtrics survey[5] that asked basic demographic questions. Next, they were shown two groups of ten microblog posts, side by side with random ordering. Two conditions were evaluated. Condition 1 showed groups of messages randomly sampled from within the 20th and 80th percentiles along the x-axis of Fig. 1. To recap, this axis represented the co-occurrence score of the best performing mechanism from Table 4. Condition 2 users were shown ten messages that were sampled from the top left portion highlighted in Fig. 1 (the "unique" and "newsworthy" messages), and ten randomly sampled from within the topic.

[5]www.qualtrics.com.

Table 5 Statistics overview across different data sets (stemming only)

Topic	# of terms in news			Avg # of *n*-grams in a tweet			Avg % of co-occurrences		
	Unigram	Bigram	Trigram	Unigram	Bigram	Trigram	Unigram	Bigram	Trigram
World cup	9274	75,036	122,573	18	17	16	77.7%	25.6%	6.3%
ISIS	2573	9764	12,724	19	18	17	63.1%	14.9%	2.4%
Earthquake	2303	7114	8772	18	17	16	64.3%	15.9%	4.1%
Occupysandy	3078	11,865	15,190	18	17	16	60.5%	10.3%	1.0%

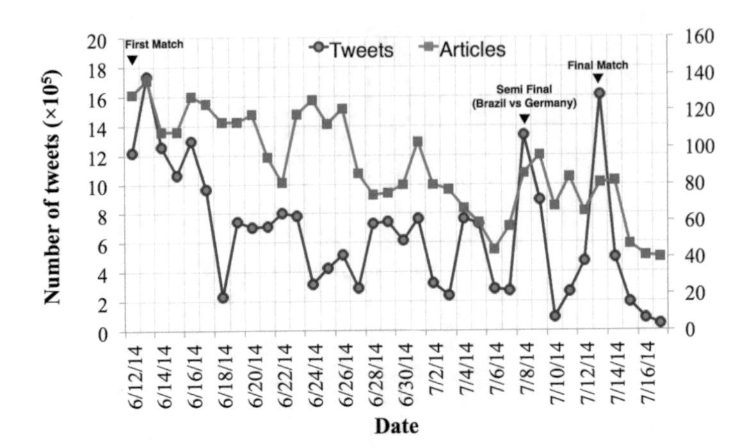

Fig. 3 Temporal distribution of the microblog messages (tweets) and news articles on the topic—*worldcup*. The time period shown in this graph corresponds to the 2014 world cup held in Brazil

This selection used both the x-axis similarity and the content-based newsworthiness score described earlier. In each case, participants were asked to rate their agreement with three statements for each group shown (total of six ratings):

1. *The messages in group x are similar to what I would find in mainstream news such as the New York Times.*
2. *The messages in group x are newsworthy*
3. *The messages in group x are personal*

3.5.3 Results

Results of the experiment are shown as box plots in Figs. 4 and 5. Our first task was to assess the effect of the co-occurrence metric chosen from the 24 options in Table 4. Two random groups of 10 tweets were sampled from the poles of this distribution (shown as the x-axis in Fig. 1) and displayed side-by-side to participants. Participants were asked to rate their agreement with the questions listed above on a Likert scale of 1–5, with 5 indicating full agreement with the statement. Responses to the above questions are shown in Fig. 4. Participants reported that the similarity to mainstream media was higher for messages with high co-occurrence, but, we did not observe a statistical significance for this result. Figure 5, however, does show a significant difference at $p < 0.05$ between the sampled messages. So, by augmenting the co-occurrence score with a content-based newsworthiness score, shown in Eq. (2), we achieved a significant shift in perception of uniqueness of content. Interestingly, the perception of newsworthiness for these messages was reasonably high and did not change significantly along the x-axis (similarity to

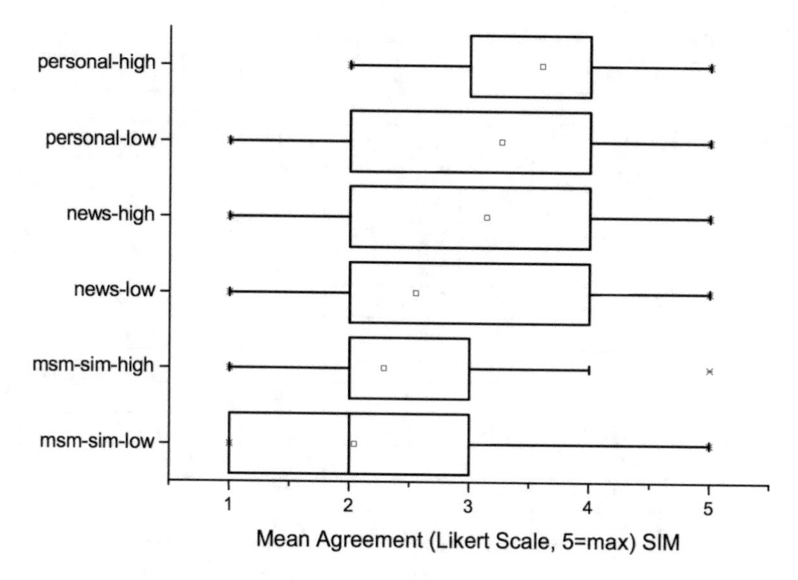

Fig. 4 Mean agreement of the responses from condition 1—`SIM`

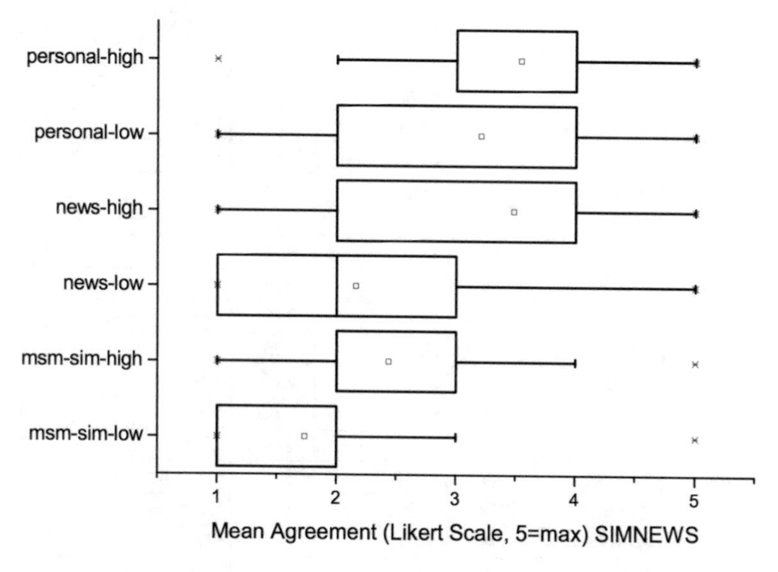

Fig. 5 Condition 2: Mean agreement of the responses from the user study—`SIMNEWS`

NYT), meaning that the approach did find messages that people felt were unique to the microblog domain and were also newsworthy.

Results of a term-based analysis are shown in Table 6 which displays three sample topics ("worldcup," "ISIS," and "Earthquake." The table shows the top $n = 10$ terms from each data set as they overlap with the source data. The left

Table 6 Top ten frequent words extracted from tweets on each topic

	Article		Common		Tweet	
	Word	#	Word	#	Word	#
Worldcup	2014	412	Worldcup	4801	Fifaworldcup	1011
	Thursday	231	World	2492	Bra	763
	Skiing	86	Cup	2363	Arg	706
	Longman	76	Soccer	1161	Ned	551
	Table	65	Brazip	1077	Joinin	418
	Association	64	Germany	873	Mesutozil1088	296
	1994	61	Ger	656	Worldcup2014	294
	Golf	60	Final	598	Gerarg	273
	Governing	60	Team	580	Fra	214
	Christopher	59	Argentina	509	Crc	211
ISIS	8217	33	Isis	4872	Amp	665
	Adeel	16	Iraq	445	Via	497
	2015	13	Syria	370	Dress	294
	Fahim	12	Obama	340	Cnn	170
	Schmitt	11	Islamic	339	Isil	162
	1973	10	Video	295	Share	134
	Fackler	8	State	281	Foxnews	126
	Corrections	6	us	274	Bokoharam	119
	Badr	6	alive	259	USA	113
	Abdurasul	5	Jordan	225	Daesh	107
Earthquake	Sniper	31	Earthquake	5165	Utc	484
	2011	22	Magnitude	835	Amp	333
	Kyle	19	Japan	515	Breaking	309
	Defense	15	Tsunami	451	Feel	274
	Former	14	California	348	Via	261
	Marine	12	Usgs	345	Newearthquake	254
	Tea	10	New	333	Mar	192
	Routh	9	Ago	295	Alert	191
	Navy	8	Strikes	256	Sismo	186
	Nations	8	Quake	245	Map	161
Occupysandy	Blackouts	49	Sandy	641	Occupysandy	5867
	Andrew	32	Help	410	Sandyaid	598
	Presidential	30	New	343	Ows	425
	Conn	29	Need	298	Sandyvolunteer	340
	Newtown	28	Hurricane	248	Please	329
	Barack	26	Relief	207	Occupywallstnyc	310
	Education	25	Nyc	205	520clintonos	269
	Connecticut	24	Volunteers	194	Today	264
	Gasoline	21	Occupy	193	Info	216
	Senate	21	Rockaway	182	Thanks	210

column (Article) shows terms that are mostly unique to news articles. The center column shows combined terms, while the rightmost column shows terms that are popular on Twitter but not overlapping with the mainstream news. From manual inspection, the combined terms in the middle column in Table 6 appear to be a good descriptor of the topic. For example, the "ISIS" topic contains "ISIS"; "IRAQ"; "SYRIA"; "OBAMA"; "ISLAMIC" as the top five terms. Terms unique to mainstream media appear to be focused more on official structures and laws, while terms unique to the microblog tend to be more personal and emotional. Interestingly, the term "BOKOHARAM" is listed in the microblog column. This is a good example of a global news phenomenon that is covered extensively in most countries, but is relatively under-reported in the United States. Now we will discuss our results in the context of the research questions presented earlier.

RQ1 How can we best detect newsworthy information in social media that is not covered by traditional media? We have examined 24 mechanisms for computing the similarity between a short microblog post and a corpus of news articles. Our findings show that a simple approach using simple unigram term matching and a porter stemming algorithm provides a better approximation of manually labeled examples than other methods tested, including POS tagging, stop-word removal, and matching on bi-grams and tri-grams. Our initial expectations were that bi-gram and tri-gram overlap would produce better matches to the manual labels. Our experimental data showed that single term overlap was a better metric. We assume that since microblog posts have a limited number of terms, overlap in bi and tri-grams was sparse, as highlighted by the statistics in Table 4. For example, unigram co-occurence for the topic "ISIS" shows 78% overlap with the news article database, while bi-gram overlap is 26% and trigram overlap is just 6.3%. For future work we plan to apply a combination of n-gram overlaps to create better mappings between microblog posts and news articles.

RQ2 How do information consumers perceive the detected information? Our online evaluation of 200 paid participants shows us that sampling messages from the distributions created by the co-occurrence computation produces a significant increase in perception of the uniqueness of messages, while not affecting perception of newsworthiness. We believe that this is a promising result for the automated detection of niche and newsworthy content in social media streams.

4 Network Based Approach

Following the previous approach, we propose another approach to capturing unique news content on microblogs using structural and functional metrics of network. In this section, we demonstrate our strategies to find differences in network structure and topic association between niche and traditional groups of tweets.

The main idea that guides our two proposed approaches is that there is a unique portion of newsworthy content in microblogs that are not covered by traditional media. The underlying assumption in the second approach is that such unique content travels from a node to its neighbors in a different fashion from those covered by traditional news outlets. Let us assume that a node u_i produces a "newsworthy" content m_i in the network and m_i becomes exposed to u_i's neighbors in a given time Δ_t. Unlike one-to-many propagations for contents directly provided by traditional media (e.g., tweets posted by @BBC), we expect arbitrary one-to-one or one-to-few type of propagations in the unique content group.

In this approach, we apply (1) network and (2) topic association analyses to our microblog datasets. First, we convert the crawled tweets and their associated users into two different graph data structures (network and topic spaces) based on the typical vertex/edge graph structure ($G = (V, E)$). Before analyzing the two spaces, for the network space, we reconstruct a retweet chain graph using our datasets. In this graph structure, every node, or a vertex, i represents a user u_i, and an edge $(i,j) \in E$ ($E \subset V \times V$) that connects nodes i and j becomes a retweet. We can say that $i \sim j$ if $(i,j) \in E$. For the topic space, we apply topic modeling to microblog messages using Latent Dirichlet Allocation (LDA) and extract associated topics from the messages. Using the topics extracted from the tweets, we construct a bipartite graph G_{LDA}. In this graph, we have a set of users $U = \{u_1, u_2, \ldots u_m\}$ and another set of topics $T = \{t_1, t_2, \ldots t_m\}$ that are associated with the users $\in U$. Afterwards, we generate the final graph G_{BPR} using Bipartite Projection via Random Walks algorithm proposed by Yildirim and Coscia [34]. The algorithmic detail of the two methods are described in Sects. 4.4 and 4.5.

4.1 Hypotheses

In this study, inspired by our motivations, we aim to answer the last research question (**RQ3**) we have in Sect. 1.

- **RQ3:** How do the niche information get propagated differently from traditional news in the network?

As a recap, in this paper, we assume that the unique and newsworthy contents on microblogs do not completely overlap with mainstream news contents. Accordingly, the following hypotheses are derived to further shape the experimental setup for our network-based approach.

Hypothesis 1 A difference in network structure can be observed between the spread of niche (unique) and traditional media content.

Hypothesis 2 A difference in topical association can be observed between the two groups.

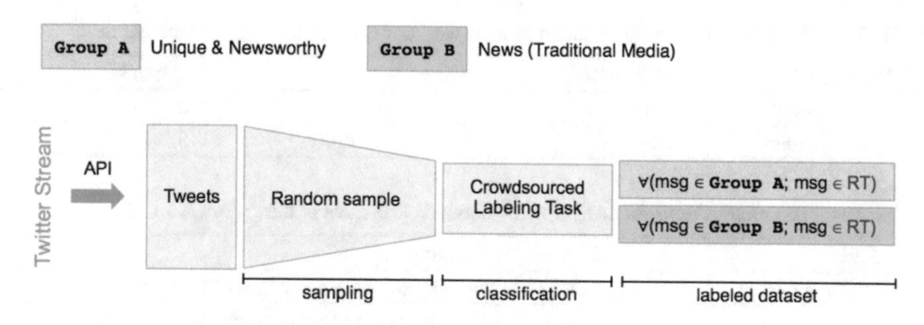

Fig. 6 A diagram that describes crawling and labeling data sets

4.2 Data Collection and Preprocessing

To utilize real data from the microblogging platform Twitter, microblog posts, or "tweets" were crawled for specific keywords. In this study, we have crawled a total of 2,353,334 tweets using Twitter REST API on three different topics: *#Calais* (86,627), *#prayforparis* (1,431,467), *#paris* (835,240). After examining all datasets, we decided to focus on the *#paris* dataset which covers most news threads and relevant discussions on related subtopics. The datasets were collected during the terrorism in Paris (Nov. 8–Nov. 15.) This crawling process is shown in Fig. 6.

Using the crawled datasets, we reconstructed retweet chain graphs in which the nodes represent users and the edges between them represent a retweet. In the data pre-processing task, the content (message text) of each tweet and corresponding metadata such as retweet count, number of friends/followers, user id and screen name, language, self-reported location are extracted using a document-oriented database[6] and parsing scripts.

4.3 Labeling Tweets

Before the comparative analysis on the two groups of contents (Group A and B), we need to classify the messages into one of the groups. Since both newsworthiness and uniqueness of content are subjective metrics, we conducted a labeling task on a crowdsourcing platform.[7] Each individual message of the 300 sampled retweets from our data collection is shown to three different participants. During the task, each user was asked to rate *newsworthiness* and *uniqueness* of the given tweet in [1–10] Likert scale and answer the foundation of their judgement on newsworthiness

[6] A NoSQL database (Mongo DB) was used.

[7] Crowdflower (http://crowdflower.com) was used for the labeling task.

Table 7 Distribution of the foundation of newsworthiness assessment in the labeling task

News type	# Responses	News type	# Responses
Usefulness	314	Timeliness	210
Interestingness	233	Novelty or rarity	143

Table 8 The metrics used for analyzing the network structures of the groups A and B

Metric	Symbol	Description
Node/edge count	N/M	Number of nodes and edges of a graph G
Average degree	$< k >$	The mean of number of edges connected to all nodes of the graph G
Closeness centrality	Cen_C	Inverse average distance to every other vertex
Betweenness centrality	Cen_B	Fraction of shortest paths that pass through the vertex
Eigenvector centrality	Cen_E	Importance of a node in a graph approximated by the centrality of its neighbors
Mean clustering coefficient	C	The mean clustering coefficient of the graph G

among usefulness, timeliness, novelty (rarity) and interestingness (see Table 7). We asked multiple participants to label on each message to avoid personal bias towards/against specific topic or information source. Thus, we only use the tweets that have high agreement on both newsworthiness and uniqueness of the content across three participants.

4.4 Network Analysis (Exp 1)

In this study, we are interested in investigating how "newsworthy and unique" content differs from other generic news contents. In particular, we want to analyze who generally produces this unique content and how this content is structured, i.e. propagated, in the network. Borrowing the perspectives from graph mining and social network analysis, we assume that each node corresponds to a message (or a user who posts/re-posts that message) and each edge to a propagation of a message from a node to its neighboring node. The list of network metrics we use are shown in Table 8.

Besides the metrics we use to indicate network structures, in this study, we examine how vertices are associated with their neighbors by looking at the structure of the graphs through graph visualizations. We will discuss our findings in Sect. 4.6.

4.5 *Topic Association* (`Exp 2`)

The second experiment seeks to explore topological differences in topic-similarity networks of users that are responsible for spreading unique versus non-unique posts. The BPR method from [34] is utilized to create a user-user content similarity network for this purpose (Figs. 7 and 8).

From the original retweet network, each user is extracted along with their 100 most recent tweets, which are aggregated into a single document. Latent Dirichlet Allocation (LDA) [9] is then performed and a document-topic matrix is produced. From this, a two-modal bipartite graph is constructed. For the *#paris* retweet network, LDA was performed with 25 topics ($K = 25$). If a user u_i's last 100 tweets

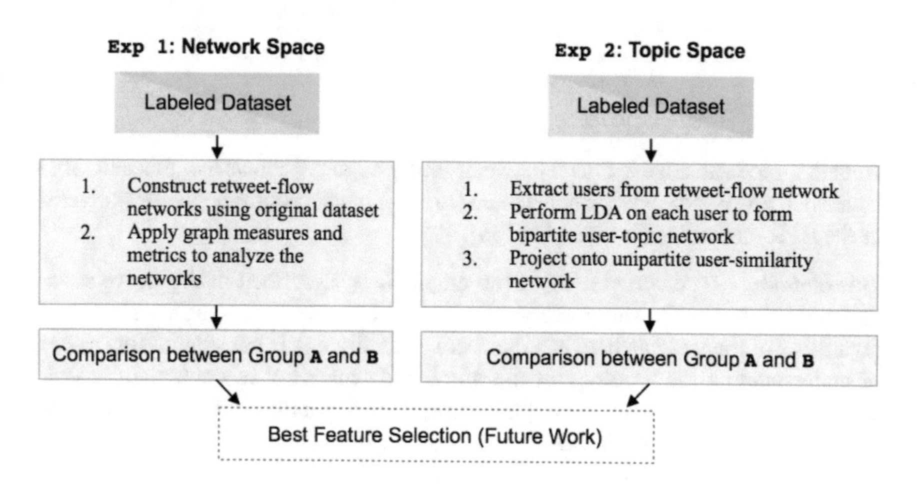

Fig. 7 A diagram that demonstrates how we process data and evaluate the model proposed in the study

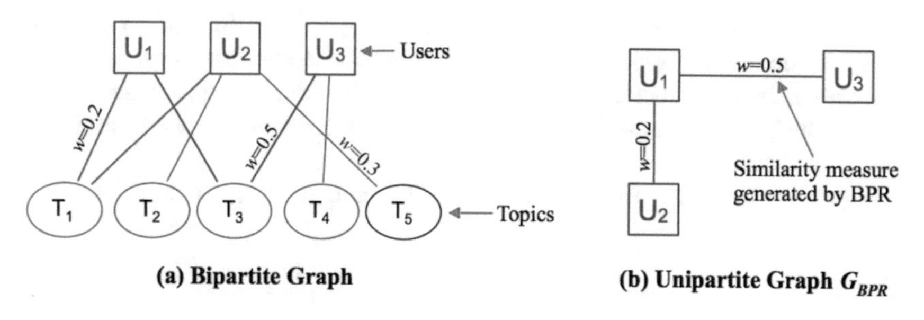

Fig. 8 Bipartite graph construction using Bipartite Projection via Random Walks. Note that although there is an inherent weight assigned to edges in the (**a**) by the document-topic matrix, (**b**) is constructed using a simple binary adjacency matrix

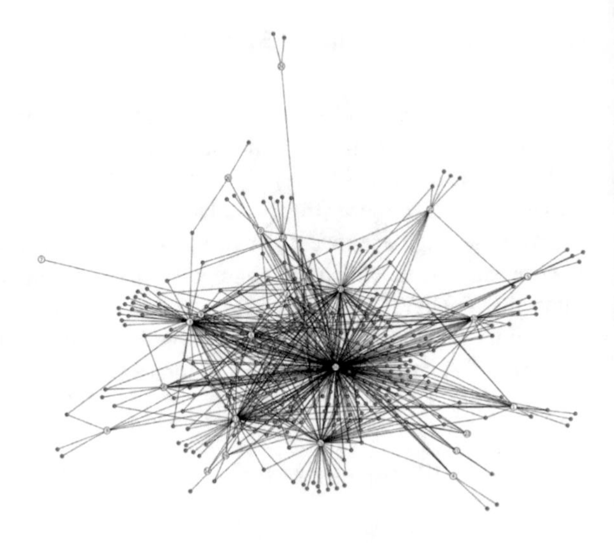

Fig. 9 *User-topic Content Similarity Network for #paris. Red nodes* represent users, and *white nodes* represent topics

contain topic t_j, an edge is drawn between i and j. Figure 9 shows that this network is connected and edges exist only between $(u_i \sim t_j)$ pairs; requirements for utilization of the BPR method can be found in [34].

Thresholding To construct a unipartite graph G_{BPR}, described in Fig. 8, we set the threshold τ, not establishing every edges when two users share at least one topic regardless of the weights between the users. This strategy is considered for the ease of understanding the topology of the graph and scalability of computation. For a given bipartite graph $\mathcal{G}$, let $\theta_{\mathcal{G}} \in [0, 1]$ denote the threshold of weight between the user u_i and the topic t_j such that

$$(u_i, t_j) \begin{cases} \text{exists} & \text{if} \quad \mathsf{weight}(u_i, t_j) \geq \theta_{\mathcal{G}} \\ \text{not exists} & \text{if} \quad \mathsf{weight}(u_i, t_j) < \theta_{\mathcal{G}} \end{cases} \tag{3}$$

The BPR [34] projection method, shown in Fig. 8, is performed on this user-topic content similarity network, and thus predicts edges in the user-user content similarity network. This technique accounts for the overall structure of the bipartite graph, which helps ensure that topic hubs do not saturate its unipartite projection with unlikely links.

Figure 7 shows the overall process of data processing and evaluation of our approach.

4.6 Results and Discussions

In this section, we will discuss the findings from our two experiments (`Exp 1` and `Exp 2`).

4.6.1 Network Analysis (`Exp 1`)

Since our primary interest is how information is produced and propagated along the connections in microblogs, we study how they differ between `Group A` and `Group B` by re-constructing retweet chains from the dataset into undirected graphs and compute the graph metrics in Table 8 on these graphs. These metrics can help us gain some insight into the structure, behavior, and dynamics of the given network. Specifically, for example, we can answer to such questions: (1) what are the dominating nodes in the propagation chain/network; (2) how densely do the nodes connected to each other; (3) can we partition this network into N different components; (4) does a giant component exist in this graph. To evaluate structural characteristic of the graphs in each group, we visualized the landscape of the entire data collection, and this is shown in Fig. 10.

Figure 10 shows the network on the topic of *#paris* with a giant component surrounded by many isolated nodes and small components. In this graph, the giant component is loosely connected with many subcomponents via single or a few edges. Intuitively, we can divide the giant component into multiple clusters (or subcomponents) through these low-connectivity edges with high betweenness centrality. Intuitively, this type of structure can be sparsified into a simplified graph structure using sparsifier graph H (d-regular Ramanujan graph). According to Benczur-karger approximation model [7], we can sample low-connectivity edges (with high probability), eliminating high-connectivity edges within densely connected components (Tables 9 and 10).

For the comparison of `Group A` and `Group B`, we sampled 2 most representative subgraphs for each group from the dataset. Structural characteristics of each set were then analyzed through visual and computational assessments.

`Group A`: *Unique and Newsworthy Contents* Our labeling task performed on the crowdsourcing platform revealed that the participants favored unique 3rd-party news

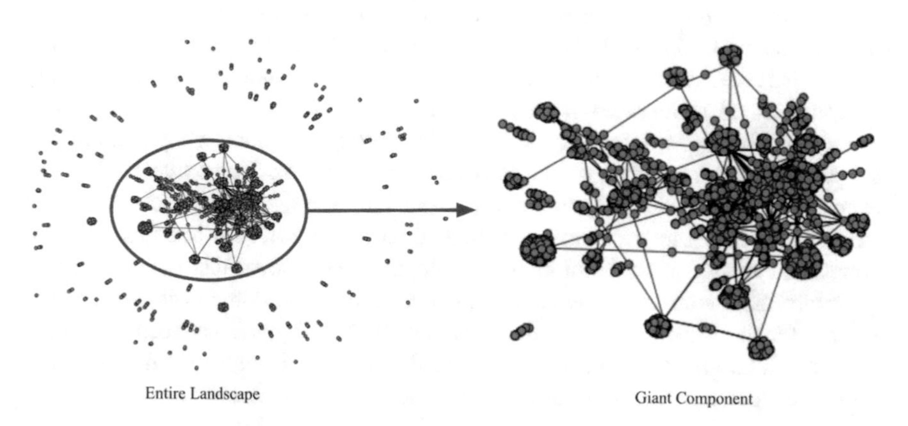

Fig. 10 The landscape of the retweet chain graph reconstructed from the dataset "#paris." One giant component and a number of small components were observed

Table 9 Basic metrics computed for the representative subgraphs (retweet chain graphs) of group A and B

Group	# Nodes	# Edges	$< k >$	C
@globalnews (Grp A)	159	151	1.8994	~ 0.0
@CNN (Grp B)	5245	6563	2.5026	0.045

$< k >$ mean degree, C mean clustering coefficient

Table 10 Network metrics computed for the bipartite (topic-user) graphs of group A and B

Group	N	M	#CComp	$< k >$	C	Cen_B	Cen_C	Cen_E
A-BrianHonan	9	8	4	0.889	0.367	0.7	1.0	0.545
A-musicnews_facts	228	6895	2	30.241	0.573	0.033	0.766	0.189
A-margotwallstrom	8	1	7	0.125	0.0	0.0	1.0	0.707
B-CNN	1743	375	1647	0.215	0.029	0.554	0.521	0.287
B-NBCNews	226	7934	7	35.106	0.551	0.025	0.777	0.181
B-FoxNews	1565	226	1494	0.144	0.029	0.382	0.769	0.322

Please note that all centrality metrics are computed on the max centrality nodes in the main connected component (#CComp: number of connected components. For other symbols, see Table 8)

providers or quotes from celebrity accounts (e.g., *@musicnews*, *@BrianHonan*) and labeled them as niche contents. For example, the tweet "*RT @BrianHonan: With the news breaking from Paris it's wise to remember this. https://t.co/bKZP5Vh46n*" was rated as highly newsworthy and unique (in other words, less likely to be seen in or covered by traditional news outlets.) Interestingly, many tweets that contain both personal opinion with sentiment and a short news headline (sometimes with a url that directs users to an external source of information) within a tweet received high newsworthy and uniqueness score.

Group B: Traditional News Most tweets that fall into this category are, expectedly, news headlines or blurbs provided by major news providers or other institutional accounts. Most of the graphs in Group B has long retweet chain that either spans across the comparatively big component or connects two neighboring components. In some cases (an example is shown in Fig. 11) one or two nodes exist(s) that bridges two small components in similar size, constructing a dumbbell-shaped graph. An example might be where the New York Times tweets about an event to its many followers, one of which is CNN News, who then retweets to its many followers. Another example of this effect that occurred in the crawled data about the Paris terrorism event involved a popular Dutch journalist who re-tweeted false information about the lights in the Eiffel Tower being turned off as a mark of respect for the victims. This created a dumbbell shaped graph between the Dutch and French communities, that also happened to contain misinformation, since the lights were actually turned off as a matter of routine (Figs. 12 and 13).

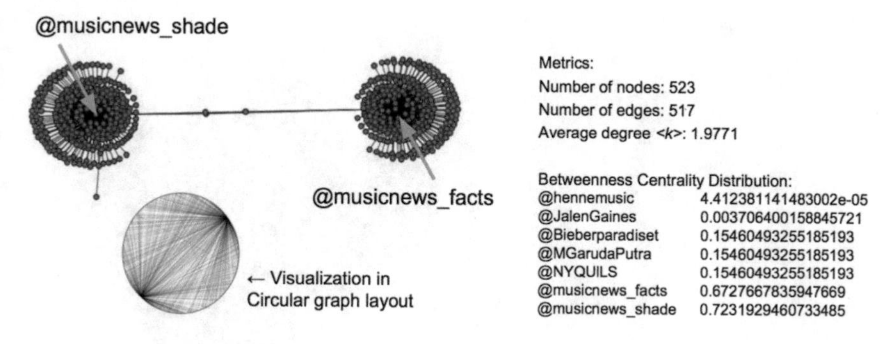

Fig. 11 An example of dumbbell type graph found in *#paris* dataset

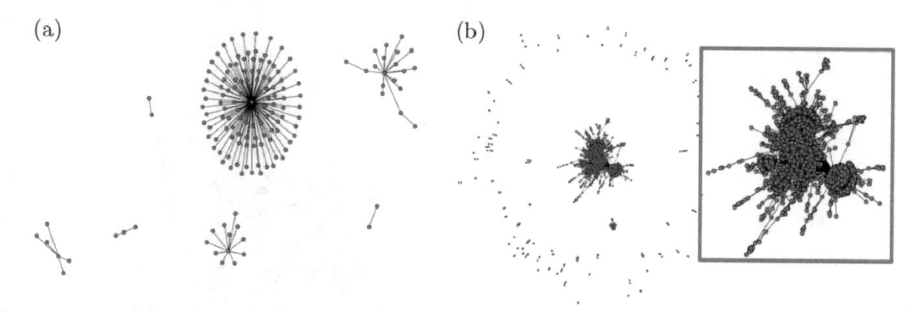

Fig. 12 Graph visualization of the unique news group `Group A` (**a**) and traditional news group `Group B` (**b**)

4.6.2 Topic Association (`Exp 2`)

The user-user content similarity network generated by the BPR is the network of interest. Figure 14 shows this network for the *#paris* example. For each projection, many possible networks can be formed based on the threshold of similarity τ between users needed to form an edge between them. Continuing the *#paris* example, the power law is reflected in Fig. 15, which plots the number of edges in the user-user network versus the similarity threshold used to form that specific network. This relationship seems to fit a power-law distribution, which would suggest that the BPR method has successfully captured scale-free decay in the number of similarities as the similarity threshold increases. Without any threshold, the giant component does not in fact grow to the entire network; the network remains unconnected. Notably, the unconnected nodes in the user-user network have an average degree of only 1.03 in the user-topic network, which explains why BPR did not predict any edges for these users.

Additionally, some user-user content similarity networks that were generated for #paris are suspected of exhibiting a power-law degree distribution themselves; an example of which is shown in Fig. 16. To corroborate this claim we will investigate further into the degree distributions of these networks as a future work.

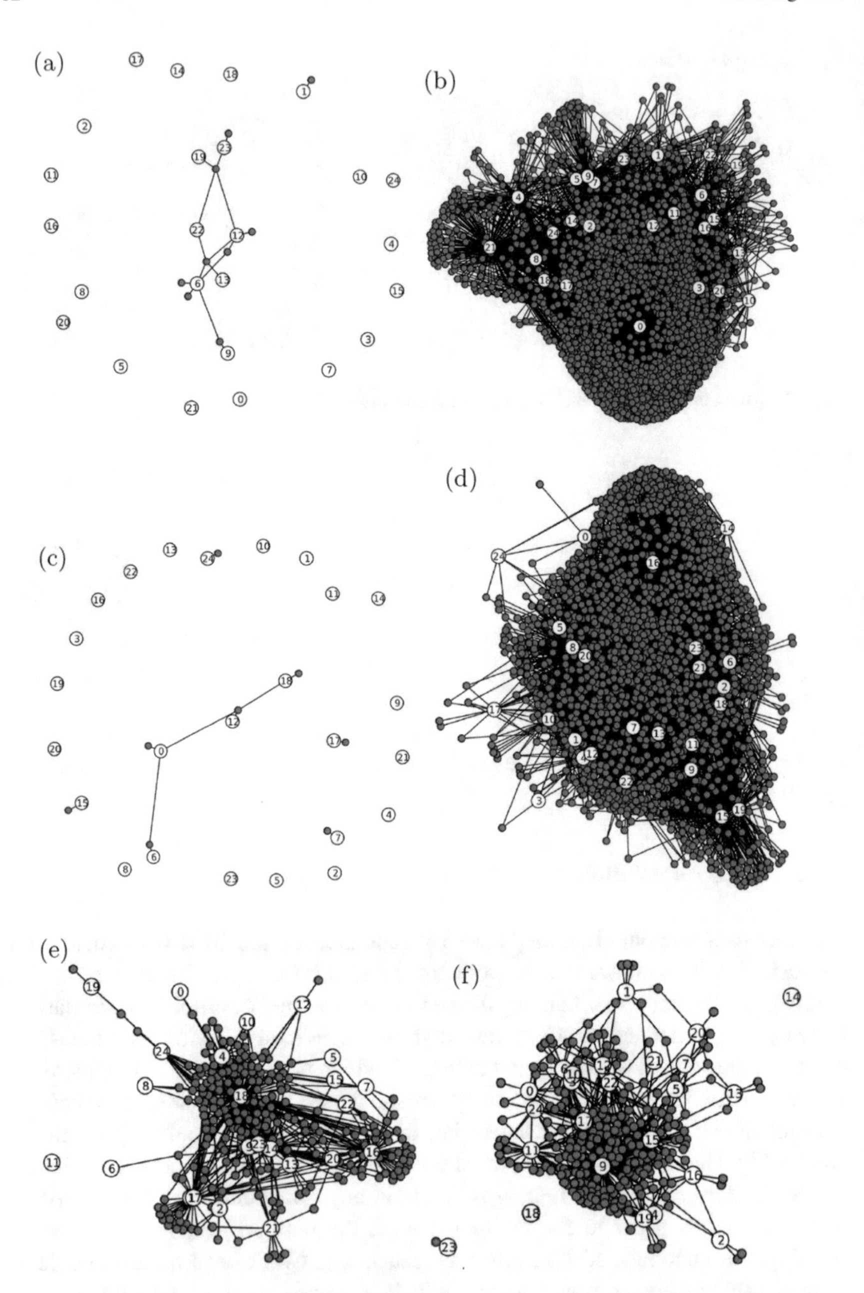

Fig. 13 Bipartite graphs of user-topic association network. Please note that LDA topic nodes are labeled with index numbers (from 1 to K; $K = 25$). Please note that (**e**) and (**f**) are the examples of crossover accounts. (**a**) `Group A` (@BrianHonan), (**b**) `Group B` (@CNN), (**c**) `Group A` (@margotwallstrom), (**d**) `Group B` (@FoxNews), (**e**) `Group A` (@musicnews_facts), (**f**) `Group B` (@RasmusTantholdt)

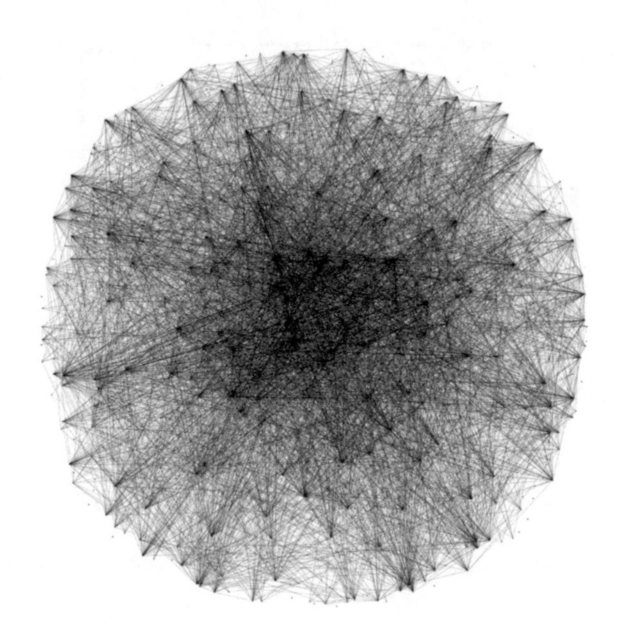

Fig. 14 *User-user Content Similarity Network for #paris.* This specific network was constructed using a similarity-threshold of 0.00001 (4295 edges)

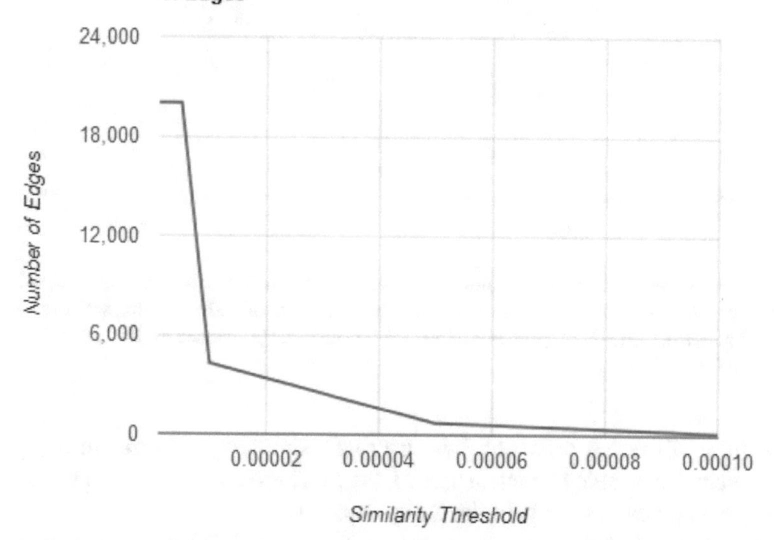

Fig. 15 Plot of similarity threshold versus number of edges generated in user–user content similarity network using the threshold τ. Calculated power-law constants using $\tau \times 1000$: alpha= -1.113, $B = 20.358$

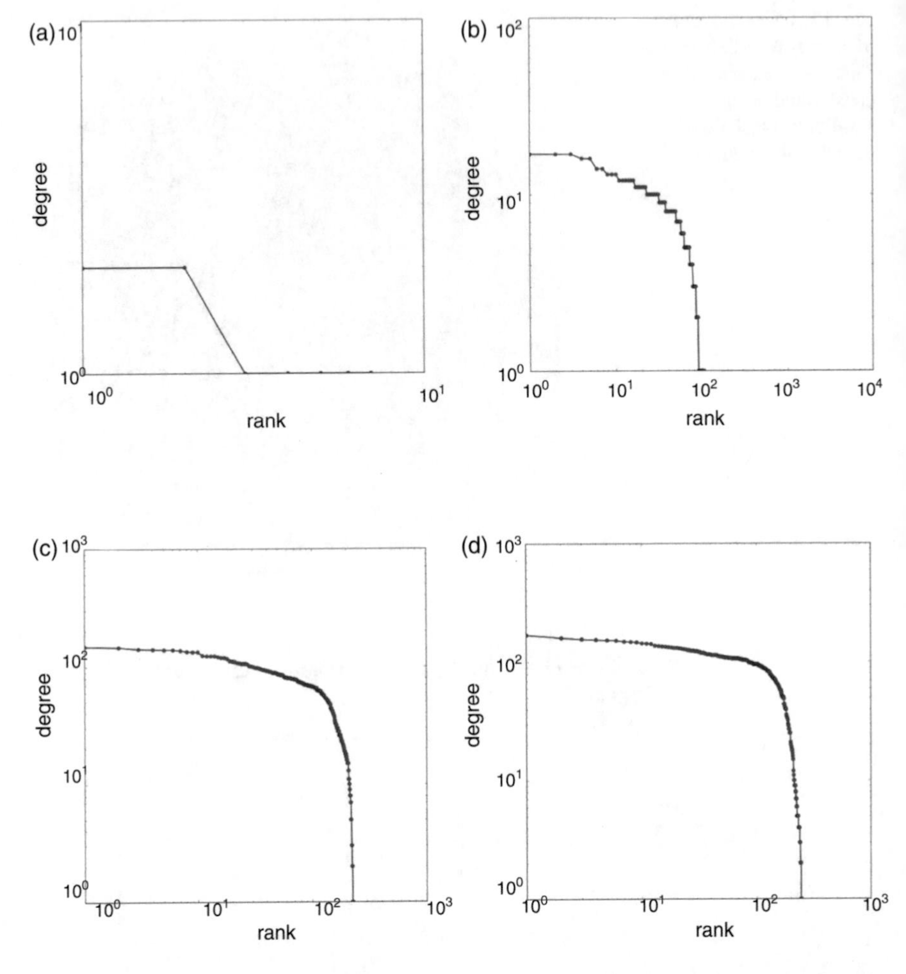

Fig. 16 Distributions of topic associations of users in group A and B. *X*-axis shows users in rank order (log scale) and *Y*-axis shows the number of topic associations, also on a log scale. (**a**) Group A (@BrianHonan), (**b**) Group B (@CNN), (**c**) Group A (@musicnews_facts, (**d**) Group B (@RasmusTantholdt)

To fully utilize the power of these user-user similarity networks in comparing unique versus non-unique content spread, the same process was carried out on a set of sampled retweet networks of the groups A and B.

It is suspected that user-user content similarity will differ between users that spread non-unique (Group B) posts versus users that spread unique (Group A) posts, as these group's corresponding retweet-chain network structures are different. Also, comparing outlying users (users that become unconnected in user-user similarity networks) to those in the giant component of the opposite group could help provide insight to any overlap in users that spread content from both groups.

5 Discussion and Future Work

In this study, a few challenges have been discussed in order to achieve our final goal: developing a reliable and automated detection algorithm for unique news content on microblogs. For future work, we will apply the salient features that we found in this study to different machine learning algorithms and find an effective way to automatically locate niche microblog contents. Moreover, a temporal analysis will be performed on retweet-chain graphs in order to reveal differences in network dynamics between the groups. Any temporal patterns, found by the analysis, may allow online learning algorithms to predict niche content across time. Specifically, by investigating multiple snapshots of each network, we can measure the temporal differences and compute related metrics over the course of development of each network. In this type of analysis, tensor and different decomposition methods such as high order SVD, PARAFAC/CANDECOMP (CP) decompositions can be applied to find out multidimensional characteristics of the given network. When we incorporate the best features into an automated algorithm, however, the algorithm might need to be optimized requiring occasional user feedback due to the ambiguity and subjectivity of newsworthiness. One of the key challenges for this work is the problem of entity detection for news. This study is limited in the sense that we focus on simple messages and compute inherent newsworthiness and similarity scores. This raises two challenges for future research. First, is a problem of understanding word meaning at the right level. Consider the classic problem of a Jaguar being either a cat or a car, for example. User-provided hashtags can help us to disambiguate, however these can lack granularity in some cases. Another challenge is diversity in spelling and form, such as "Ph.D," "PHD," "ph.d" for example. Approaches such as Bostandjiev et al's study on LinkedVis [10] use entity resolution methods to resolve these problems. For example, by resolving every mention of PhD to the Wikipedia page for PhD. Such content-based approaches could help for some news event detection and disambiguation, but of course would not guarantee a comprehensive solution to these difficult problems.

5.1 *Scalability and Real-Time News Gathering*

Many information filtering methods that rely on underlying similarity models, such as automated collaborative filtering, for example, are, or have components that are computationally complex. In contrast to these approaches, which typically employ a modeling phase with complexity of $O(n^2)$ on the number of users, our method for computing inherent newsworthiness can theoretically be applied as a linear time function of number of terms in the message, and further, with efficient indexing structures for n-gram access in the news corpus (an offline process, which, lets say using a quicksort algorithm, would run in $O(nlog(n))$ time), results in a linear time search that is a function of the number of n-grams per message. This allows for a highly scalable, real-time news search experience for the end user. However, aside from the theoretical analysis, there are practical rate limitations on the Twitter end-points that would require commercial agreements to achieve the full potential of our approach.

6 Conclusion

This paper evaluated novel approaches for automatic detection of unique and newsworthy content in microblogs, using a comparative analysis between a corpus of curated news articles from traditional media and collections of "uncurated" microblog posts. Our initial approach examined differences in content similarity between the two. 24 combinations of simple NLP techniques were evaluated to optimize a similarity score between a short Twitter post and a corpus of news articles about a target topic. Next, a user study was described that gathered human annotations of newsworthiness for use as groundtruth to evaluate our filtering method. Results showed general agreement between predicted scores from our approach and the human annotations.

We extend our news detection method to include information about the underlying network and dynamics of the information flow within it. LDA and BPR algorithms were used to explore structural and functional network metrics for the purpose of predicting newsworthiness and uniqueness of content. Primarily, we have studied the structure of various subgraphs underlying multiple topic-specific collections of microblog posts. Moreover, we have proposed a method to explore the topical association between different nodes in a graph, i.e. the vertices that tend to belong to either unique or traditional news groups. The results of our empirical analysis show that structural differences are observed between the unique and traditional news groups in microblogs. For example, the majority of subgraphs in the traditional group have long retweet chains and exhibit a giant component surrounded by a number of small components, unique contents typically propagate from a dominating node with only a few multi-hop retweet chains observed. Furthermore, results from LDA and BPR algorithms indicate that strong and dense topic associations between users are frequently observed in the graphs of the traditional group, but not in the unique group.

Acknowledgements This work was partially supported by the U.S. Army Research Laboratory under Cooperative Agreement No. W911NF-09-2-0053; The views and conclusions contained in this document are those of the authors and should not be interpreted as representing the official policies, either expressed or implied, of ARL, NSF, or the U.S. Government. The U.S. Government is authorized to reproduce and distribute reprints for Government purposes notwithstanding any copyright notation here on.

References

1. Agichtein, E., Castillo, C., Donato, D., Gionis, A., Mishne, G.: Finding high-quality content in social media. In: Proceedings of the 2008 International Conference on Web Search and Data Mining, pp. 183–194. ACM, New York (2008)
2. Allison, L., Dix, T.I.: A bit-string longest-common-subsequence algorithm. Inf. Process. Lett. **23**(5), 305–310 (1986)
3. Anderson, A., Huttenlocher, D., Kleinberg, J., Leskovec, J.: Effects of user similarity in social media. In: Proceedings of the Fifth ACM International Conference on Web Search and Data Mining, WSDM'12, pp. 703–712. ACM, New York (2012)
4. André, P., Bernstein, M., Luther, K.: Who gives a tweet?: evaluating microblog content value. In: Proceedings of the ACM 2012 Conference on Computer Supported Cooperative Work, CSCW'12, pp. 471–474. ACM, New York (2012)
5. Bär, D., Biemann, C., Gurevych, I., Zesch, T.: Ukp: Computing semantic textual similarity by combining multiple content similarity measures. In: Proceedings of the 1st Joint Conference on Lexical and Computational Semantics, SemEval'12, pp. 435–440. Association for Computational Linguistics, Stroudsburg (2012)
6. Becker, H., Naaman, M., Gravano, L.: Learning similarity metrics for event identification in social media. In: Proceedings of the Third ACM International Conference on Web Search and Data Mining, WSDM'10, pp. 291–300. ACM, New York (2010) .
7. Benczúr, A.A., Karger, D.R.: Approximating st minimum cuts in õ (n 2) time. In: Proceedings of the Twenty-Eighth Annual ACM Symposium on Theory of Computing, pp. 47–55. ACM, New York (1996)
8. Biemann, C.: Creating a system for lexical substitutions from scratch using crowdsourcing. Lang. Resour. Eval. **47**(1), 97–122 (2013)
9. Blei, D.M., Ng, A.Y., Jordan, M.I.: Latent Dirichlet allocation. J. Mach. Learn. Res., **3**, 993–1022 (2003)
10. Bostandjiev, S., O'Donovan, J., Höllerer, T.: Tasteweights: A visual interactive hybrid recommender system. In: Proceedings of the Sixth ACM Conference on Recommender Systems, RecSys'12, pp. 35–42, ACM, New York (2012)
11. Budak, C., Goel, S., Rao, J.M.: Fair and balanced? quantifying media bias through crowdsourced content analysis. In: Proceedings of the Nineth International Conference on Weblogs and Social Media, Oxford, UK. AAAI, Palo Alto (2015)
12. Canini, K., Suh, B., Pirolli, P.: Finding credible information sources in social networks based on content and social structure. In: *Privacy, Security, Risk and Trust (PASSAT) and 2011 IEEE Third Inernational Conference on Social Computing (SocialCom), 2011 IEEE Third International Conference on*, pp. 1–8, Oct 2011
13. Castillo, C., Mendoza, M., Poblete, B.: Information credibility on twitter. In: Proceedings of the 20th International Conference on World Wide Web, pp. 675–684. ACM, New York (2011)
14. Gao, W., Li, P., Darwish, K.: Joint topic modeling for event summarization across news and social media streams. In: Proceedings of the 21st ACM International Conference on Information and Knowledge Management, CIKM'12, pp. 1173–1182. ACM, New York (2012)
15. Gusfield, D.: Algorithms on Strings, Trees and Sequences: Computer Science and Computational Biology. Cambridge University Press, Cambridge (1997)
16. Guy, I., Jacovi, M., Perer, A., Ronen, I., Uziel, E.: Same places, same things, same people?: mining user similarity on social media. In: Quinn, K.I., Gutwin, C., Tang, J.C. (eds.) CSCW, pp. 41–50. ACM, New York (2010)
17. Herdağdelen, A.: Twitter n-gram corpus with demographic metadata. Lang. Resour. Eval. **47**(4), 1127–1147 (2013)
18. Hermida, A., Fletcher, F., Korell, D., Logan, D.: Share, like, recommend: decoding the social media news consumer. Journal. Stud. **13**(5–6), 815–824 (2012)
19. Holcomb, J., Gottfried, J., Mitchell, A.: News use across social media platforms (2013)

20. Horan, T.J.: softversus hardnews on microblogging networks: semantic analysis of twitter produsage. Inf. Commun. Soc. **16**(1), 43–60 (2013)
21. Hurlock, J., Wilson, M.: Searching twitter: separating the tweet from the chaff. In: International AAAI Conference on Web and Social Media (2011). Retrieved from http://www.aaai.org/ocs/index.php/ICWSM/ICWSM11/paper/view/2819
22. Kang, B., Höllerer, T., O'Donovan, J.: The full story: Automatic detection of unique news content in microblogs. In: Proceedings of the 2015 IEEE/ACM International Conference on Advances in Social Networks Analysis and Mining, ASONAM 2015, Paris, France, August 25–28, 2015, pp. 1192–1199 (2015)
23. Koehn, P.: Europarl: a parallel corpus for statistical machine translation. In: MT Summit, vol. 5, pp. 79–86 (2005)
24. Kothari, A., Magdy, W., Darwish, K., Mourad, A., Taei, A.: Detecting comments on news articles in microblogs. In: Proceedings of the Seventh International Conference on Weblogs and Social Media (ICWSM), Cambridge, Massachusetts, USA, July 8–11 (2013). http://www.aaai.org/ocs/index.php/ICWSM/ICWSM13/paper/view/6011
25. Kouloumpis, E., Wilson, T., Moore, J.D.: Twitter sentiment analysis: the good the bad and the omg!. In: Proceedings of the Fifth International Conference on Weblogs and Social Media, Barcelona, Catalonia, Spain, July 17–21 (2011). http://www.aaai.org/ocs/index.php/ICWSM/ICWSM11/paper/view/2857
26. Mihalcea, R., Corley, C., Strapparava, C.: Corpus-based and knowledge-based measures of text semantic similarity. In: AAAI, vol. 6, pp. 775–780 (2006)
27. Nagarajan, M., Purohit, H., Sheth, A.P.: A qualitative examination of topical tweet and retweet practices. In: ICWSM, 2010 (2010)
28. Rosen-Zvi, M., Griffiths, T., Steyvers, M., Smyth, P.: The author-topic model for authors and documents. In: Proceedings of the 20th Conference on Uncertainty in Artificial Intelligence, UAI'04, pp. 487–494, AUAI Press, Arlington (2004)
29. Shoemaker, P.J.: News and newsworthiness: A commentary. Communications **31**(1), 105–111 (2006)
30. Sikdar, S., Kang, B., O'Donovan, J., Höllerer, T., Adalı, S.: Understanding information credibility on twitter. In: IEEE/ASE SocialCom, pp. 19–24 (2013)
31. Stamatatos, E.: Plagiarism detection using stopword n-grams. J. Am. Soc. Inf. Sci. Technol. **62**(12), 2512–2527 (2011)
32. Willnat, L., Weaver, D.H.: The American journalist in the digital age. Technical Report, School of Journalism, Indiana University (2014)
33. Yang, J., Counts, S.: Predicting the speed, scale, and range of information diffusion in twitter. In: ICWSM, vol. 10, pp. 355–358 (2010)
34. Yildirim, M.A., Coscia, M.: Using random walks to generate associations between objects. PLoS ONE **9**(8) (2014)
35. Zhao, W.X., Jiang, J., Weng, J., He, J., Lim, E.-P., Yan, H., Li, X.: Comparing twitter and traditional media using topic models. In: Advances in Information Retrieval, pp. 338–349. Springer, New York (2011)

Prediction of Elevated Activity in Online Social Media Using Aggregated and Individualized Models

Jimpei Harada, David Darmon, Michelle Girvan, and William Rand

1 Introduction

For a wide variety of organizations, companies, and individuals there is a growing interest in using social media to get their message out. For instance, brand managers are often tasked with launching promotions that raise the awareness of their brand among users of social media. However, the signal that a brand is trying to convey can easily get lost in the "noise" produced by other brands, individuals, bots, etc. While good content is important to engage an audience, it is also important to know when users will pay attention to the content in order to increase the chance that the message is spread. Therefore, a brand manager must consider not only what they want to say, but also *when* they want to say it.

In order to effectively spread a message on a social media platform, an important first step is to understand the patterns of user engagement. After receiving and becoming aware of information, users on a social media platform then evaluate the content of the information and decide whether it should be retransmitted or not. Previous research has examined different criteria for this decision, including

J. Harada
Center for Complexity in Business, University of Maryland, College Park, MD, USA
e-mail: jimpei.harada@gmail.com

D. Darmon (✉)
Department of Mathematics, University of Maryland, College Park, MD, USA
e-mail: ddarmon@math.umd.edu

M. Girvan
Department of Physics, University of Maryland, College Park, MD, USA
e-mail: girvan@umd.edu

W. Rand
Poole College of Management, North Carolina State University, Raleigh, NC, USA
e-mail: wmrand@ncsu.edu

R. Missaoui et al. (eds.), *Trends in Social Network Analysis*, Lecture Notes in Social Networks, DOI 10.1007/978-3-319-53420-6_7

the sender's level of activity and the freshness of the information [1], as well as the user's benefit from spreading the information [2]. In this research, instead of exploring criteria related to evaluation of content and the decision to retransmit, we focus on timing when users on social media are engaged in retransmission behavior. A key assumption of our approach is that information is most likely to be retransmitted during the highest activity periods. In particular, in this paper, we study the task of predicting user engagement on Twitter, and we measure engagement in terms of the number of users actively issuing retweets.

It is well known that user activity on social media services follows both diurnal and weekly patterns [3, 4]. For example, Fig. 1 demonstrates the number of users active on Twitter out of a collection of 2145 over a 4 week period, starting on Mondays at 9 am EST. At the daily level, the number of users actively retweeting

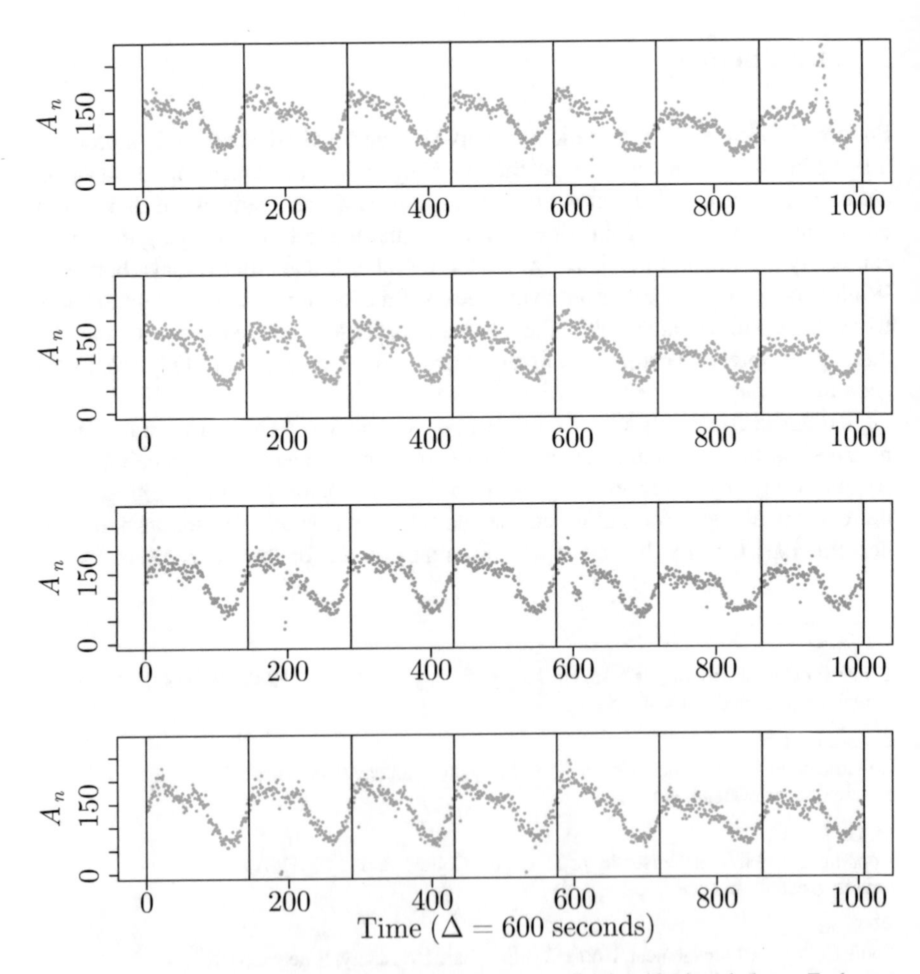

Fig. 1 The number of users actively retweeting during disjoint 10 min windows. Each row corresponds to a week, and each column corresponds to a day of the week, starting from Monday

increases over the course of a day and then decreases at night. However, the times of peak activity also fluctuate from week to week. Such fluctuations in social systems have been attributed to the fact that observed aggregate social behavior, driven by individual human actions, can be described as mixtures of Poisson and non-Poisson processes, where these processes can be seen as modeling individual decision making [5]. Thus, we expect the aggregate behavior of a collection of users who have different decision making processes to exhibit significant temporal fluctuation from seasonality from week to week. In order to effectively reach a large number of activated users, it is therefore important to determine when they are the most engaged by tracking such fluctuations while controlling for the diurnal and weekly seasonality. Moreover, recent work studying the attention of users on Twitter has found that retweets of a given tweet typically occur on the time scale of minutes [6, 7]. Given this observation, it is also important that we track seasonal fluctuations at a fine temporal resolution.

In order to model the number of active retweeters on Twitter at any given time, we propose three approaches: a seasonality model that assumes the overall retweet activity on Twitter is fully explained by the time-of-day and day-of-week, an autoregressive model that explicitly models deviations from the day-to-day seasonality, and an aggregation-of-individuals approach that models the activity patterns of each individual user and then aggregates these models to describe the overall activity pattern.

The seasonality model is based on the assumption that user engagement over time can be explained by seasonal patterns at the daily and weekly level. Thus, in order to predict the time when users are engaged at a certain level using this model, we consider only engagement patterns from the past.

The autoregressive model seeks to describe the population-level fluctuations about the seasonality using a simple linear autoregressive model. We assume that the deviations from seasonality have memory where we can think of this memory in terms of activation/deactivation of the users on Twitter. For example, a certain topic might become popular over the course of several hours, leading to activity greater than expected by the baseline seasonality. Such bursts of activity have been observed on both Twitter and blogging platforms [8]. By noting when and how such bursts occur, we can better predict the number of users active on Twitter compared to using seasonality alone.

The aggregation-of-individuals model explicitly views the overall activity as the accumulation of the activity patterns of all of the users under consideration. In particular, we model each user-to-be-aggregated as a point process with memory [9]. In this case, each individual user can become activated/deactivated, depending on their own previous behavior and the behavior of their inputs. By viewing the user as a computational unit, we can build a predictive model of how they interact with Twitter. This approach has been successfully applied to individual-level prediction [10] on Twitter, where many high volume users were found to be well-described by such a model. We can then aggregate these individual-level models to produce a global prediction of activity levels that accounts for individual-level activation.

In the rest of this paper, we explore the problem of identifying periods of high activation on a social media platform. We begin by describing our three models and relevant literature. Then we describe the data sets used to test the predictive ability of these models for the proposed problem. Next, we review the predictive ability of the various models, and compare the benefits and tradeoffs of each approach. Finally, we conclude with the limitations of the present work and future directions to extend and improve it.

This paper is an extended version of our previous paper *Forecasting High Tide: Predicting Times of Elevated Activity in Online Social Media* from the proceedings of ASONAM 2015 [11]. In addition to expanding on the results presented there, this paper explores in much greater detail the construction and interpretation of the individualized models, including a simple method to adjust for associations in the behavior amongst users and a demonstration of how the individual-level models map to behavioral profiles of user behavior.

2 Related Work

A large body of work has investigated the dynamics of technology-mediated human interaction. Relevant to our work, Goh et al. [12] found that human behavior on email services is dominated by bursty-type behavior, with periods of high activity separated by long stretches of inactivity. The authors of [13] found stereotypical temporal patterns in the interaction between blogs and mainstream media news. Studies of Twitter have found similar stereotypical aggregate behavioral patterns for the popularity of particular hashtags over time [13, 14]. More recent work has sought to develop first principle mathematical models explicitly geared towards human behavior on social media [15, 16]. In the field of information systems, a related body of work considers the prediction of network traffic [17–19]. However, in this context the goal is to predict elevated activity in order to balance between the available network resources and the expected network demand.

A great deal of work has been done on the problem of predicting the future popularity of individual tweets and hashtags based on their features. As a very recent example, in [20], the authors performed an experiment to investigate how the wording of a tweet impacts whether it is retweeted, controlling for both the author and the topic of the tweet. In [21], the authors predict the volume of tweets about a hashtag day-to-day using features extracted from a corpus of tweets containing that hashtag on previous days. Similar studies can be found in [22–25]. The problem of predicting individual tweet, hashtag, and topic popularity has been well-studied, and these references are only meant to give a sampling of the much larger literature on the subject.

The problem of predicting the total volume of tweets over time has attracted much less attention from the research community. Notable exceptions include [26–28]. In [26], the authors build a predictive model for the overall volume of tweets related to a particular hashtag. Similar to one of our approaches, the authors do this by aggregating individual predictive models for a universe of users, where the users were chosen if they previously tweeted on a topic and followed a user who also tweeted on that topic. They then identified predictive models for each user at the resolution of days, where predictions were made based on previous activity of a user and their local network structure. The goal of predicting day-resolution volume from users on a particular topic differs greatly from predicting high volume times from a collection of users determined based on their network properties, which is the goal of this paper. In [27], the authors seek to determine the 1 h period in which the followers of a given collection of users are most likely to be active. However, their investigation is purely sociological in nature, in that they make no predictions, and the data used in their analysis only covered a single week of activity. Thus, their approach is not directly applicable to forecasting retweet volume from streaming data. Finally, in [28], the authors use a two state Hidden Markov modeling framework, where the hidden states correspond to when the user is either in an active mode or an inactive state. Using these models, they predict the expected interarrival time for a user given their observed previous behavior by filtering their hidden state, and make predictions based on this time. Thus, this approach is similar in spirit to our aggregation-of-individuals approach. However, they assume a particular hidden state model architecture that is homogeneous across users, while our approach, as we will see, allows for model heterogeneity across users. Moreover, while their approach could theoretically be used to predict total retweet volume by aggregating their individual model predictions, they do not do this, focussing instead on individual-level predictions.

3 Methodology

Here we define our exact problem and the proposed solutions. Consider a set $\mathcal{U} = \{u_1, u_2, \ldots, u_U\}$ of U users. Each user in $\mathcal{U}$ has an individual retweet history. Let Δ be a time interval; here we take $\Delta = 10$ min. Then for each user u in the set of users $\mathcal{U}$, we specify their retweet activity during any window of length Δ by

$$X_n(u) = \begin{cases} 1 : \text{user } u \text{ retweeted between times} \\ \qquad (n-1)\Delta \text{ and } n\Delta \\ 0 : \text{otherwise} \end{cases}. \tag{1}$$

That is, $\{X_n(u)\}_{n=1}^{N}$ specifies the retweet activity of the user during each of the N time intervals $[0, \Delta), [\Delta, 2\Delta), \ldots, [(N-1)\Delta, N\Delta)$.

The total number of users active during any time interval $[(n-1)\Delta, n\Delta)$ is then given by

$$A_n = \sum_{u \in \mathcal{U}} X_n(u). \tag{2}$$

This is the value we seek to predict.

3.1 Seasonality

For the seasonality model, we assume that retweet activity shows day-to-day variability, but regularity from week-to-week. We assume that the seasonality repeats every T time steps,

$$s_n = s_{n+jT}, \qquad j = 1, 2, \ldots \tag{3}$$

and that the observed number of users retweeting A_n is given by

$$A_n = s_n + \epsilon_n \tag{4}$$

where ϵ_n can be thought of as the deviation from the seasonality at any given time n. Under the assumption of seasonality, we infer the seasonal component by averaging across W weeks [29],

$$\hat{s}_n = \frac{1}{W} \sum_{j \in \{0,1,\ldots,W-1\}} A_{n+jT}, \qquad n = 1, \ldots, T. \tag{5}$$

Figure 2 shows the aggregate retweet activity across the 4 weeks from Fig. 1 with the estimated seasonality superimposed.

If we assume that $\{\epsilon_n\}_{n=1}^{N}$ is a realization from a white noise process, the optimal predictor under mean-squared loss for A_n is s_n, the seasonality. Thus, we use our estimator for the seasonality as the predictor for the seasonality model,

$$A_n^{\mathrm{S}} = \hat{s}_n. \tag{6}$$

3.2 Aggregate Autoregressive Model

In the seasonality model, we have assumed that the residuals $\{\epsilon_n\}_{n=1}^{N}$ are white noise. More explicitly, we have assumed that they show no autocorrelation: $E[\epsilon_t \epsilon_s] = \sigma_\epsilon^2 \delta_{st}$, where σ_ϵ^2 is the variance of the white noise process and δ_{st} is the Kronecker delta. A more reasonable model for the residual would incorporate memory, since aggregate social systems are known to exhibit such memory [30]. Thus, a simple refinement of the previous model allows for memory in the deviations from seasonality. More explicitly, we consider the model

$$A_n = s_n + Y_n \tag{7}$$

where we now take $\{Y_n\}_{n=1}^{N}$ to be a realization from an autoregressive process of order p, an AR(p) model [29]. That is, we consider the dynamics of Y_n to be governed by

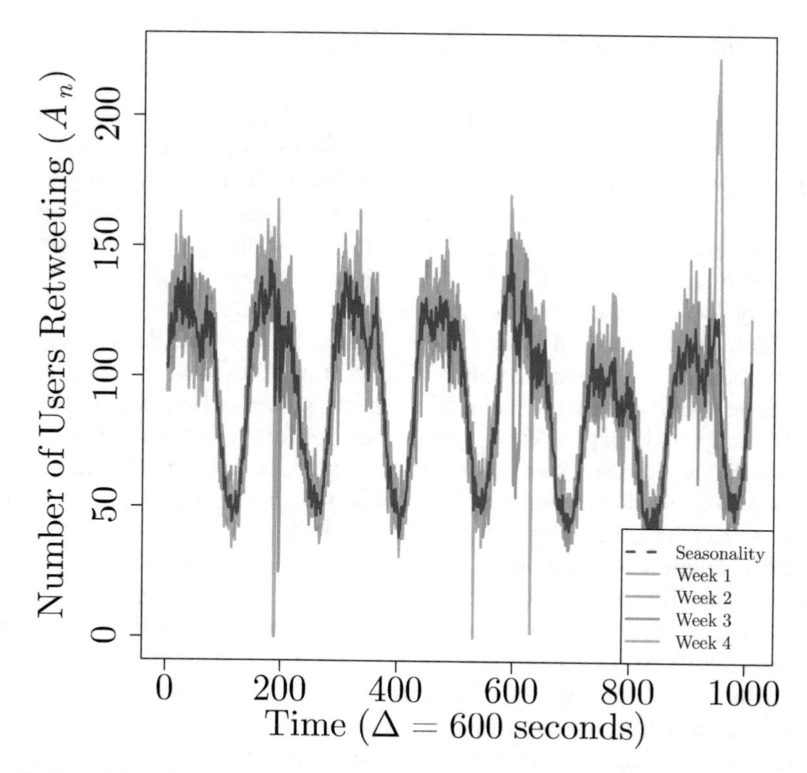

Fig. 2 The number of users retweeting A_n over four consecutive weeks. The estimated seasonality $\hat{s}_n$ is shown in *blue*

$$Y_n = \sum_{j=1}^{p} b_j Y_{n-j} + \epsilon_n \tag{8}$$

where $\{\epsilon_n\}$ is again a white noise process with mean 0 and variance σ_ϵ^2.

The predictor for the aggregate autoregressive model is

$$A_n^{\text{AR}} = \hat{s}_n + \sum_{j=1}^{\hat{p}} \hat{b}_j \hat{Y}_{n-j}, \tag{9}$$

where $\hat{Y}_n = A_n - \hat{s}_n$ is the deviation of the observed aggregate retweeting activity from the estimated seasonality at time n. We choose the autoregressive order $\hat{p}$ by minimizing the Akaike information criterion on the training set [31].

3.3 Aggregation of Causal State Models

Before describing the aggregation procedure, we briefly review computational mechanics, which is our basic modeling approach for the individual-level models. Shalizi and Crutchfield [32] provides a more in-depth introduction to computational mechanics, and [10] describes an application of computational mechanics to modeling individual user activity on Twitter. Computational mechanics provides a framework for describing stationary [33] (and more generally, *conditionally* stationary [34]), discrete-time, discrete-alphabet stochastic processes by linking the observed process to a hidden state process. In this way, the formalism of computational mechanics is closely related to Hidden Markov Models and other state-based models of discrete-alphabet stochastic processes [35]. In particular, any conditionally stationary stochastic process $\{X_n\}$ naturally induces a hidden state process $\{S_n\}$, where the transition structure of the hidden state process is determined by the predictive distribution of $\{X_n\}$. The hidden state process $\{S_n\}$ is always Markov, and the combination of its Markov chain representation and the state conditional emission probabilities $P(X_n = x \mid S_{n-1} = s)$ is called the *causal state model* or *ϵ-machine* for the stochastic process $\{X_n\}$. In the case where the predictive distribution for $\{X_n\}$ is unknown, machine reconstruction algorithms can be used to automatically infer the ϵ-machine that best describes the observed data $\{X_n\}_{n=1}^{N}$. We use the Causal State Splitting Reconstruction (CSSR) algorithm [36] to infer an ϵ-machine for each user's observed retweeting activity (Fig. 3).

As with the autoregressive model, the CSSR algorithm requires a maximum history length $L_{\max}$ to look into the past in order to reconstruct the ϵ-machine associated with a user u's behavior. While theory exists for choosing the largest $L_{\max}$ such that we can consistently infer the one-step-ahead predictive distributions used in CSSR [37], we take the practical approach of choosing $L_{\max}$ based on cross-validation. In particular, we perform fivefold cross-validation using the log-likelihood of the held out data as our objective function [38, 39]. The form of the log-likelihood associated with a realization from a stochastic process under an ϵ-machine model may be found in [40].

For each user u, we reconstruct their associated ϵ-machine. We then perform prediction as follows: at time $n-1$, we determine the current causal state $S_{n-1}(u)$ for each user u based on their activity pattern $X_1^{n-1}(u) = (X_1(u), X_2(u), \ldots, X_{n-1}(u))$. The causal state $S_{n-1}(u)$ specifies the one-step-ahead predictive distribution for each user, $P(X_n(u) = 1 \mid S_{n-1}(u) = s(u))$. We then aggregate these probabilities to form our prediction for the number of active users at the next time step,

$$A_n^{\mathrm{CSM}} = \sum_{u \in \mathcal{U}} P(X_n(u) = 1 \mid S_{n-1}(u) = s(u)). \tag{10}$$

This can be seen to be the expected number of users active at time n given the causal states of the users at time $n-1$, under the assumption that the behavior of a user u at time n is independent of the causal states of all others users at time $n-1$ given the causal state of u at time $n-1$.

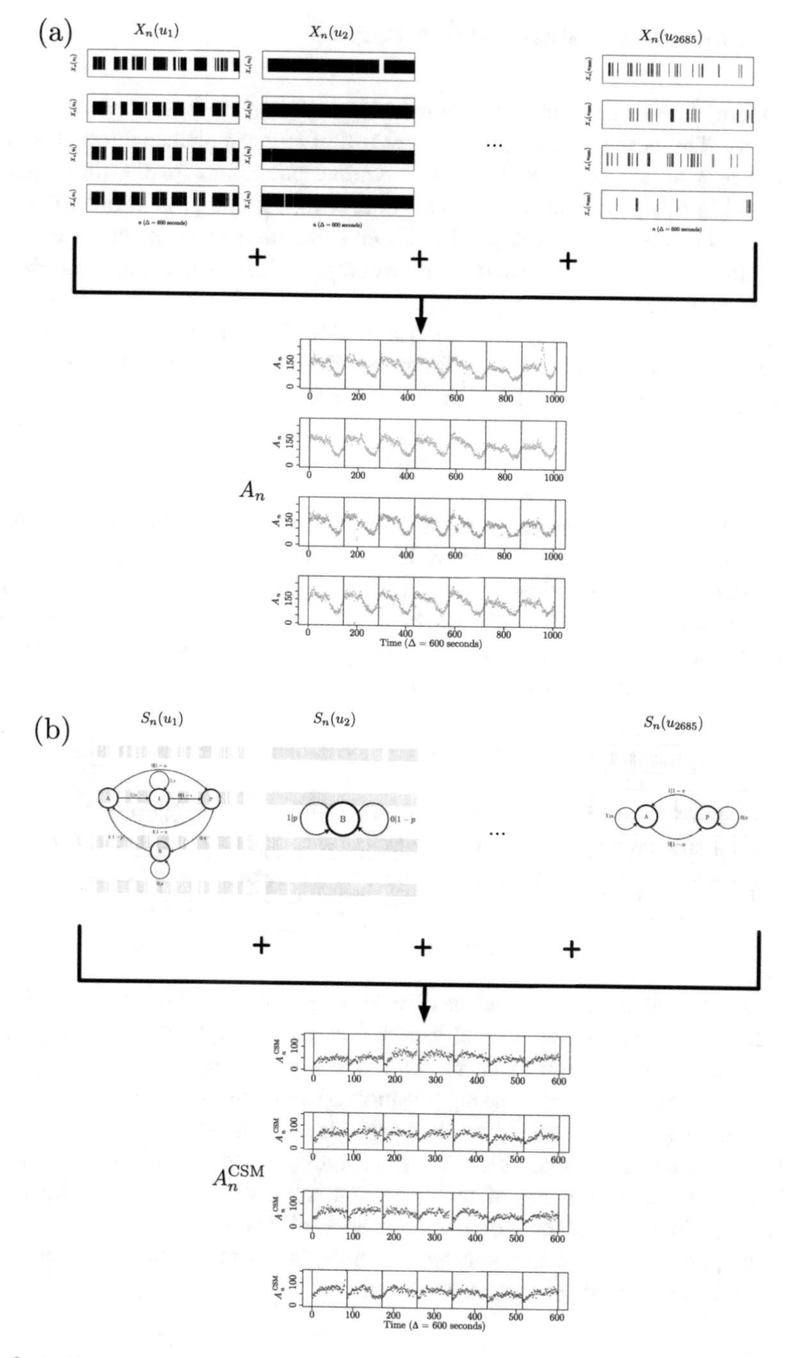

Fig. 3 A demonstration of how (**a**) the retweet volume A_n results from the summation of the individual retweet behavior $\{X_n(u)\}_{u \in \mathcal{U}}$ of the users in $\mathcal{U}$ and (**b**) the aggregation-of-individuals prediction A_n^{CSM} is formed via filtering through each user u's ϵ-machine

3.4 Time Complexity of Each Model

An important consideration in the practical application of these methods is their scalability. The best possible model is not useful if its implementation cannot be realized in a reasonable amount of time. Online prediction for the three models developed in this paper occurs in linear time, with a single pass through the data. However, the time complexity of the offline estimation of each model can vary greatly. In this section, we present the time complexities of each model and discuss their consequences.

Estimating the seasonality across $|\mathcal{U}|$ users for W weeks with T time units per week requires estimating the overall activity from the users per each week and averaging across weeks, which has the time complexity

$$O(|\mathcal{U}|WT). \tag{11}$$

The aggregate autoregressive model is a linear regression of the past activity on the future, and can thus be solved by least squares using, for example, an SVD decomposition. An individual linear regression of order p can be performed in $O(p^2 WT)$ time [41]. Thus, with the estimation of the appropriate model order amongst model orders up to $p_{\max}$, the least squares estimation is $O(p_{\max}^3 WT)$. The initial aggregate activity must also be computed, so the overall complexity is

$$O(p_{\max}^3 WT + |\mathcal{U}|WT). \tag{12}$$

The computational complexity of CSSR for estimating a single ϵ-machine on a binary alphabet using K-fold cross-validation with a maximum history length of $L_{\max}$ is $O(KL_{\max}(2^{L_{\max}+1} + WT))$ [36]. Thus, for $|\mathcal{U}|$ users, the overall complexity is

$$O(|\mathcal{U}|KL_{\max}(2^{L_{\max}+1} + WT)). \tag{13}$$

Clearly the seasonality model has the smallest time complexity, linear in the number of users and the number of weeks. The time complexity of the aggregate autoregressive model is not much worse, especially with modern implementations of linear algebra packages. The aggregation-of-individuals model has the worst time complexity, as is expected since it depends on first estimating the individual models and then aggregating them. Thus, depending on the number of users under consideration, the aggregation-of-individuals model may be infeasible. However, we note that the inference of the individual models is also completely parallelizable, so depending on the computing resources available, the increased burden of inferring the individual models can be greatly reduced.

4 Data Collection and Selection of $\mathcal{U}$

We begin with a collection of 15,000 Twitter users whose statuses (Tweet text) were collected over two disjoint 5 week intervals: from 25 April 2011 to 29 May 2011 and from 1 October 2012 to 5 November 2012. The users are embedded in a 15,000 node network collected by performing a breadth-first expansion of the active followers of a random seed user. In particular, the network was constructed by considering the followers of the seed user, and including those followers considered active (i.e., users who tweeted at least once per day over the past 100 days). The collection of users continued from the followers of these followers, etc., until 15,000 users were included. From this network of users, the subset of users $\mathcal{U}$ was chosen to account for 80% of the retweet volume for the first 4 weeks in the 5 week period under consideration. That is, we take u_1 to be the user issuing the greatest number of retweets, then u_2 to be the user issuing the second greatest number of retweets, etc., until we reach the user u_U such that the total number of retweets issued by the users in $\mathcal{U}$ account for 80% of the retweet volume. This results in $U = 2145$ users for the 2011 collection and $U = 1610$ users for the 2012 collection. Because we are interested in predicting times of greatest retweet activity, for each day we only consider the retweet activity from 6 AM EST to 10 PM EST. The data used in our analysis can be made available upon request by the corresponding author.

5 Results

In the following results, we use the first 4 weeks of the 5 week periods from 2011 and 2012 for inference of the three model types, and leave the last weeks from each year for testing. As described in the methodology section, we choose the parameters of each model as follows. The seasonality model has no tuning parameter, and we use the full 4 weeks to infer the seasonality component. We choose the model order p of the autoregressive model to maximize the Akaike information criterion on the 4 week training period. For each causal state model in the aggregation-of-individuals model, we infer the user-specific history length L by fivefold log-likelihood cross-validation over the 28 days in the training sets.

5.1 *Adjustment to the Aggregation-of-Individuals Model*

As described in the methodology section, the predictor for the aggregation-of-individuals model (10) is equivalent to the expected number of users in $\mathcal{U}$ who are active at time step n given their causal states at $n - 1$ under a certain independence assumption. In particular, we have taken the predictor to be

$$A_n^{\mathrm{CSM}} = E\left[\sum_{u\in\mathcal{U}} X_n(u) \,\middle|\, S_{n-1}(u_1),\ldots,S_{n-1}(u_U)\right] \tag{14}$$

$$= \sum_{u\in\mathcal{U}} E[X_n(u) \mid S_{n-1}(u_1),\ldots,S_{n-1}(u_U)] \tag{15}$$

$$= \sum_{u\in\mathcal{U}} E[X_n(u) \mid S_{n-1}(u)] \tag{16}$$

$$= \sum_{u\in\mathcal{U}} P(X_n(u) = 1 \mid S_{n-1}(u)), \tag{17}$$

where going from (15) to (16) we make the assumption that for all $u \in \mathcal{U}$,

$$X_n(u) \perp\!\!\!\perp \{S_{n-1}(u'), u' \neq u\} \mid S_{n-1}(u). \tag{18}$$

That is, we assume that the observed behavior of user u at time n is independent of the causal states of all other users u' at time $n-1$, given the causal state of user u at time $n-1$. While such an independence relationship holds when conditioning on the local causal states of a time-varying random field [42], it need not be true when conditioning on the marginal causal states.

Motivated by the form of the deviation of (10) from the predicted value (see Fig. 4), we define the adjusted aggregation-of-individuals predictor as

$$A_n^{\mathrm{CSM}^*} = \beta_0 + \beta_1 A_n^{\mathrm{CSM}}, \tag{19}$$

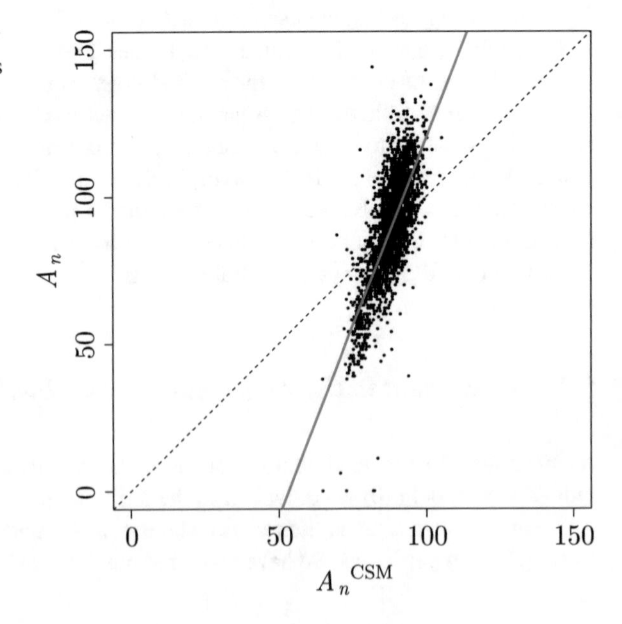

Fig. 4 The transformation of the aggregation-of-individuals model used to adjust for associations in user behavior. The *red line* corresponds to the linear least squares fit from regressing the true values A_n from the training set on the unadjusted aggregation-of-individuals predictions A_n^{CSM} from the 2011 data

where the parameters β_0 and β_1 were estimated by regressing the true values A_n from the training set on the unadjusted aggregation-of-individuals predictions A_n^{CSM} from the training set. We will use this predictor for the remainder of this work, and address alternative corrections in the conclusion.

5.2 Predicting Activation Level at Varying Thresholds

We next present an experiment to test the predictive capability for each of the three proposed models. As mentioned in the introduction, ideally a potential influencer would like to choose the optimal time(s)-of-day to send out a message such that the largest number of users will be active around those times. As a proxy for this goal, we consider the task of identifying whether or not the activity level over an interval of length Δ will fall into the 100pth percentile for that day. As an example, how well can we predict whether the number of activated users falls within the 80th percentile for a given day?

Let N_Δ be the number of time points to predict on in a day ($N_\Delta = 86$ for this analysis). For a given day $d \in \{1, 2, \ldots, 7\}$ in the testing set, the true distribution of the activity levels is given by

$$F_d^{\mathrm{True}}(a) = \frac{1}{N_\Delta} \sum_{n=n_{\mathrm{train}}+N_\Delta(d-1)+1}^{n_{\mathrm{train}}+N_\Delta d} \mathbb{1}\left[A_n \le a\right]. \tag{20}$$

We then define the historical distribution of the activity levels for a day d in terms of the estimated seasonality for that day from the training set

$$F_d^{\mathrm{Hist}}(a) = \frac{1}{N_\Delta} \sum_{n=N_\Delta(d-1)+1}^{N_\Delta d} \mathbb{1}\left[A_n^S \le a\right]. \tag{21}$$

We will use $F_d^{\mathrm{Hist}}(\hat{A}_n)$ to predict whether or not a predicted activity level $\hat{A}_n$ exceeds the quantile p^* of activity for a given day, where $\hat{A}_n$ is one of the A_n^{S}, A_n^{AR}, or $A_n^{\mathrm{CSM}^*}$. That is, for a threshold p, we predict the indicator for whether the activity at time n will exceed some quantile p^* as

$$\hat{I}_n(p) = \begin{cases} 1 : F_{d(n)}^{\mathrm{Hist}}(\hat{A}_n) > p \\ 0 : \text{otherwise} \end{cases}. \tag{22}$$

Whether or not the activity at time n exceeded the quantile p^* is then given in terms of the true distribution as

$$I_n = \begin{cases} 1 : F_{d(n)}^{\mathrm{True}}(A_n) > p^* \\ 0 : \text{otherwise} \end{cases}. \tag{23}$$

As we vary the threshold value p, the true positive rate is given by

$$\mathrm{TPR}(p) = \frac{\sum_{n=n_{\mathrm{train}}+1}^{n_{\mathrm{test}}} \mathbb{1}\left[\hat{I}_n(p) = 1, I_n = 1\right]}{\sum_{n=n_{\mathrm{train}}+1}^{n_{\mathrm{test}}} \mathbb{1}\left[I_n = 1\right]} \tag{24}$$

and the false positive rate is given by

$$\mathrm{FPR}(p) = \frac{\sum_{n=n_{\mathrm{train}}+1}^{n_{\mathrm{test}}} \mathbb{1}\left[\hat{I}_n(p) = 1, I_n = 0\right]}{\sum_{n=n_{\mathrm{train}}+1}^{n_{\mathrm{test}}} \mathbb{1}\left[I_n = 0\right]}. \tag{25}$$

We show the ROC curves associated with the fixed quantiles $p^* = 0.70, 0.75, 0.80$, along with their AUCs, for the test weeks from 2011 and 2012 in Fig. 5 and Table 1. The true and false positive rates are computed using the last 86 of the 96 time points in each day, since both the autoregressive and aggregation-of-individuals models require up to ten time points to begin prediction depending on the model order p or largest history length L, respectively.

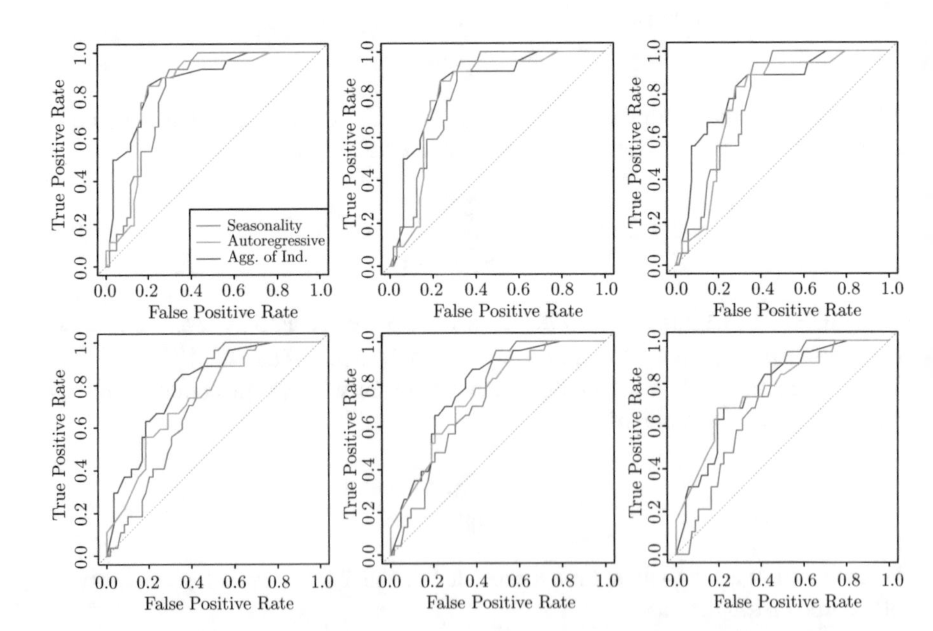

Fig. 5 The ROC curves associated with the seasonality (*red*), aggregate autoregressive (*green*), and aggregation of individuals (*blue*) approaches for the testing week in 2011 (*top*) and 2012 (*bottom*) with p^* fixed at 0.70 (*left*), 0.75 (*middle*), and 0.80 (*right*). The AUC values for each ROC curve are given in Table 1

Table 1 The AUC for each of the ROC curves in Fig. 5

Year	p^*	Seasonality	Autoregressive	Agg. of ind.
2011	0.70	0.817	0.829	**0.865**
	0.75	0.816	0.812	**0.841**
	0.80	0.778	0.773	**0.825**
2012	0.70	0.715	0.738	**0.797**
	0.75	0.731	0.751	**0.782**
	0.80	0.720	**0.773**	0.771

Bold values indicate the method with the largest AUC

Overall, based on the AUC values, the aggregation-of-individuals model performs best in 2011 over all values of p^*, and all but $p^* = 0.80$ in 2012 where the aggregate autoregressive model is best. However, inspection of the ROC curves indicates that based on the desired balance between true and false positives, each of the models may outperform the others, with no model strictly dominating. For example, if a high false positive rate is acceptable, the seasonality model achieves the lowest false positive rate to give a 100% true positive rate on the testing sets in 2011 and 2012 across all values of p^*. However, the seasonality model generally underperforms when the desired false positive rate is low, in which case both the aggregation-of-individuals model and the autoregressive model are competitive.

5.3 *Utility of Individual-Level Models Beyond Aggregate Prediction*

Though we do not focus on individual-level prediction in this paper, we wish to highlight some of the possible advantages offered by the aggregation-of-individuals approach not immediately evident from the ROC analysis above. In particular, as demonstrated in Fig. 3, the aggregation-of-individuals generates individual level, behavioral models for each user u. These models have the advantage of being interpretable. Consider the four models in Fig. 6. The models can be represented as directed graphs, where each vertex corresponds to a causal state, and each arrow corresponds to an allowed emission from that state. The arrows are decorated with the emission symbol $x \in \{0, 1\}$ (i.e., user u either retweets or does not during a time interval) and the causal state conditioned emission probability $P(X_n(u) = x \mid S_{n-1}(u) = s)$ of transitioning from state s while emitting symbol x. That is, each arrow is decorated as $x|P(X_n(u) = x \mid S_{n-1}(u) = s)$.

These models allowed for user-specific targeting. Consider the model represented by (b). Users of this type tend to retweet in a bursty manner, with an active state A and a passive state P. This corresponds to a simple order-1 Markov model. For such users, it is sufficient to target them when they have recently retweeted. Users exhibiting behavior like models (c) or (d) require more subtle targeting. Model (c) has the same active and passive states as in (b), but with an additional refractory state

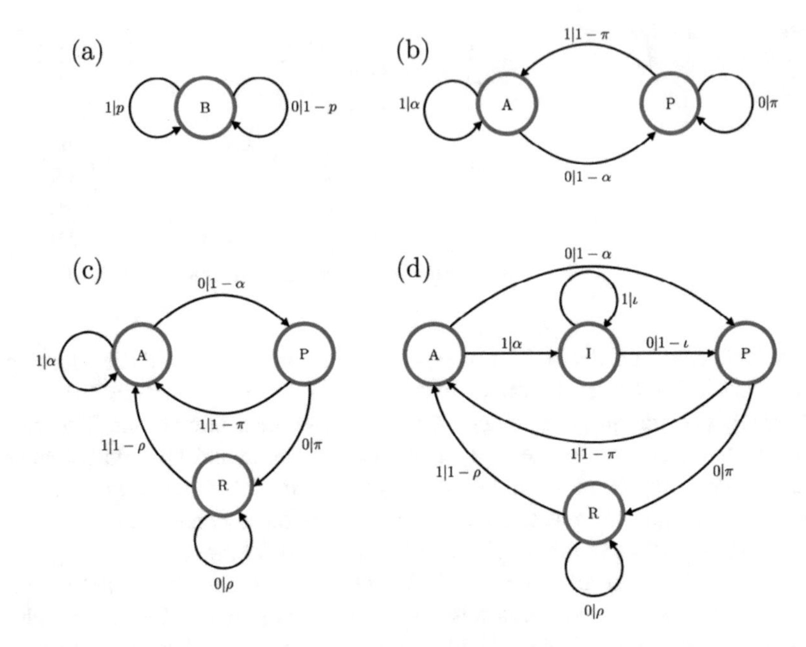

Fig. 6 Four example ϵ-machines inferred from the users. (**a**) A user who retweets at random with bias p. (**b**) A user who retweets in a bursty manner, with an active state A and a passive state P. (**c**) A user who retweets in a bursty manner, with a refractory state R. (**d**) A user who retweets in a bursty manner with both a refractory state R and an intermittent state I

R that occurs after the user is quiescent while in the passive state P. Depending on the balance between $1-\rho$ and α, which corresponds to the probabilities of retweeting from the active and refractory states, it may be more beneficial to target the user when they are currently active or when they are "resting" in the refractory state. Model (d) is similar to model (c), with an additional intermediate state I that occurs after the user has issued a retweet from the active state A. Again, depending on the balance between α, $1 - \rho$, and ι, the user can be targeted for when they are most likely to retweet. Many of the users have simple ϵ-machines similar to (a), (b), (c), and (d), which allow for this sort of user-specific targeting.

6 Conclusion

We have found that while user retweet activity clearly exhibits seasonality from day-to-day and week-to-week, seasonality alone does not explain the times of high user activity on social media. We have presented three models for the observed engagement of users on social media: a seasonal model that assumes engagement is dictated entirely by time-of-day and day-of-week, an aggregate autoregressive model that captures the memory present in the deviations of engagement from seasonality, and an aggregation-of-individuals model that leverages the individual-level behavior patterns of users to forecast overall engagement. The aggregate autoregressive and aggregation-of-individuals models capture the excitation of users in contrasting ways, the former by modeling system-wide deviations from the seasonal baseline, and the latter by modeling individualized deviations from quiescence/activity. By incorporating additional information about either the deviations from seasonality or the behavioral patterns of individual users, more accurate prediction of times of high volume are possible, especially when a low false positive rate is desired. If we return to our motivating example of identifying times that would maximize engagement by an audience with a particular tweet, then overexposure to a message may lead to reduced user engagement (content fatigue) due to the repetitive nature of the message. Therefore, in this scenario, it can be said that having a low false positive is important, which indicates that a seasonality model alone will not suffice.

In future work, we will explore more sophisticated models that should provide even greater predictive power. For example, the individual models used in the aggregation-of-individuals method did not incorporate social inputs to the users beyond their own previous behavior. The computational mechanics framework allows for the incorporation of inputs via either dynamic random field-based [42] or transducer-based [43, 44] models of a user's behavior. Such an extension could eliminate the need for the adjustment of the aggregation-of-individuals predictor to translate the model's output to a prediction.

This work highlights that in building predictive models for complex social systems, a multi-level view of the system under consideration often leads to improved predictive ability. Thus, in the predictive problem considered in this paper, influencers who track potential user engagement can use complexity modeling to make better informed decisions.

References

1. Stephen, A.T., Dover, Y., Muchnik, L., Goldenberg, J.: Fresh is best: the effect of source activity on the decision to retransmit content in social media. Available at SSRN 1609611 (2014)
2. Toubia, O., Stephen, A.T.: Intrinsic vs. image-related utility in social media: why do people contribute content to twitter? Mark. Sci. **32**(3), 368–392 (2013)
3. Golder, S.A., Macy, M.W.: Diurnal and seasonal mood vary with work, sleep, and daylength across diverse cultures. Science **333**(6051), 1878–1881 (2011)
4. Grinberg, N., Naaman, M., Shaw, B., Lotan, G.: Extracting diurnal patterns of real world activity from social media. In: ICWSM (2013)
5. Barabasi, A.-L.: The origin of bursts and heavy tails in human dynamics. Nature **435**(7039), 207–211 (2005)
6. Hodas, N.O., Lerman, K.: How visibility and divided attention constrain social contagion. In: Privacy, Security, Risk and Trust (PASSAT), 2012 International Conference on and 2012 International Conference on Social Computing (SocialCom), pp. 249–257. IEEE, New York (2012)

7. Hodas, N.O., Lerman, K.: Attention and visibility in an information-rich world. In: 2013 IEEE International Conference on Multimedia and Expo Workshops (ICMEW), pp. 1–6. IEEE, New York (2013)
8. Yang, J., Leskovec, J.: Patterns of temporal variation in online media. In: Proceedings of the Fourth ACM International Conference on Web Search and Data Mining, pp. 177–186. ACM, New York (2011)
9. Ver Steeg, G., Galstyan, A.: Information transfer in social media. In: Proceedings of the 21st International WWW Conference, pp. 509–518. ACM, New York (2012)
10. Darmon, D., Sylvester, J., Girvan, M., Rand, W.: Predictability of user behavior in social media: bottom-up v. top-down modeling. In: ASE/IEEE Int'l Conference on Social Computing, pp. 102–107 (2013)
11. Harada, J., Darmon, D., Girvan, M., Rand, W.: Forecasting high tide: predicting times of elevated activity in online social media. In: Proceedings of the 2015 IEEE/ACM International Conference on Advances in Social Networks Analysis and Mining 2015, pp. 504–507. ACM, New York (2015)
12. Goh, K.-I., Barabási, A.-L.: Burstiness and memory in complex systems. EPL (Europhys. Lett.) **81**(4), 48002 (2008)
13. Leskovec, J., Backstrom, L., Kleinberg, J.: Meme-tracking and the dynamics of the news cycle. In: Proceedings of the 15th ACM SIGKDD International Conference on Knowledge Discovery and Data Mining, pp. 497–506. ACM, New York (2009)
14. Lehmann, J., Gonçalves, B., Ramasco, J.J., Cattuto, C.: Dynamical classes of collective attention in twitter. In: Proceedings of the 21st International Conference on World Wide Web, pp. 251–260. ACM, New York (2012).
15. Bauckhage, C., Kersting, K., Hadiji, F.: Mathematical models of fads explain the temporal dynamics of internet memes. In: ICWSM (2013)
16. Bauckhage, C., Kersting, K., Rastegarpanah, B.: Collective attention to social media evolves according to diffusion models. In: Proceedings of the Companion Publication of the 23rd International Conference on World Wide Web, pp. 223–224 (2014)
17. Doulamis, A., Doulamis, N., Kollias, S.D.: An adaptable neural-network model for recursive nonlinear traffic prediction and modeling of MPEG video sources. IEEE Trans. Neural Netw. **14**(1), 150–166 (2003)
18. Chang, B.R., Tsai, H.F.: Improving network traffic analysis by foreseeing data-packet-flow with hybrid fuzzy-based model prediction. Expert Syst. Appl. **36**(3), 6960–6965 (2009)
19. Dalmazo, B.L., Vilela, J.P., Curado, M.: Predicting traffic in the cloud: a statistical approach. In: 2013 Third International Conference on Cloud and Green Computing (CGC), pp. 121–126. IEEE, New York (2013)
20. Tan, C., Lee, L., Pang, B.: The effect of wording on message propagation: topic-and author-controlled natural experiments on twitter. arXiv preprint arXiv:1405.1438 (2014)
21. Ma, Z., Sun, A., Cong, G.: Will this #hashtag be popular tomorrow? In: Proceedings of the 35th International ACM SIGIR Conference on Research and Development in Information Retrieval, pp. 1173–1174. ACM, New York (2012)
22. Hong, L., Dan, O., Davison, B.D.: Predicting popular messages in twitter. In: Proceedings of the 20th International Conference Companion on World Wide Web, pp. 57–58. ACM, New York (2011)
23. Petrovic, S., Osborne, M., Lavrenko, V.: Rt to win! predicting message propagation in twitter. In: ICWSM (2011)
24. Yang, J., Counts, S.: Predicting the speed, scale, and range of information diffusion in twitter. ICWSM, vol. 10, pp. 355–358 (2010)
25. Suh, B., Hong, L., Pirolli, P., Chi, E.H.: Want to be retweeted? Large scale analytics on factors impacting retweet in twitter network. In: 2010 IEEE Second International Conference on Social Computing, pp. 177–184. IEEE, New York (2010)
26. Ruan, Y., Purohit, H., Fuhry, D., Parthasarathy, S., Sheth, A.: Prediction of topic volume on twitter. WebSci (short papers) (2012)

27. Alwagait, E., Shahzad, B.: Maximization of tweet's viewership with respect to time. In: 2014 World Symposium on Computer Applications & Research (WSCAR), pp. 1–5. IEEE, New York (2014)
28. Raghavan, V., Steeg, G.V., Galstyan, A., Tartakovsky, A.G.: Modeling temporal activity patterns in dynamic social networks. IEEE Trans. Comput. Soc. Syst. **1**, 89–107 (2014)
29. Fan, J., Yao, Q.: Nonlinear Time Series. Springer, New York (2002)
30. Mathiesen, J., Angheluta, L., Ahlgren, P.T.H., Jensen, M.H.: Excitable human dynamics driven by extrinsic events in massive communities. Proc. Natl. Acad. Sci. **110**(43), 17259–17262 (2013)
31. Hyndman, R.J., Khandakar, Y.: Automatic time series for forecasting: the forecast package for R. J. Stat. Softw. **27**(3), 1–22 (2008)
32. Shalizi, C.R., Crutchfield, J.P.: Computational mechanics: pattern and prediction, structure and simplicity. J. Stat. Phys. **104**(3–4), 817–879 (2001)
33. Grimmett, G., Stirzaker, D.: Probability and Random Processes, vol. 2. Oxford University Press, Oxford (1992)
34. Caires, S., Ferreira, J.A.: On the nonparametric prediction of conditionally stationary sequences. Probability, Networks and Algorithms, pp. 1–32 (2003)
35. Littman, M.L., Sutton, R.S., Singh, S.P.: Predictive representations of state. In: NIPS, vol. 14, pp. 1555–1561 (2001)
36. Shalizi, C.R., Klinkner, K.L.: Blind construction of optimal nonlinear recursive predictors for discrete sequences. In: M. Chickering, J.Y. Halpern, (eds.) Uncertainty in Artificial Intelligence: Proceedings of the Twentieth Conference (UAI 2004), pp. 504–511, Arlington, Virginia, 2004. AUAI Press, Arlington
37. Marton, K., Shields, P.C.: Entropy and the consistent estimation of joint distributions. Ann. Probab. **22**, 960–977 (1994)
38. Breiman, L., Spector, P.: Submodel selection and evaluation in regression. The X-random case. Int. Stat. Rev. **60**, 291 (1992)
39. Kohavi, R.: A study of cross-validation and bootstrap for accuracy estimation and model selection. In: Proceedings of the 14th International Joint Conference on Artificial Intelligence (1995)
40. Haslinger, R., Klinkner, K.L., Shalizi, C.R: The computational structure of spike trains. Neural Comput. **22**(1), 121–157 (2010)
41. O'Leary, D.P.: Scientific Computing with Case Studies. SIAM, Philadelphia (2009)
42. Shalizi, C.R: Optimal nonlinear prediction of random fields on networks. Discret. Math. Theor. Comput. Sci. **AB**, 11–30 (2003); DMTCS Proceedings of the Discrete Models for Complex Systems (DMCS)
43. Shalizi, C.R.: Causal architecture, complexity and self-organization in the time series and cellular automata. PhD Thesis, University of Wisconsin–Madison (2001)
44. Barnett, N., Crutchfield, J.P.: Computational mechanics of input-output processes: structured transformations and the ϵ-transducer. arXiv preprint arXiv:1412.2690 (2014)

Unsupervised Link Prediction Based on Time Frames in Weighted–Directed Citation Networks

Mehmet Kaya, Mujtaba Jawed, Ertan Bütün, and Reda Alhajj

1 Introduction

The advances in internet has given facility to people and organizations to interact and collaborate more and more which provide the basis for the emergence of social networks in internet. Social networks are the outcome of nodes and edges that can be shown as a graph, in which the nodes act as people or organizations and edges act as different forms of social relationships (such as citation, in which two authors are connected if they have cite each other) between them. In social networks, connections tend to appear, become stronger, become weaker, and disappear along time, which makes them very dynamic and complex systems.

Social Network Analysis (SNA) is a wide area of researches that tries to overpass such kind of problems [1]. Several tasks can be related to SNA. In this paper our aim is to consider the dynamic of links in Author-Author social network, thus we want to predict those citation links that will form or become stronger along the different frames of time based on previous state of the network. SNA deals with this well-known problem called link prediction.

Lots of studies have been done to treat the link prediction problem [2–4]. The application of proximity measures [5–8] on the non-connected pairs of nodes at the current time in the network has been considered by most of the studies in the past to predict new links at future time. These kinds of metrics give scores to any pair of nodes, and then the given scores used for performing the prediction task by an unsupervised or a supervised method. In unsupervised method, the non-connected

M. Kaya (✉) • M. Jawed • E. Bütün
Department of Computer Engineering, Fırat University, Elazığ, Turkey
e-mail: kaya@firat.edu.tr; mujtabajawed786@gmail.com; ebutun@firat.edu.tr

R. Alhajj
Department of Computer Science, University of Calgary, Calgary, AB, Canada
e-mail: alhajj@cpsc.ucalgary.ca

R. Missaoui et al. (eds.), *Trends in Social Network Analysis*, Lecture Notes in Social Networks, DOI 10.1007/978-3-319-53420-6_8

pairs of nodes are ranked by a chosen metric and then the top ranked pairs are specified as a predicted links. In supervised method the link prediction problem is considered as a classification task. In this method the connected pairs of nodes assign to positive class while non-connected ones assign to negative class. Also, the similarity scores which are chosen from a set of topological metrics are accepted as features and then used by a classifier for performing the prediction task. In the previous studies [2, 9–14] the proximity score calculation usually was done without taking the evolution of the network into account. This can be seen as a limitation in the previous works. The proximity metrics were computed using all network data up to the current time (i.e., present state of the network) without considering when links were created. Moreover, some of real world problems can be only modeled by directed networks. But, most studies of link prediction assumed that links of networks are undirected except for a few attempts [15, 16].

As a remedy to the above mentioned problems, in our previous paper, we presented a time-frame based link prediction method for directed networks [17]. The method used unsupervised learning strategy. The experiments conducted on unweighted–directed citation network denoted that the method outperforms the traditional predictors. In the present paper, we improve our time-frame based link prediction method for weighted–directed networks. For this purpose, we define new temporal events for weighted networks. In order to check the performance of the proposed approach, we performed experiments on the weighted citation networks extracted from different parts of DBLP (Digital Bibliography & Library Project). DBLP is a bibliographic dataset that provides a vast amount of data about various research publications in the Computer Science area. In order to compare our method with the traditional approaches, we used the Common Neighbors, Jacquard's Coefficient, Preferential Attachment and Adamic/Adar [5–8] traditional similarity measures in an unsupervised method [9–11].

The rest of the paper is organized as follows: Sect. 2 brings the related works, Sect. 3 briefly discusses about link prediction. Section 4 describes the traditional proximity metrics. Section 5 presents the citation network and triad patterns. Section 6 gives the proposed time-frame based link prediction and time-frame based score on triad patterns. Section 7 presents the experimental results. Finally, Sect. 8 shows the conclusion of this work and future works.

2 Related Works

In the literature of link prediction the most developed algorithms are based on topological information of network. According to the structural properties of the network, link prediction approaches can be categorized into three groups: (1) Common neighbor based, (2) Global information based, and (3) Paths based. Common neighbor based methods take only information of first order neighbors into account namely, Common Neighbor Index [6, 9], Jaccard Index [8], Preferential Attachment Index [11, 12], Adamic–Adar Index [7], and Resource Allocation Index [13].

Although most of the link prediction studies did not take weights of links into account, here we remark some previous studies in which the use of temporal information, weighted networks and citation networks in link prediction are the matters of importance. For instance, the weighted versions of some proximity measures have been proposed by De Sa and Prudencio in [14]. They record a good prediction performance with weighted proximity measures. Lin et al. [18] proposed a method based on Markov Chain system in weighted networks with the use of Resource Allocation into it. Nodes with various degrees and weights along various order neighbors were considered by their method. A weighted PageRank algorithm based on weighted–directed author citation was proposed by Radicchi et al. in [19], aiming to rank the scientists by taking their scientific publication credits into account. In [20] a weighted graph modeled as a network by authors. Link's weights was acting as the age of the most recent activities between nodes and after that the extended version of Adamic–Adar for weighted networks has deployed to accomplish the link prediction task. Mining of network data completed with temporal information for discovering the association rules which explain the network in a best way was proposed by authors in [21]. Juszczyszyn et al. [22] acknowledged an approach in which the probabilities of transmissions between triads of nodes derived by using the history of the network (records of network during the past frames of time). Even if comprising results can be achieved in this approach, frequent subgraphs mining is a very expensive task [23]. In [24] time series for each pair of nodes was built by authors, in which the frequency of happening of links among the nodes during a particular cycle of time is the observation of each series. Potgieter et al. in [25] and Soares and Prudêncio in [26] adopted a similar idea, but according to their studies for prediction the proximity score, time series models were used. In this approach, they used a chosen proximity metric, and for each pair of nodes time series were built by calculating the score in an array of time periods (i.e., various snapshots of the network used over the time). As a result, a final proximity score for each pair was obtained by forecasting its corresponding time series.

3 Link Prediction

Link prediction focuses on finding hidden connections or predicting links that tend to appear in the future time by considering the previous states of network. It is a very well-known task applicable to a bored of areas, such as citation networks, recommending systems, bibliographic domain, criminal investigations, and molecule biology [2, 4]. A classic definition of the link prediction problem is expressed by: "Given a snapshot of a social network at time t, we seek to accurately predict the edges that will be added to the network during the interval from time t to a given future time t'" [27]. Several approaches proposed to deal with this problem. Some well-known ones listed as: (1) Node-wise based approach, (2) Topological/structural patterns based approach.

Extracting the values of measures that show the similarity between pairs of nodes is the basic point for node-wise based approach. Nodes act as a vector of features, in order to find their closeness similarity metrics applied to each pair of nodes. After that the scores can be used by unsupervised method [11, 27–30]. In the unsupervised strategy, a proximity measure is selected and applied to pairs of nodes in the network aiming to rank them. The top ranked ones are predicted to be linked.

The topological patterns based approach [9, 11, 31, 32] extracts the scores from non-connected nodes of the network by using topological metrics. By tracking structural patterns of the network, these metrics provide a similarity degree between two nodes [9]. Then the scores are used as a basis to build models for performing the prediction. The topological based approach is the most widespread one. It is also easy to implement and presents good performance.

In the most of the previous studies, the classic definition of link prediction has been followed, the link prediction task performed by analyzing the complete structure of network at the current time. In other words, creation time of existing links or temporal information was not considered. However, temporal information (e.g., the interaction moments of nodes in the past or the first observation time of a connection) is a meaningful perspective that should be focused during the link prediction task [33, 34]. For instance, it would be interesting to consider not only how many but also when the links are formed between neighbors during the neighborhood based proximity score calculation. Common neighbor's recent activities can be more valuable than old activities. In order to predict new connections they use the static structure of the network and changes in the network over the time are not considered in these approaches. As a result they cannot model its evolution as such. Also, static approaches are suitable for investigating the occurrence of a certain link in a network but they are not so much useful, as an example, if the prediction of repeated link occurrences be a point of interest.

4 Traditional Proximity Metrics

In this section, we explain the traditional metrics applied as predictor and comparator attributes in our unsupervised link prediction approach. For this purpose, we first give some definitions and notations which will be useful to understand the descriptions below. $\Gamma(x)$ is the set of neighbors of x, let $|\Gamma(x)|$ be the degree of node x, and let $w(x, y)$ be the link weight between node x and y.

4.1 Number of Common Neighbors (CN)

The most widespread and simplest metric acknowledged in link prediction problem is the common neighbor (CN) metric. Also, a high number of common neighbors make it possible that a connection between two nodes x and y will create in future [6]. The CN measure for unweighted networks is defined as:

$$\mathrm{CN}\,(x,y) = |\Gamma(x) \cap \Gamma(y)| \tag{1}$$

The weighted version of common neighbor (CN) metric can be extended as below for weighted networks.

$$\mathrm{CN}\,(x,y) \sum_{z\in\Gamma(x)\cap\Gamma(y)} w\,(x,z) + w\,(y,z) \tag{2}$$

4.2 *Jacquard's Coefficient*

For comparing sets in data mining the JC metric is well explored [35]. The JC presumes the likelihood of creating a new link between pairs of nodes that share a high ratio of common neighbors comparative to the total number of their neighbors. Jacquard's coefficient for unweighted networks is given as:

$$\mathrm{JC}\,(x,y) = \frac{|\Gamma(x) \cap \Gamma(y)|}{|\Gamma(x) \cup \Gamma(y)|} \tag{3}$$

The weighted version of Jacquard's coefficient (JC) metric can be extended as below for weighted networks.

$$\mathrm{JC}\,(x,y) = \sum_{z\in\lceil(x)\cap\lceil(y)} \frac{w\,(x,z) + w\,(y,z)}{\sum_{a\,\in\,\lceil(x)} w\,(a,x) + \sum_{b\in\lceil(y)} w\,(b,y)} \tag{4}$$

4.3 *Preferential Attachment (PA)*

The PA measure supposes that the occurrence probability of a future connection between two nodes is rational to their degrees (i.e., nodes with a high number of relationships at the present tend to form new links in the future). Barabasi and Bonabeau [12], and Newman [6] have proposed that production of collaborators number between two nodes can express the probability of a future link between them. For unweighted networks, the PA measure is defined as:

$$\mathrm{PA}\,(x,y) = |\Gamma(x)| \times |\Gamma(y)| \tag{5}$$

The weighted version of PA metric can be extended as below for weighted networks.

$$\mathrm{PA}\,(x,y) = \sum_{a\in\lceil(x)} w\,(a,x) \times \sum_{b\in\lceil(y)} w\,(b,y) \tag{6}$$

4.4 Adamic–Adar Coefficient (AA)

Adamic and Adar [7] formulated this metric related to the Jaccard's coefficient. The common neighbors with fewer neighbors weighted/carried a higher importance in this metric. Moreover, AA evaluates the exclusiveness (or strength) of relationship between an evaluated pair of nodes and a common neighbor. The AA measure is given for unweighted networks as:

$$\mathrm{AA}\,(x,y) = \sum_{z\in\Gamma(x)\cap\Gamma(y)} \frac{1}{\log\left(\Gamma(z)\right)} \tag{7}$$

The weighted version of Adamic and Adar (AA) metric can be extended as below for weighted networks.

$$\mathrm{AA}\,(x,y) = \sum_{z\in\lceil(x)\cap\lceil(y)} \frac{w\,(x,z) + w\,(y,z)}{log\left(1 + \sum_{c\in\lceil(z)} w\,(z,c)\right)} \tag{8}$$

5 Citation Network and Triad Patterns

A directed weighted graph in which each node indicates an author and each edge indicates a citation link from an author to another one, and weights show the citation numbers between two nodes. This can be shown by an arrow (link) going from the node showing A_i to the node showing A_j. In this case the nodes from a collection A (authors) form a directed weighted citation network. Figure 1 represents a citation network over the time.

Dyad is the basic unit of analysis in social network theory. In undirected networks, a dyad is a pair of nodes who may share a social relation with one another. In directed networks, a dyad is a pair of nodes who may share a social relation through mutual links, a nonreciprocal relation, or no relation. Nonreciprocal means that one node is interested in the other node but the other node is not interested. A set of three parties, which includes three dyads, is called triad. A triad is "closed" if all nodes are linked with each other in some manner. A closed triad is also called

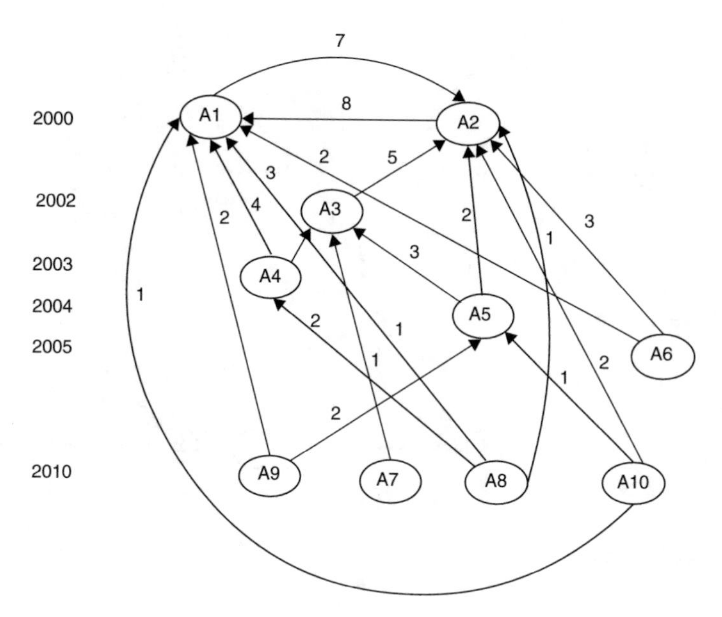

Fig. 1 Citation network

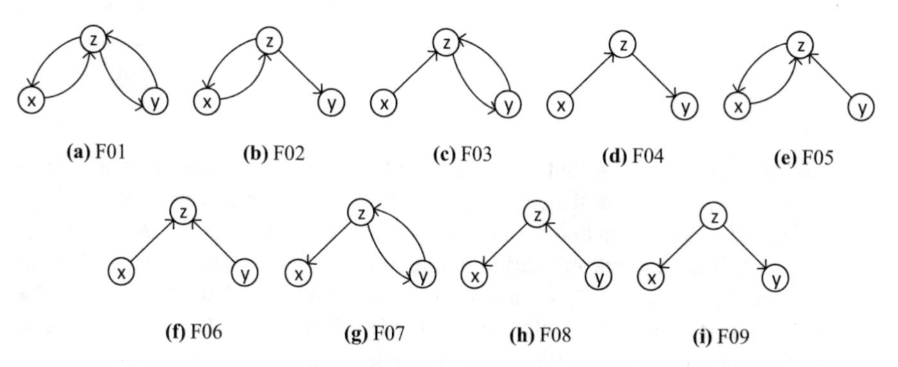

Fig. 2 Triad graph patterns

triangle. Figure 2 shows triad patterns of the actors *x*, *z*, and *y*. Edges are directed because our aim is to model patterns in directed social networks (e.g., citation networks). All patterns are open triads with *z* being the common neighbor of *x* and *y* [16].

Figure 2 shows all possible connectivity configurations between *x*, *z*, and *y* with the condition that *x* and *y* are not directly connected. Open triads are labeled as F0*X* where *X* is a changing index between [1, 9]. The pattern F01 shows the case where *x* and *z* as well as *y* and *z* are mutually connected. According to the theory of triadic closure, the chances are high that *x* will also connect to *y* (i.e., *x* and *z* as well as *y* and *z* mutually cited each other, *x* and *y* will possibly cite each other also). In F02, only *x* and *z* are mutually cited by each other. The node *y* is cited by *z* but the

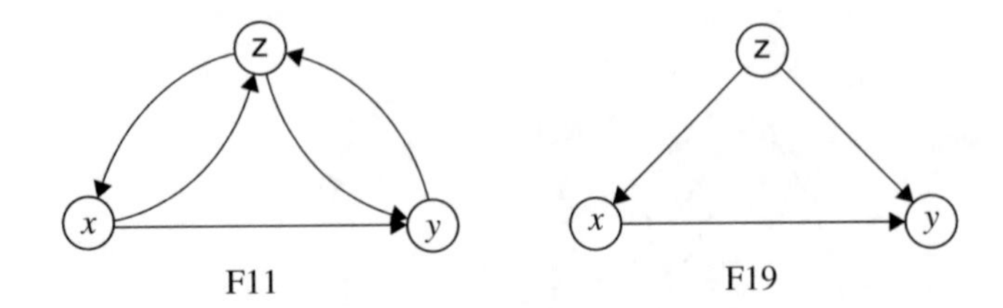

Fig. 3 Closed F1X triads

Fig. 4 Closed F2X triads

Fig. 5 Closed F3X triads

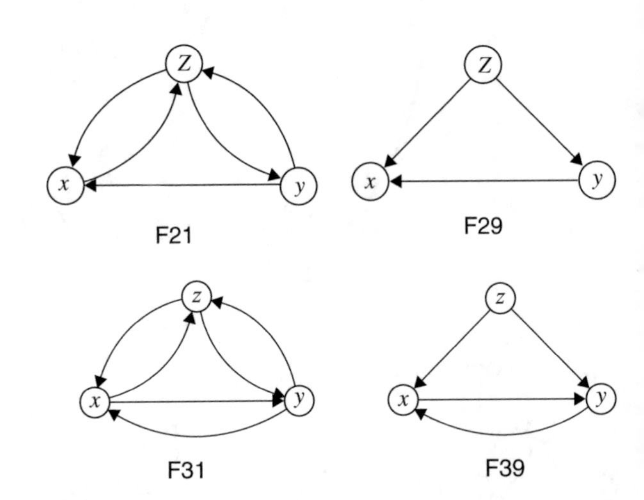

relationship is not reciprocated. F03, F05, and F07 indicate complementary cases where a mutual citation among one dyad exists. The other cases labeled as F04, F06, F08, and F09 show patterns without any mutual citation among the dyads. The goal which follows by link prediction is to determine which one of the triads is or will be closed (i.e., becoming a triangle). A triad can be closed as follows, if x cites y, y cites x, or if x and y mutually cite each other. The figures below show closed triads based on F01 and F09 (the first and the last pattern of Fig. 1 are shown for brevity). Figure 3 shows the patterns where the triads are closed from x to y.

The triads which are closed via x to y are labeled as F1X with $X = [1, 9]$. In the same manner, the triads that are closed via y to x are labeled as F2X with $X = [1, 9]$ (Fig. 4). Finally, the triads which are closed by mutual connections of x and y (Fig. 5), labeled as F3X with $X = [1, 9]$. To summarize the discussions about triad patterns, triads that are related to link prediction may have 36 different configurations with respect to how nodes are connected to each other through directed links. Open triads have 2 connected dyads and 2–4 links. Closed triads have 3 connected dyads and 3–6 links.

6 The Proposed Time-Frame Based Link Prediction in Directed Networks

6.1 Temporal Structure

Generally the prediction of new links with a particular type between a pair of nodes $x, y \in V$ in the future is the link prediction task in social networks. Temporal link prediction considers the dynamic evolvement of network as different from static link prediction. Let's observe the network G at time t that should be split into numerous time-segment snapshots, which present the network at different times in the past. Then a prediction window is defined that shows how further we want to make the prediction in the future. Afterwards, small sets which we named them frames consist of from sequential snapshots with length of prediction window.

Let G_t represent the network at time t. The frame created from the union of the graphs from time 1 to T shows by $[G_1, G_2, \ldots, G_T]$. The number of periods (frames) in the series is given by n. w represents the prediction window.

In this paper we focused on a network which indicates user information up to the year $T = 2012$ with a 4 year length of prediction window (i.e., new links prediction from 2009 up to 2012). We extracted $k = 3$ frames from the network structure and N was obtained as ($N = F1, F2, F3$).

$N = \{[1997\text{–}2000], [2001\text{–}2004], [2005\text{–}2008], [2009, 2012]\}$.

6.2 Unweighted and Weighted Temporal Events

A temporal event occurs when a pair of nodes changes from a state to another state (connected or non-connected) between consecutive time frames. There are three type events. These are protective, inventive, and regressive. The simple cases of these temporal event types were also used in the study in [26] for undirected and unweighted networks. We extend these temporal event types and scores for weighted and directed networks. A weighted temporal event occurs when the state of a link changes from a state to another state (connected, non-connected) or the weight of link becomes stronger or weaker between consecutive time frames. Three types of the weighted events are categorized into W-Protective, W-Inventive, and W-Regressive.

6.2.1 Protective

A protective event occurs when a dyad's relation is not dropped with the evolvement of the network that is when a dyad shares a link in a frame and the connection is preserved in the next frame. With respect to the above mentioned graph patterns protective event can be redefined as, a protective event occurs when a triad state doesn't change along the evolvement of the network from a frame to the next frame. For each pair of nodes (u, v), a $p(u, v, k)$ reward related to frame F_k is defined in order to take into account a protective event during the transition from the $(k-1)$-th to the k-th frame and formally:

$$p(u, v, k) = \begin{cases} p, & \text{if } n_{k-1}(u, v) = n_k(u, v) \\ 0, & \text{otherwise} \end{cases} \tag{9}$$

In Eq. (9), $n_{k-1}(u, v)$ is the number of edges between u and v nodes in the $(k-1)$-th frame. If there is a mutual relation between the related nodes, this number equals to 2. The constant p shows the reward for protective event, as the link between two nodes is stable and the event's value should be a positive.

6.2.2 W-Protective

A weighted protective event occurs when total weight of relations between two nodes does not decrease less than an event changing rate (e) pre-determined or does not increase more than this rate through the evolvement of network from a time frame to another one.

In order to take a weighted protective event into account during the transition from the $(k-1)$-th to the k-th frame, $w\text{-}p(u, v, k)$ reward for each pair of nodes (u, v) related to frame F_k is defined as follows:

$$w - p(u, v, k) = \begin{cases} p \text{ if } (w_{k-1}(u, v) + w_{k-1}(v, u)) - e \times (w_{k-1}(u, v) + w_{k-1}(v, u)) < \\ \qquad (w_k(u, v) + w_k(v, u)) < \\ \quad (w_{k-1}(u, v) + w_{k-1}(v, u)) + e \times (w_{k-1}(u, v) + w_{k-1}(v, u)) \\ 0 \text{ otherwise} \end{cases} \tag{10}$$

In (10), the constant p shows the reward for weighted protective event, $w_{k-1}(u, v)$ and $w_{k-1}(v, u)$ are the weights of edges between u and v, and v and u in the frame F_{k-1}, respectively. e $(0 < e \leq 1)$ is the event changing rate pre-determined, the events between frames change with respect to this rate.

6.2.3 Inventive

The formation of a new link between two nodes in different frame of time is showed by inventive event. This happens when there is no connection between two nodes in a frame and a link is observed in the next frame. For each pair of nodes (u, v) an $i(u, v, k)$ reward related to frame F_k is defined in order to take into account an inventive event during the transition from the $(k-1)$-th to the k-th frame and formally:

$$i(u, v, k) = \begin{cases} i, & \text{if } n_k(u, v) > n_{k-1}(u, v) \\ 0, & \text{otherwise} \end{cases} \tag{11}$$

In the above equation the constant i indicates the reward for inventive events. Since the tie between two nodes becomes stronger, its value should be positive.

6.2.4 W-Inventive

A weighted inventive event occurs when the total weight of relations between two nodes increases more than the event changing rate (e) in the next frame. For each pair of nodes (u, v), w-$i(u, v, k)$ reward related to frame F_k is defined in order to take a weighted inventive event into account during the transition from the (k − 1)-th to the k-th frame and formally:

$$w - i(u, v, k) = \begin{cases} i \text{ if } (w_k(u, v) + w_k(v, u)) > \\ \quad (w_{k-1}(u, v) + w_{k-1}(v, u)) + e \times (w_{k-1}(u, v) + w_{k-1}(v, u)) \\ 0 \text{ otherwise} \end{cases} \tag{12}$$

In the above equation the constant i indicates the reward for weighted inventive events. Since the tie between two nodes becomes stronger, its value should be positive.

6.2.5 Regressive

Regressive events are the opposite form of inventive events. The removal of an existing link between two nodes from a frame to another frame is representing by regressive event. For each pair of nodes (u, v) an $r(u, v, k)$ reward related to frame F_k is defined in order to take into account a regressive event during the transition from the (k − 1)-th to the k-th frame and formally:

$$r(u, v, k) = \begin{cases} r, & \text{if } n_{k-1}(u, v) > n_k(u, v) \\ 0, & \text{otherwise} \end{cases} \tag{13}$$

In the above equation the constant r indicates the reward for regressive events. Since the tie between two nodes tends to decrease, then its value should be negative.

6.2.6 W-Regressive

A weighted regressive event occurs when the total weight of relations between two nodes decreases more than the event changing rate (e) in the subsequent frame. In

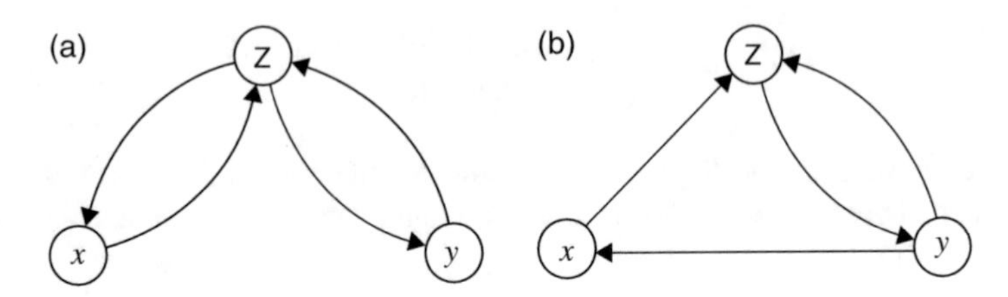

Fig. 6 Protective, inventive, and regressive events. (**a**) Frame $k-1$. (**b**) Frame k

this case, W-regressive score is computed as follows:

$$w-r(u,v,k)=\begin{cases} r \text{ if } \left(w_k(u,v)+w_k(v,u)\right)< \\ \quad \left(w_{k-1}(u,v)+w_{k-1}(v,u)\right)-e.\left(w_{k-1}(u,v)+w_{k-1}(v,u)\right) \\ 0 \text{ otherwise} \end{cases} \tag{14}$$

In (14) the constant r indicates the reward for weighted regressive events. Since the tie between two nodes tends to decrease r should assume a negative value.

Figure 6 shows an example of the above mentioned three events (protective, inventive, and regressive) in which we can see that a protective event occurred between node z and y after the evolvement of the network from frame $k-1$ to frame k. A link reduction can be seen between x and z which is the representation of the regressive event. On the other hand, there is no connection between x and y in frame $k-1$ while a new link is observed in frame k between them which presents the inventive event.

6.3 Time Frame Based Score

There are different kinds of approaches for link prediction that calculate scores to pair of nodes by applying proximity metrics, to determine how similar are the nodes and is there any connection tend to appear or form between them in the future.

In the proposed method we have combined the scores of primary cases with the scores of secondary cases which are the events observed in the node's neighborhood.

$$\text{Score }(u,v)=\sum_{k=2}^{n}P(u,v,k)+\alpha\times S(u,v,k) \tag{15}$$

$$P(u,v,k)=\begin{cases} p(u,v,k) & \text{if protective event observed} \\ i(u,v,k) & \text{if inventive event observed} \\ r(u,v,k) & \text{if regressive event observed} \end{cases} \tag{16}$$

$$S(u,v,k) = \sum_{z \in \Gamma(x) \cap \Gamma(y)} P(u,z,k) + P(v,z,k) \quad (17)$$

In Eq. (15) the parameter α is an amortization factor that indicates how strong secondary events affect the tie between u and v. $P(u, v, k)$ represents the reward of the primary event for (u, v) pairs observed in the transition from the frame $k - 1$ to the frame k. In Eq. (17) $S(u, v, k)$ shows the collected rewards of secondary events appointed to the pairs (u, v).

In our experiment, the proposed measures were compared with commonly used measures in the literature for link prediction. For this purpose, four proximity scores were considered: Common Neighbors (CN) [6], Jaccard's Coefficient (JC) [8], Preferential Attachment (PA) [6], and Adamic–Adar Coefficient (AA) [7].

The steps of the proposed method are as follows:

1. The weak connections with a weight less than *threshold* pre-determined or non-connected pairs are selected from the frame (validation set) before the prediction frame. The non-connected pairs that did not meet with any events during the network evolution were eliminated.
2. The proximity scores of five different predictors (CN, JC, PA, AA, and our method) are computed for the chosen pairs.
3. The ranked top n-pair are taken into account as the future links or strong links that are more likely to appear.
4. If the weight of a pair with a weight less than *threshold* pre-determined increases more than the even changing rate (e) from the frame under consideration to the prediction frame, the connection of this pair turn from week to strong and vice versa.
5. The performance measures of all the predictors considered are compared.

7 Experimental Results

In order to evaluate the performance of the proposed method we conducted a set of experiments using DBLP-Citation-Network-V7 dataset. We downloaded the dataset from Arnetminer (https://aminer.org) [36]. The dataset contains 2,244,021 papers and 4,354,534 citation relationships. We used the validation set to empirically evaluate and determine the most appropriate values of p, i, r, and α parameters in the link prediction task. We showed with some tests that the best performance was achieved at $p = 0.40$, $i = 0.80$, $r = -0.37$, and $\alpha = 0.1$. These results indicate that protective (p) and regressive (r) values almost balance with each other. The best performance values were achieved when $p < i$. The *threshold* value is the proportion of weight of the node pair under consideration to the sum of weights of all the links in the related frame. Throughout this study, the *threshold* value is pre-specified as 0.4%.

Table 1 Precision values (%) of five methods on the unweighted network

Method	$P@10$	$P@50$	$P@100$	$P@500$
CN	0	4	21	38
JC	10	12	29	41
PA	10	21	28	53
AA	50	59	64	68
Our method	70	78	81	84

Table 2 Precision values (%) of five methods on the weighted network

Method	$P@10$	$P@50$	$P@100$	$P@500$
CN	5	7	28	44
JC	14	17	33	46
PA	13	23	32	57
AA	59	63	68	73
Our method ($e = 20\%$)	75	82	86	89
Our method ($e = 40\%$)	72	79	82	85

The first experiment is dedicated to evaluate the precision values of the proposed method and the traditional proximity scores on the unweighted network with {[1997–2000], [2001–2004], [2005–2008], [2009–2012]} frames. In this structure, the frame size is four and the snapshot at [2009–2012] time interval is used as the prediction frame. Precision values found by five methods are reported in Table 1. Precision can be evaluated at different points in a ranked list of extracted citations. Mathematically, precision at rank n ($P@n$) is defined as the proportion of relevant citations and extracted citations.

$$P@n = \frac{\text{number of relevant citations with rank } n}{n} \tag{18}$$

As can be easily seen from Table 1, our approach outperforms the other well-known predictors for every four $P@n$. Our method also performs best at Precision@500. CN delivers worst in terms of $P@n$.

The second experiment finds the precision values of the proposed method and the other predictors on the weighted network. This experiment also investigates the effect of the event changing rate e for the performance of the time frame based score on the weighted citation network. Table 2 reports the results of this experiment. As can be seen from Table 2, the best prediction performances are achieved by our link predictor over first weighted version of the network with $e = 20\%$. Since the event changes of the links (protective, inventive, or regressive) between the frames are determined by this rate, it is important to choose the most appropriate event changing rate. This is because while the size of frame remains constant, if the event changing rate increases, less number of the event changes between frames is observed. The observations in Table 2 are similar to those described above for the frames on the unweighted network. However, it should be noted that all values are larger than those of the unweighted network.

Table 3 AUC values of the proposed method for different values of α

Network	$\alpha=0$	$\alpha=0.05$	$\alpha=0.1$	$\alpha=0.2$
Unweighted	0.68	0.82	0.92	0.87
Weighted ($e=20\%$)	0.72	0.86	0.95	0.91
Weighted ($e=40\%$)	0.70	0.83	0.89	0.88

The third experiment deals with investigating the effect of the amortization factor α for the performance of the time frame based score. As mentioned before, this parameter emphasizes the importance of the second events in relation to primary events in the proposed method. For this purpose, we evaluated the time frame based score for four different values of α. It should be noted that when $\alpha=0$, secondary events are not considered. However, for higher values possible, an extreme importance is assigned to secondary events. The AUC values obtained as measure of performance for different α values are given in Table 3. Our method performs best with an AUC value of 0.95 at $\alpha=0.1$ on the weighted network with $e=20\%$. Second ranked is again the weighted network having $e=20\%$ with an AUC of 0.91 at $\alpha=0.2$.

8 Conclusions

Generally, link prediction problem is a difficult task in directed and weighted networks. This paper proposed a time-frame based unsupervised link prediction method for directed and weighted networks. For this purpose, we first defined weighted temporal events in such networks. Then, we proposed a novel approach to calculate the time-frame based score of nodes. The method is based on the common neighbor based metrics and does not consider the global information based metrics and path based metrics.

In the experiments, we made a comparison between weighted and unweighted proximity metrics. According to the experimental results we can highlight that the weighted proximity metrics outperform the unweighted metrics. Experimental results applied on real dataset show that the proposed method gives accurate prediction and promising results.

References

1. Wasserman, S., Faust, K.: Social Network Analysis: Methods and Applications (Structural Analysis in the Social Sciences). Cambridge University Press, Cambridge (1994)
2. Getoor, L., Diehl, C.P.: Link mining: a survey. ACM SIGKDD Explor. Newsl. **7**(2), 3–12 (2005)
3. Wang, C., Satuluri, V., Parthasarathy, S.: Local probabilistic models for link prediction. In: Proceedings of the 2007 Seventh IEEE International Conference on Data Mining, pp. 322–331 (2007)

4. Wang, P., Xû, B., Wu, Y., Zhou, X.: Link prediction in social networks: the state-of-the-art. Sci. China Inf. Sci. **58**, 011101(38) (2015)
5. Barabâsi, A.L., Jeong, H., Néda, Z., Ravasz, E., Schubert, A., Vicsek, T.: Evolution of the social network of scientific collaborations. Phys. A Stat. Mech. Its Appl. **311**(3–4), 590–614 (2002)
6. Newman, M.E.J.: Clustering and preferential attachment in growing networks. Phys. Rev. E **64**(2), 025102 (2001)
7. Adamic, L.A., Adar, E.: Friends and neighbors on the web. Soc. Netw. **25**(3), 211–230 (2003)
8. Salton, G., McGill, M.J.: Introduction to Modern Information Retrieval. McGraw-Hill, New York, NY (1986)
9. Liben-Nowell, D., Kleinberg, J.: The link-prediction problem for social networks. J. Am. Soc. Inf. Sci. Technol. **58**, 1019–1031 (2007)
10. Lü, L., Zhou, T.: Link prediction in weighted networks: the role of weak ties. EPL (Europhys. Lett.). **89**(1), 18001 (2010)
11. Murata, T., Moriyasu, S.: Link prediction based on structural properties of online social networks. N. Gener. Comput. **26**(3), 245–257 (2008)
12. Barabasi, A.L., Bonabeau, E.: Scale-free networks. Sci. Am. **288**(5), 60–69 (2003)
13. Zhou, T., Lü, L., Zhang, Y.C.: Predicting missing links via local information. Eur. Phys. J. B. **71**(4), 623–630 (2009)
14. De Sa, H., Prudencio, R.B.C.: Supervised link prediction in weighted networks. In: International Joint Conference on Neural Networks, pp. 2281–2288 (2011)
15. Yu, Y., Wang, X.: Link prediction in directed network and its application in microblog. **2014**, 509282 (2014)
16. Schall, D.: Link prediction in directed social networks. Soc. Netw. Anal. Min. **4**, 157 (2014)
17. Jawed, M., Kaya, M., Alhajj, R.: Time frame based link prediction in directed citation networks. In: Proceedings of the 2015 IEEE/ACM International Conference on Advances in Social Networks Analysis and Mining (2015)
18. Lin, Z., Xiong, L., Zhu, Y.: Link prediction using BenefitRanks in weighted networks. In: IEEE/WIC/ACM International Conferences on Web Intelligence and Intelligent Agent Technology (2012)
19. Radicchi, F., Fortunato, S., Makines, B., Vespignani, A.: Diffusion of scientific credits and the ranking of scientists. Phys. Rev. E. **80**(5), 056103 (2009)
20. Tylenda, T., Angelova, R., Bedathur, S.: Towards time-aware link prediction in evolving social networks. In: Proceedings of the Third Workshop on Social Network Mining and Analysis, SNA-KDD'09, pp. 1–9 (2009)
21. Bringmann, B., Berlingerio, M., Bonchi, F., Gionis, A.: Learning and predicting the evolution of social networks. IEEE Intell. Syst. **25**(4), 26–35 (2010)
22. Juszczyszyn, K., Musial, K., Budka, M.: Link prediction based on subgraph evolution in dynamic social networks. In: Privacy security risk and trust (PASSAR) 2011 IEEE Third International Conference on Social Computing, pp. 27–34 (2011)
23. Jiang, C., Coenen, F., Zito, M.: A survey of frequent subgraph mining algorithms. Knowl. Eng. Rev. **28**(1), 75–105 (2013)
24. Huang, Z., Lin, D.K.J.: The time-series link prediction problem with applications in communication surveillance. INFORMS J. Comput. **21**(2), 286–303 (2009)
25. Potgieter, A., April, K.A., Cooke, R.J.E., Osunmakinde, I.O.: Temporality in link prediction: understanding social complexity. J. Emerg. Complex. Organ. **11**(1), 83–96 (2009)
26. Soares, P.R.S., Prudencio, R.B.C.: Proximity measures for link prediction based on temporal events. Expert Syst. Appl. **40**(16), 6652–6660 (2013)
27. Liben-Nowell, D., Kleinberg, J.: The link prediction problem for social networks. In: Proceedings of the 2003 International Conference on Information and Knowledge Management, pp. 556–559 (2003)
28. Lü, L., Zhou, T.: Link prediction in complex networks: a survey. . **390**(6), 1150–1170 (2011)
29. Kaya, B., Poyraz, M.: Age-series based link prediction in evolving disease networks. Comput. Biol. Med. **63**, 1–10 (2015)

30. Kaya, B., Poyraz, M.: Unsupervised link prediction in evolving abnormal medical parameter networks. Int. J. Mach. Learn. Cybern. **7**(1), 145–155 (2016)
31. Xiang, E.W.: A survey on link prediction models for social network data. Technical Report, Department of Computer Science and Engineering, The Hong Kong University of Science and Technology (2008)
32. Huang, Z.: Link prediction based on graph topology: the predictive value of the generalized clustering coefficient. In: Proceedings of 12th ACM SIGKDD International Conference on Knowledge Discovery and Data Mining (2006)
33. Kaya, B., Poyraz, M.: Supervised link prediction in symptom networks with evolving case. Measurement. **56**, 231–238 (2014)
34. Kaya, B., Poyraz, M.: Finding relations between diseases by age-series based supervised link prediction. ASONAM. **2015**, 1097–1103 (2015)
35. Tan, P.N., Steinbach, M., Kumar, V.: Introduction to Data Mining. (Vol. 1). Pearson Addison Wesley, Boston (2006)
36. Tang, J., Zhang, J., Yao, L., Li, J., Zhang, L., Su, Z.: ArnetMiner: extraction and mining of academic social networks. In: Proceedings of the Fourteenth ACM SIGKDD International Conference on Knowledge Discovery and Data Mining (SIGKDD'2008), pp. 990–998 (2008)

An Approach to Maximize the Influence Spread in the Social Networks

Ibrahima Gaye, Gervais Mendy, Samuel Ouya, and Diaraf Seck

1 Introduction

The social networks, such as *Facebook*, *Twitter*, *viadeo*, and *Linkedin*, become more and more popular and diverse in recent decades. Many people begin to integrate them. Noting the very important number of users, the social networks can change the nature of communication and information. Thus, the Social Network Analysis (*SNA*) attracts many attention, thanks to its varied fields of applications. For example, in marketing, the online social networks show a big potential. They are much more effective than traditional marketing techniques. To make visible a product, organizations can use the advertising technique, word of mouth in social networks for good visibility [1, 2]. This approach is known under influence maximization expression in social networks [3]. Seen the importance of this problem, the research community in data mining was interested. The problem is to find a small set of k individuals or organizations (i.e., seeds) in the social network that will begin and maximize the influence propagation in a small delay.

As an example of implementation, we can consider the case of politicians who wish to popularize their programs with the electors during the electoral campaigns. They can search and communicate their programs to the most influential individuals in the social network. These targeted people will influence their neighbors, who in turn will influence their neighbors, and so on. We can take another example in the field of marketing: A company can find in the social network the most influential

I. Gaye (✉) • G. Mendy • S. Ouya
UCAD-ESP-LIRT of Sénégal, Dakar, Fann, Senegal
e-mail: gaye.ibrahima@esp.sn; gervais.mendy@ucad.edu.sn; samuel.ouya@gmail.com

D. Seck
UCAD-FST-LMDAN of Sénégal, Dakar, Fann, Senegal
e-mail: diaraf.seck@ucad.edu.sn

R. Missaoui et al. (eds.), *Trends in Social Network Analysis*, Lecture Notes in Social Networks, DOI 10.1007/978-3-319-53420-6_9

customers by granting them facilities in order to better dispose of its products. So these customers can trigger the process of influencing as many individuals as possible in the social network.

For a long time, the social network has been used to represent the interactions between individuals (or organizations) in different contexts. The social network makes it possible to represent individuals (or organizations) and the links between them (individuals or organizations). Formally, a social network is modeled by an undirected graph or a digraph $G(V, E)$, where $V = (v_1, \ldots, v_n)$ is the set of nodes and $E \subseteq (V \times V)$ the set of edges (or links). The nodes represent the individuals (or organizations) and the edges denote the relationships between the individuals (or organizations). Influence maximization problem is to find a subset of $k-nodes$ (i.e., seed set) in a social network that could maximize the influence spread. Mathematically we can define the problem as follows: find S_k^* such as:

$$S_k^* = \text{argmax}_{S \subseteq V,\ |S|=k}\ \sigma(S) \tag{1}$$

where $\sigma(S)$ denotes an influence or objective function that gives the number of influenced nodes according to the set of seeds S.

The first to deal with this $NP-hard$ problem are Kempe et al. [4]. It is very difficult to choose the $k-nodes$ that maximize the $\sigma(S)$ function. In this paper, we propose a new approach to determine the seeds, which approach prevents the information feedback toward a node chosen as a seed. We will propose to extract a particular spanning graph. This extraction uses the existing centrality measure (like *closeness* and *degree*). Three extraction algorithms are proposed. The two first are called *SCG*-algorithm (**S**panning **C**onnected **G**raph-algorithm) and they are only applicable with connected graphs. The *SCG*-algorithm is proposed in two versions. The second version is just an improvement of the first version. Then we proceeded to a generalization of the *SCG*-algorithm which is proposed. The general algorithm is called *SG*-algorithm (*S*panning *G*raph-algorithm) and it is applicable to an arbitrary graph. After we determined the spanning graph, we use the existing heuristics, like degree, *degree discount* [5, 6], *PageRank* [7], *diffusion degree* ℓ-th [8], etc., to find the seeds. The three algorithms are effective and each one has an $O(mn)$ complexity. This paper is organized as follows. First we will develop the related works and will show the limits of these works in order to propose our contribution. Then, we will propose three algorithms to determine a particular spanning graphs and we discuss the benchmark propagation models and finally some simulations will be made to show the pertinence of our approach.

2 Related Works

To determine the seeds, several works have been done. Kempe et al. [4] proposed centrality measures, *High Degree Heuristic* Model, based on the notion of neighborhood. It is the most classic approach to solve the influence maximization problem.

The seeds are determined according to their degree. Thus, the k nodes having the highest degree are selected as seed nodes. In this same vein, Chen et al. [9] propose the *Degree Discount Heuristic* Model. The general idea of the *Degree Discount algorithm* is that if a node u is chosen as a seed, for each neighbor v of u that should be considered as a new seed, the edge uv should not be counted in the calculation of its degree. Chen et al. [10] propose scalable heuristics to estimate the coverage of a set by considering Maximum Influence Paths (*MIP*) based on Independent Cascade Model (*ICM*). An *MIP* between a pair of nodes (u, v) is the path with the maximum spread probability from u to v. They propose also in [11], a scalable heuristic for Linear Threshold Model (*LTM*). They use the Directed Acyclic Graph (*DAG*) construction property. In recent works, Suman et al. [12] proposed the *Diffusion Degree*. In this heuristic the authors consider the centrality degree define in [4] which is based on the propagation probability and degree centrality. Recently, we proposed a generalization [8] of the previous heuristic for that, the ℓ-th neighbors participate in information propagation process. We have done [13] some treatments on the initial graph before to determine the seeds by using the existing centrality heuristics. They eliminate the information feedback (the fact to try to activate a node that already has been it) toward a considered user as seed, by extracting a spanning social network.

3 Our Contribution

The first works on the influence maximization problem were posed by Domingos et al. [3] as an algorithmic problem. They modeled the problem by using Markov random fields and proposing heuristic solutions. Then, Kempe et al. [4] reformulated it as a discrete optimization problem and presented an extensive of their study. They demonstrate that the greedy hill climb algorithm gives good approximation if the activation function is sub-modular and monotone. There are also Chen et al. [9] that consider the *MIP*. In recent works Suman et al. [12] use the degree centrality measure and the diffusion probability based on the Independent Cascade Model (ICM) [4] to determine the initial diffusers. The works of Suman K. et al. are ameliorated in [8]. They generalize until the ℓ-th neighbors with the propagation probability. To maximize the information propagation in a social network, the choice of the diffusion model is very important. In the two benchmark models, (Independent Cascade Model (ICM) [4, 18] and Linear Threshold Model (LT M) [4]) and their derivatives, the nodes are activated for ever. In these propagation models, the information can return toward an already active node (information feedback) even if it would be ignored. In [13], the authors propose to prevent this information feedback under the cascade models. They extract a particular spanning graph and use the existing centrality measures to determine the seeds from this latter. In the construction of the spanning graph, the authors choose randomly the children of a node. In this paper, we propose an extension of the works done in [13]. We develop a second version of the algorithm that chooses the children by basing on an analysis of node neighborhood. We also show that this approach can be based not only on the cascade models but also on the threshold models.

4 Spanning Graph

To determine initial diffusers in the influence maximization problem, we follow the methodology of Fig. 1. To obtain the seeds, firstly, we determine a particular spanning graph from the initial graph. If the input graph is connected, we use one version of *SCG*-algorithms. If the input graph is not connected, we use the *SG*-algorithm. This latter takes an arbitrary graph (connected or unconnected). Thereafter, we use the existing centrality measures to determine these seeds from this spanning graph given by one of these algorithms. The real challenge of construction of the spanning graph is to select the initial node. In this part, we show the importance of the centrality measure by designing and developing the two versions of *SCG*-algorithms and the *SG*-algorithm.

4.1 Centrality Measure

The centrality measure in [5, 6, 14–17] is a fundamental concept in the social network analysis. It gives a strict indication of how a node is connected in a network. We can apply these centrality measures in several types of networks like the social networks, the information networks, the biological networks. Several centrality measures have been proposed and they have not the same importance in the network. For example, the *degree* centrality is based on the neighborhood notion; the *closeness* centrality is based on geodesic distance with all accessible nodes; the *betweenness* centrality is the degree that a node contributes to sum of maximum path between all pairs of nodes; the *degree discount* centrality is also based on the neighborhood notion by modifying the degree of the neighbors of a node chosen

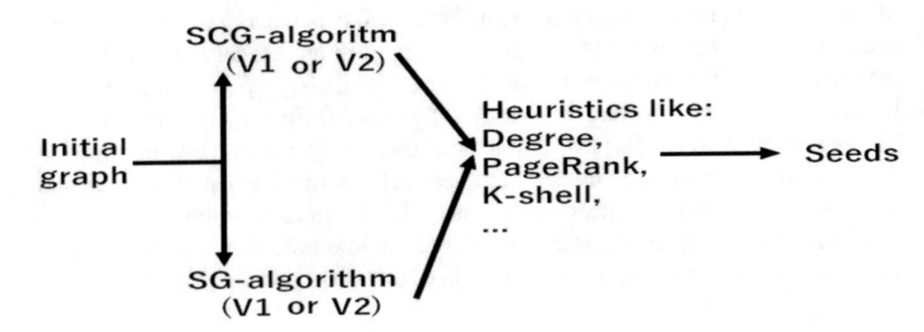

Fig. 1 Methodology

as seed; the *Diffusion* degree [12] is based on the neighborhood and the diffusion probability; the *Diffusion* degree ℓ-th [8] is a generalization of the *Diffusion* degree, the *PageRank* [7] based on the importance of the neighborhood, etc.

Our works consist to determine a particular spanning graph to prevent the information feedback. The challenge is to choose the first node that begins this construction. Like centrality measures, we have the *closeness* centrality that gives the position of a node in the network. The node which has the smallest measure is considered as the center of the network. If we begin the construction of the spanning graph from this node, we will have a balanced spanning graph. So, we do less iterations to visit all nodes of the graph. It is given by Eq. (2). To calculate it, we follow the next steps:

- Calculate the shorter path distance, called "the geodesic distance," between the node u and all other nodes different of u.
- Calculate the sum of all geodesic distance.

$$C_c(u) = \sum_{v \in V, v \neq u} d(u, v) \tag{2}$$

where $d(u, v)$ denotes the geodesic distance between u and v.

In certain algorithms of this paper, the children are not chosen randomly. We use the neighborhoods of each node. The number of neighbors is given by the *degree centrality*. It is defined by the following equation:

$$C_d(u) = \sum_{v \in V} \lambda(u, v) \tag{3}$$

where $\lambda(u, v) = 1$ if $\{u, v\} \in E$ and 0 else.

4.2 SCG-Algorithm

In this paragraph, we develop two versions of *SCG*-algorithm. Both algorithms use a connected graph as input data and provide a spanning tree. In the first version (Algorithm 1), we use closeness centrality measure to determine the first node which begins the construction of the spanning graph. This node is given by Eq. (4). After the determination of this node, its children are built randomly. Suppose that, we are in the iteration i-th, the algorithm must build the neighbors of the nodes u and v. Let the node w be a neighbor of u and v in the initial graph. In the Algorithm 1, the node w will be randomly a child u or v in the spanning graph. This issue will be corrected in the second version (Algorithm 2) that builds the children by analyzing the neighborhood of the nodes u and v. We use the degree centrality measure to choose the neighbor of w between u and v. This choice is no more random.

$$BeginNode = \text{argmin}_{v, v \in V}\ C_c(v) \tag{4}$$

where $C_c(v)$ denotes the closeness centrality measure of v.

Algorithm 1: SCG_{v1}-algorithm

1 Important variables used in algorithm 1
2 SG: Spanning graph of output 3 E_{SG}: Set of edges of the graph SG
4 V_{SG}: Set of node of the graph SG
5 $Level^{v}(u)$: Level of u in relation to v

Data: A connected graph $G(V,E)$
Result: Spanning Graph $SG(V_{SG}, E_{SG})$

6 **for** *all node* $v \in V$ **do**
7 | Calculate $C_c(v)$;
8 **end**
9 $BeginNode \longleftarrow argmin_{v,v \in V}\ C_c(v)$;
10 $V_{SG} \longleftarrow \{BeginNode\}, E_{SG} \longleftarrow \phi, level \longleftarrow 0$;
11 $Level^{BeginNode}(BeginNode) \longleftarrow level$;
12 **while** $(| V_{SG} | \neq | V |)$ **do**
13 | **for** *all* $u \in V_{SG}$ *and* $Level^{BeginNode}(u) = level$ **do**
14 | | **for** *all node* $z \in N(u)$ *and* $z \notin V_{SG}$ **do**
15 | | | $E_{SG} \longleftarrow E_{SG} \bigcup \{u, z\}$;
16 | | | $V_{SG} \longleftarrow V_{SG} \bigcup \{z\}$;
17 | | | $Level^{BeginNode}(z) \longleftarrow level + 1$;
18 | | **end**
19 | **end**
20 | $level \longleftarrow level + 1$;
21 **end**
22 return SG;

4.2.1 SCG_{v1}-Algorithm

We develop the first version of SCG-algorithm that has an $O(mn)$ complexity, where m denotes the number of edges and n the number of nodes. The goal of this algorithm is to prevent the information feedback toward the seeds before their determination. This algorithm extracts a spanning graph of the initial connected graph. The general idea of Algorithm 1 is to choose the node that begin the construction of the spanning graph. It is given by Eq. (4) that gives the node which has the smallest closeness centrality measure. The children of the nodes are building randomly without considering their neighborhood in the initial graph.

The input graph of Algorithm 1 is a connected graph and the output a spanning graph (a spanning tree). The choice of the first node is very important. In [5, 6, 14–17], several centrality measures are defined. Among them, we have the *closeness* centrality which measures the node importance on in relation to others in the network (i.e., the geodesic distance of a node in relation to others). So, if we consider the node which has the smallest closeness centrality measure, then we have the center node of the network. From this node, we can visit all other nodes with a small number of iterations. At line 6–8 of Algorithm 1, we calculate the C_c of all node. At line 9, we determine the first node ($BeginNode$) that begin the construction of the spanning graph. Between the lines 14–18, each non-explored neighbor z of u

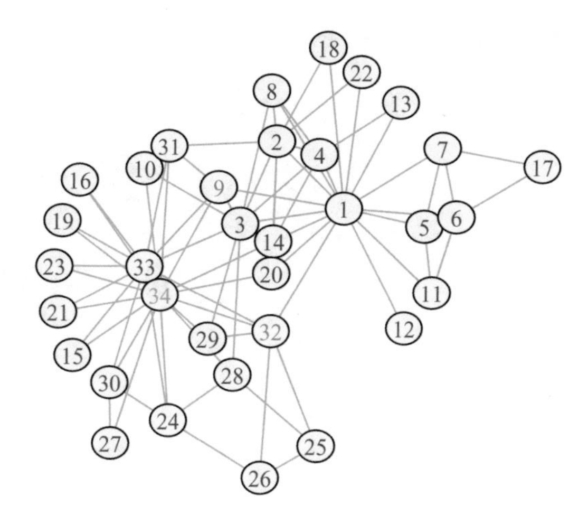

Fig. 2 A connected graph

in G will be consider as neighbor of u in SG. The level of z becomes level + 1, the edge (u, z) will be in the set E_{SG} and z in the set V_{SG}. Algorithm 1 chooses randomly the edge (u, z) without considering the number of neighbors of u (Fig. 2).

For example, when we take the graph of Fig. 2, the node 1 has the smallest closeness centrality measure than the other nodes. So, we consider it as the beginning node (*BeginNode*). Algorithm 1 begins the construction of the spanning graph from this node. The level of the beginning node is zero. For the first iteration, it builds the children of the first node. The set of the children of node 1 will be $children(1) = \{2, 3, 4, 5, 6, 7, 8, 9, 11, 12, 13, 14, 18, 20, 22, 32\}$. For the second iteration (i.e., current level is 1), we build the children of each node of the children(1) set. Algorithm 1 builds at first the children of node 2, secondly, the children of node 3, etc. Finally, it ends by the node 32. In Fig. 2, node 34 is neighbor of node 9 and node 32 in G. The level of 9 and 32 is 1 in relation to *BeginNode*. Algorithm 1 chooses randomly between 9 and 32 the neighbor of 34 in the spanning graph. If Algorithm 1 takes the graph of Fig. 2, it gives the spanning graph of Fig. 3. The node 9 is chosen by Algorithm 1 as children of 34. The seeds given by the heuristics like degree centrality, PageRank, diffusion degree, in the spanning graph give better results than the seeds given by the initial graph with the same centrality measure under the cascade models and the linear threshold models.

4.2.2 SCG_{v2}-Algorithm

Here, we develop a second version of the SCG-algorithm with an $O(mn)$ complexity. The goal is to eliminate the information feedback toward the seeds. The general idea of Algorithm 2 is first to choose the beginning node. This latter is simply the node which has the smallest closeness centrality measure. Then its children are built by doing an analysis of the neighborhood of nodes in G. Algorithm 1 builds the children randomly. This inconvenience is corrected in Algorithm 2. In the latter the neighbors

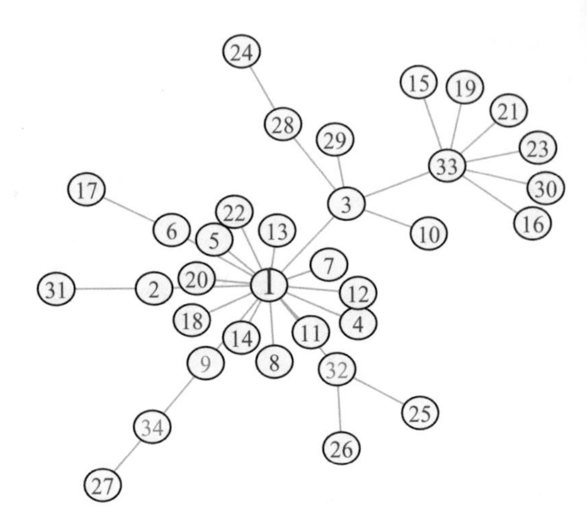

Fig. 3 Spanning graph given by Algorithm 1

are built in relation to the number of neighbors in *G*.

$$u = \text{argmax}_{v, v \in V \text{ and } v \notin V_{SG}} \; C_d(v) \tag{5}$$

Algorithm 2 corrects the disadvantage noted in Algorithm 1. In this latter, we note that the construction of children of nodes of same level is done randomly. Algorithm 2 gives output of a spanning graph by taking a connected graph as input data. In the second version, we lose less information on the initial graph than the first version. The methodology does not change. At line 7–9 of Algorithm 2, we calculate the C_c of all node and take the node (*BeginNode*) which has the smallest *closeness* centrality measure(*BeginNode*) at line 10. At line 16 we consider all nodes u of the current level in relation to *BeginNode*. Between the line 17 and 21, let u a node of the current level, each non-explored neighbor z of u in G will be consider as neighbor of u in SG. The level of z will be level + 1, and the edge (u, z) will be put in E_{SG} and z in V_{SG}. If we take one level, there are several nodes in SG which have the same neighbor w in G. There will be considered like the children of the node which has the most neighbors in G. For example, let u and v, two nodes of level α in SG. Let w be a not visited common neighbor of v and u in G. In SG, w will be considered as a neighbor of the node which has the most neighbors in G. At line 14 of Algorithm 2, we build a set of nodes whose level is the current level by the instruction named $SortNodeSet^{BeginNode}(level)$. At next line, we sort this set ($SortNodeSet^{BeginNode}(level)$) of nodes according to the number of their neighbors in G. The construction of neighbors of the nodes of the same level begins by the node respecting Eq. (5).

For example, Algorithm 2 takes the graph of Fig. 2 as input and returns the spanning graph of Fig. 4. At line 07 of Algorithm 2, we obtain the set: $children(1) = \{2, 3, 4, 5, 6, 7, 8, 9, 11, 12, 13, 14, 18, 20, 22, 32\}$. With the sorting done at the line 08, we have the follow result

Algorithm 2: SCG_{v2}-algorithm

1 important variables used in algorithm 2
2 SG: Output spanning graph
3 E_{SG}: Set of the edges of the graph SG
4 V_{SG}: Set of the nodes of the graph SG
5 $Level^{v}(u)$: Level of u in relation to v
6 $SortNodeSet^{v}(1)$: Set of the nodes of level 1 in relation to v

Data: A connected graph $G(V, E)$
Result: Spanning Graph $SG(V_{SG}, E_{SG})$

7 **for** *all node* $v \in V$ **do**
8 | Calculate $C_c(v)$
9 **end**
10 $BeginNode \longleftarrow argmin_{v,v \in V}\ C_c(v)$;
11 $V_{SG} \longleftarrow \{BeginNode\}, E_{SG} \longleftarrow \phi, level \longleftarrow 0$;
12 $Level^{BeginNode}(BeginNode) \longleftarrow level$;
13 **while** $(| V_{SG} | \neq | V |)$ **do**
14 | $SortNodeSet^{BeginNode}(level) \longleftarrow$ All nodes where level is $level$;
15 | Sort $SortNodeSet^{BeginNode}(level)$ according to their neighbors number;
16 | **for** *all* $u \in SortLevelNodeSet$ **do**
17 | | **for** *all node* $z \in N(u)$ *and* $z \notin V_{SG}$ **do**
18 | | | $E_{SG} \longleftarrow E_{SG} \bigcup \{u, z\}$;
19 | | | $V_{SG} \longleftarrow V_{SG} \bigcup \{z\}$;
20 | | | $Level^{BeginNode}(z) \longleftarrow level + 1$;
21 | | **end**
22 | **end**
23 | $level \longleftarrow level + 1$;
24 **end**
25 return SG;

$children(1)' = [3, 2, 4, 32, 9, 14, 8, 6, 7, 11, 20, 5, 18, 13, 22, 12]$. All children of the set $children(1)'$ will be built in the order. First it builds the children of node 2, then those of node 3, etc., and finally those of node 12. The noted conflict in the Algorithm 1 between the node 32 and the node 9 is regulated by Algorithm 2. The node 34 will be a child of node 32 which has the bigger centrality degree than node 9.

After we determine the particular spanning graph, we can use the existing heuristic like: *degree* centrality, *PageRank*, *diffusion* degree, etc., to determine the set of seeds. With the independent cascade and linear threshold methods, the seeds given by the spanning graph give a better result than the seeds given by the initial graph.

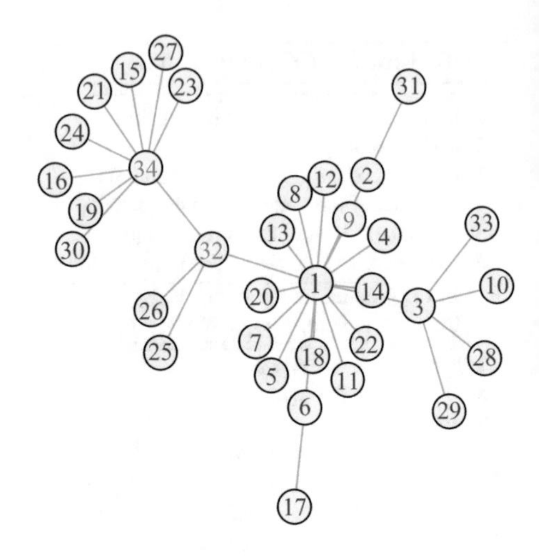

Fig. 4 Spanning graph given by Algorithm 2

4.3 SG-Algorithm

In this part, we propose a generalization of Algorithm 1 called *SG*-algorithm. For a generalization of Algorithm 2, you integrate the line 15 of Algorithm 2 in Algorithm 3 after the line 16. Algorithm 3 gives a spanning graph that is exactly a spanning forest if the input graph is unconnected. The complexity is O(mn). Before to propose a generalization of Algorithm 1, we use the following definition of a connected digraph in relation to a given node.

definition A digraph $G(V, E)$ is connected in relation to node $u \in V$ if and only if from this latter, we can visit all nodes $v \in V$ with the orientation.
If not, it is unconnected in relation to $u \in V$ if and only if it exists one node $v \in V$ which is not accessible from u.

The general idea is that from the node (*BeginNode*) which has the smallest closeness centrality measure, we build a spanning tree. This tree will contain all accessible nodes from the node *BeginNode*. If there are remaining nodes, we determine again the non-exploited node which has the smallest closeness centrality measure and another spanning tree will be created. The process continues until all nodes in the initial graph are exploited.

From line 7–8, Algorithm 3 calculates the closeness centrality measure of all nodes of the initial graph *G*. From line 10 to line 33, the spanning graph which is built is a spanning forest. At line 11, it determines the not exploited node that has the smallest *closeness* centrality measure. The latter is the node that begins the *i*-th spanning tree of the spanning forest. So, *BeginNode*(i) represents the node that begins the *i*-th spanning tree of the spanning forest. For each spanning graph, the level is initialized to zero. In the line 16, Algorithm 3 determines a set which represents all nodes of level the current level. This set is put in *leafnode*. For each node u in *leafnode*, their non-exploited neighbors in the graph G will be considered

Algorithm 3: *SG*-algorithm

```
1  Important variables used in algorithm 3
2  i   Connected component i
3  E_SG(i)   Set of edges of i-th Connected component
4  V_SG(i)   Set of nodes of i-th Connected component
5  Level^v(u)   Level of u in relation to v
   Data: A graph G(V,E)
   Result: Spanning Graph SG(V_SG, E_SG)
6  for all node v ∈ V do
7  |  Calculate C_c(v);
8  end
9  E_SG ← φ, V_SG ← φ, i ← 1;
10 while (| V_SG |≠| V |) do
11    BeginNode(i) ← argmin_{v,v∈V and v∉V_SG} C_c(v);
12    V_SG(i) ← {BeginNode(i)}, E_SG(i) ← φ, level ← 0;
13    Level^BeginNode(i)(BeginNode(i)) ← level;
14    CC(i) ← true;
15    while CC(i) do
16       leafNode ← all nodes of level level;
17       if leafNode not empty then
18          for all u ∈ leafNode do
19             for all node z ∈ N(u) and z ∉ V_SG do
20                E_SG(i) ← E_SG(i) ⋃{u,z} ;
21                V_SG(i) ← V_SG(i) ⋃{z} ;
22                Level^BeginNode(i)(z) ← level + 1 ;
23             end
24          end
25          level ← level + 1 ;
26       else
27          CC(i) ← false;
28       end
29    end
30    V_SG(i) ← V_SG ⋃ V_SG(i);
31    E_SG(i) ← E_SG ⋃ E_SG(i);
32    i ← i + 1 ;
33 end
34 return SG;
```

as of neighbors of u in the i-th spanning tree. If several nodes of the same level have the same neighbors in the initial graph G, the algorithm begins to build their neighbors randomly. If we consider a generalization of Algorithm 2, a sorting based on the number of neighbors in G is done and the children of the node that has the biggest centrality degree will be built. After the construction of all neighbors of nodes in leaf node, the current level is incremented at line 25 of Algorithm 3. The *leafnode* set will be updated: if it is not empty, another level will be created; if it is empty, another spanning tree will be built. This process continues until all nodes of G are exploited.

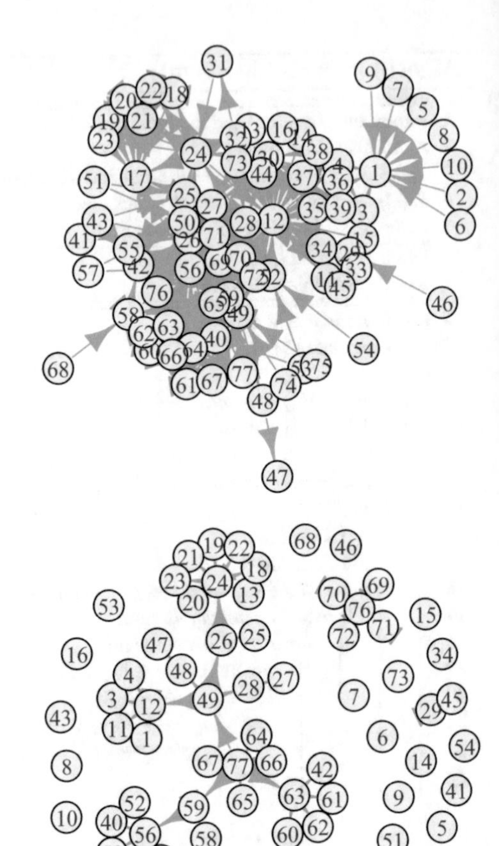

Fig. 5 An unconnected digraph

Fig. 6 Spanning graph given by Algorithm 3

For example, let's consider a social network that is represented by the digraph of Fig. 5. It is used as input graph of Algorithm 3 which returns the spanning forest of Fig. 6. Here, the output graph is a forest because from the node which has the smallest centrality closeness, there are several not accessible nodes. So, the Algorithm 3 returns a spanning forest. In this graph, several nodes are isolated, for example: 7, 9, 15. If we look at the graph of Fig. 5, these nodes have only one link toward another node or toward itself. They are not fundamental in the influence spread processes (Figs. 7 and 8).

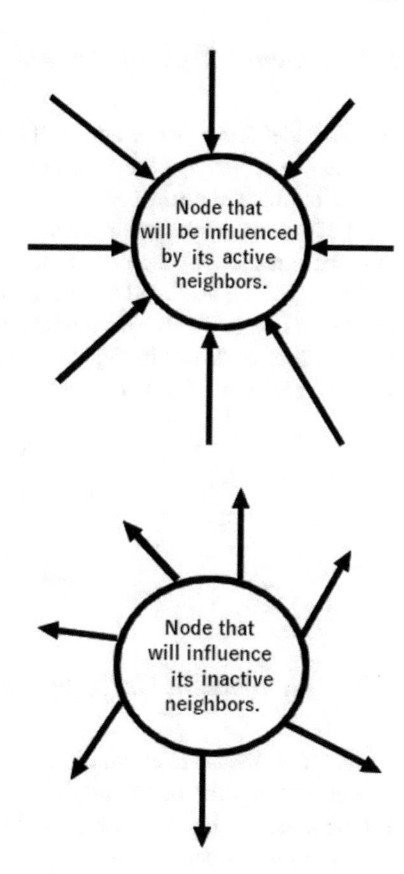

Fig. 7 LT model: node is activated by its active neighbors

Fig. 8 IC model: node activates its inactive neighbors

5 Propagation Model

In the process of influence maximization, after the selection of the seeds, it is very important to have a spread model. Several works have been done. Among them, we have two benchmark heuristics: Linear Threshold Model (*LTM*) proposed by Kempe et al. [4] and the Independent Cascade Model (*ICM*) proposed by Kempe et al. [4], Lopez-Pintado [18]. But other studies have been done and they are an evolution of the two benchmark heuristics cited above. For example, in [19], they propose a generalization of ICM. The probability of a node u is modified according to the influenced nodes. In the *E*xtended *I*ndependent *C*ascade (*EIC*) model proposed by Jingyu et al. [20], the activation probability is not uniform for all iterations. There is also the *C*ontinuous-*T*ime *M*arkov *C*hain into the *I*ndependent *C*ascade Model ($CTMC - ICM$) which is also an evolution of the ICM proposed by Zhu et al. [21].

5.1 Linear Threshold Model

In this model, a non-active node u at time t can be activated by these active neighbors at time $t + 1$. Let $p_{v,u}$ the social pressure of the node v on the node u. Let θ_u the activation threshold (the social resistance) of the randomly chosen node u between [0,1]. If the sum of the influence factor of all active neighbors of the node u is bigger than the threshold of the node u and smaller than 1, then the node u begins active and forever. The activated node can also participate in the activation of its inactive neighbors at time $t + 1$. This model gives importance to the node which will receive information, as shown in Fig. 7. Mathematically we can define the *LTM* by Eq. (7) with the condition of Eq. (6).

$$\sum_{v \text{ active and } v \in N(u)} p_{(v,u)} \prec 1 \tag{6}$$

$$\sum_{v \text{ active and } v \in N(u)} p_{(v,u)} \succ \theta_u \tag{7}$$

5.2 Independent Cascade Model

In the *ICM* the node u which influences its neighbors is also called contagious model. Let $p_{u,v}$ the probability that the node u will influence the node v. At time t if the node v becomes active, it can speed up the influence of its neighbors at time $t + 1$. In this propagation model also, an active node u remains activated forever. It is called Cascade Independent Model because it does not handle the historic. The probability is independent of time. This model gives importance to the node which will spread information, as shown in Fig. 8.

6 Experiment and Results

To show the pertinence to prevent information feedback toward the seed before their determination, we follow the following steps:

- first, we determine a spanning graph from initial graph by using one of these proposed algorithms,
- second, we use the same heuristic (like *PageRank*) to determine the seeds in the two graphs (initial graph and spanning graph),
- finally, we measure the number of influenced nodes by the two seed sets with a spread model.

In our simulation, to select the initial diffusers, we use the following heuristics: *degree* centrality, *discount* degree centrality, and *PageRank* centrality. After selecting the seeds, we use the two propagation model benchmarks: Independent Cascade Model (*ICM*) and Linear Threshold Model (*LTM*).

6.1 Data

The networks *com* – *Amazon* Communities [22][1] and *Eron Email* [23][2] communication networks are used in [13]. While in this extension, we use the networks: *DBLP* [24] collaboration network and *Eron Email* communication network. In this extension of [13], we changed of networks because it does not impact the model. The com-Amazon bibliography information gives a comprehensive list of computer research documents. They build a network of co-authors, where two authors are connected if they publish at least one paper together. The place of publication, such as a newspaper or a conference, defines a community. The authors are considered like the edges and if the author has published with the author, then the edge is present in the graph. This graph has 317,080 authors and 1,049,866 relations. This network is used to simulate the two versions of SCG-algorithm.

The *Eron Email* communication network covers all communication emails. Its data were originally made public, and posted to the web, by the Federal Energy Regulatory Commission during its investigation. The nodes of the network are email addresses. Let two email users α and β, if α sends a message to β, we add a directed edge from α to β. We have a digraph that contains Emails and their relationships. This graph contains 36,692 email users and 183,831 communications. This network is used to simulate the *SG*-algorithm.

The *com* – *Amazon* bibliography information gives a comprehensive list of computer research documents. They build a network of co-authors, where two authors are connected if they publish at least one paper together. Place of publication, such as a newspaper or a conference, defines a community. The authors are considered the edges and if the author α is published with the author β, then the edge α, β is present in the graph. This graph has 317,080 authors and 1,049,866 relations. This network is used to simulate the two versions of *SCG*-algorithm.

6.2 Benchmark Heuristics and Parameters

For simulation parameters, we use the following heuristics to determine the seeds: *degree* centrality, *discount degree* centrality, and *PageRank* centrality. These heuristics are applied in the spanning graphs given by these algorithms and the initial graph to determine the seeds for each one. In our simulation, the activation probability of each node is chosen randomly between [0.01 and 0.05]. The parameter *p* of *discount degree* heuristic is fixed at 0.01. The damping factor *d* of *PageRank* heuristic is fixed at 0.85. We use the *IC* and *LT* benchmark propagation models to spread the information. The activation threshold of *LT* model is fixed at 0.04 for each node.

[1]http://snap.stanford.edu/data/com-Amazon.html.

[2]http://snap.stanford.edu/data/email-Enron.html.

6.3 Results

To calculate the influence propagation, we denote the following sets:

- s_0^k: set of k seeds obtained from the initial graph,
- s_1^k: set of k seeds obtained from the graph given by SCG_{v1}-algorithm,
- s_2^k: set of k seeds obtained from the graph given by SCG_{v2}-algorithm,
- s_3^k: set of k seeds obtained from the graph given by SG-algorithm.

In Figs. 9 and 10, we simulate SCG_{v1}-algorithm and SCG_{v2}-algorithm. We show the number of nodes influenced by S_0^k, S_1^k, and S_2^k in relation to the k values that vary between 5 and 30 by step of 5. These seed sets are determined by using the degree centrality heuristic. The iteration number is fixed at 4. The *ICM* is used in Fig. 9 and the *LTM* in Fig. 10.

In Figs. 11 and 12, we simulate the SCG_{v1}-algorithm and SCG_{v2}-algorithm. We show the number of nodes influenced by S_0^k, S_1^k, and S_2^k in relation to the k values that vary between 5 and 30 by step of 5. These seed sets are determined by using the *degree discount* heuristic. The iteration number is fixed at 4. The *ICM* is used in Fig. 11 and the *LTM* in Fig. 12.

In Figs. 13 and 14, we simulate the SCG_{v1}-algorithm and SCG_{v2}-algorithm. We show the number of nodes influenced by S_0^k, S_1^k, and S_2^k in relation to the k values that vary between 5 and 30 by step of 5. These seed sets are determined by using the *PageRank* heuristic. The iteration number is fixed at 4. The *ICM* is used in Fig. 13 and the *LTM* in Fig. 14.

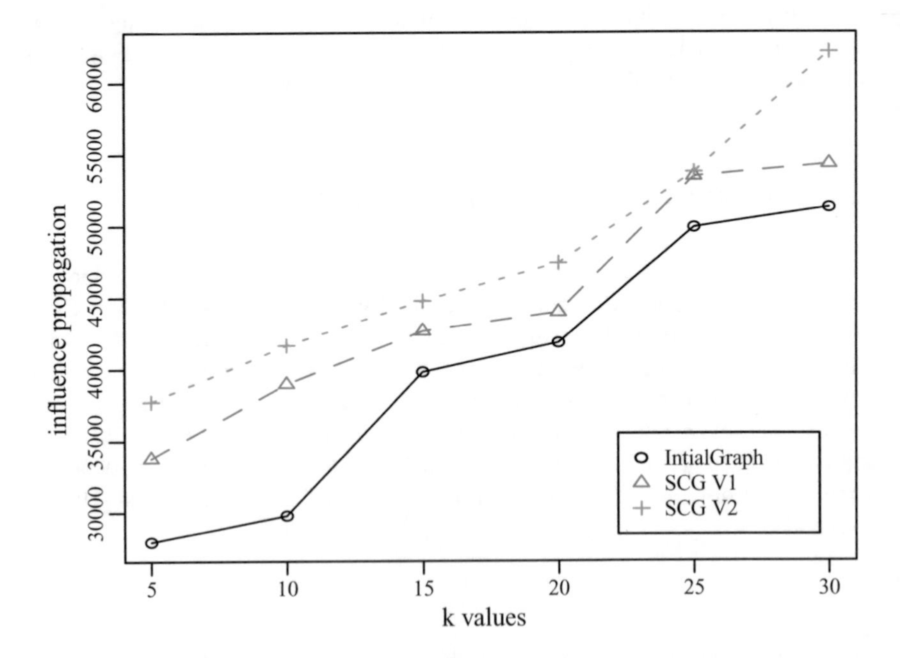

Fig. 9 Influence propagation with centrality degree heuristic under *ICM*

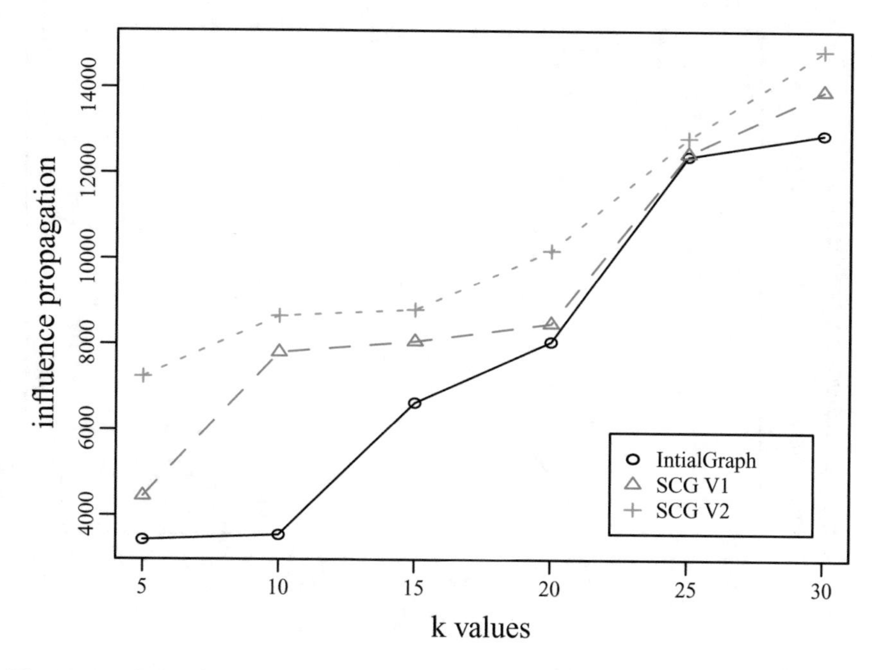

Fig. 10 Influence propagation with centrality degree heuristic under *LTM*

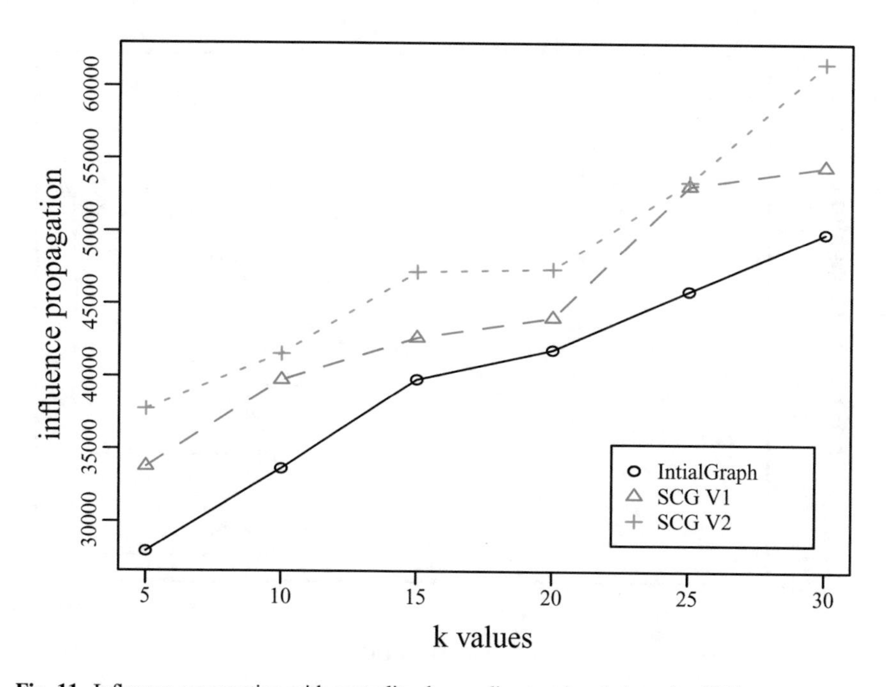

Fig. 11 Influence propagation with centrality degree discount heuristic under *ICM*

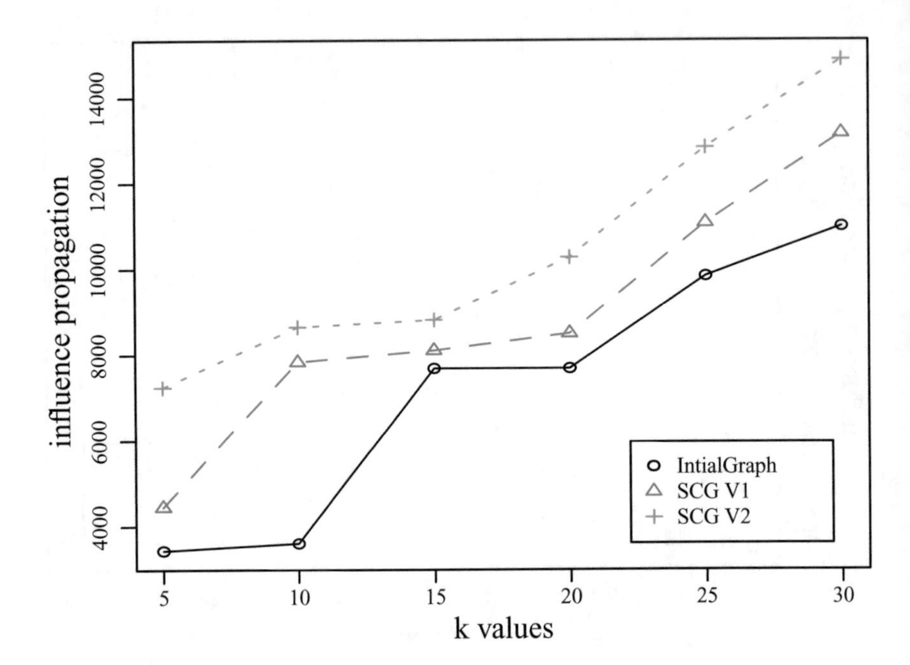

Fig. 12 Influence propagation with centrality degree discount heuristic under *LTM*

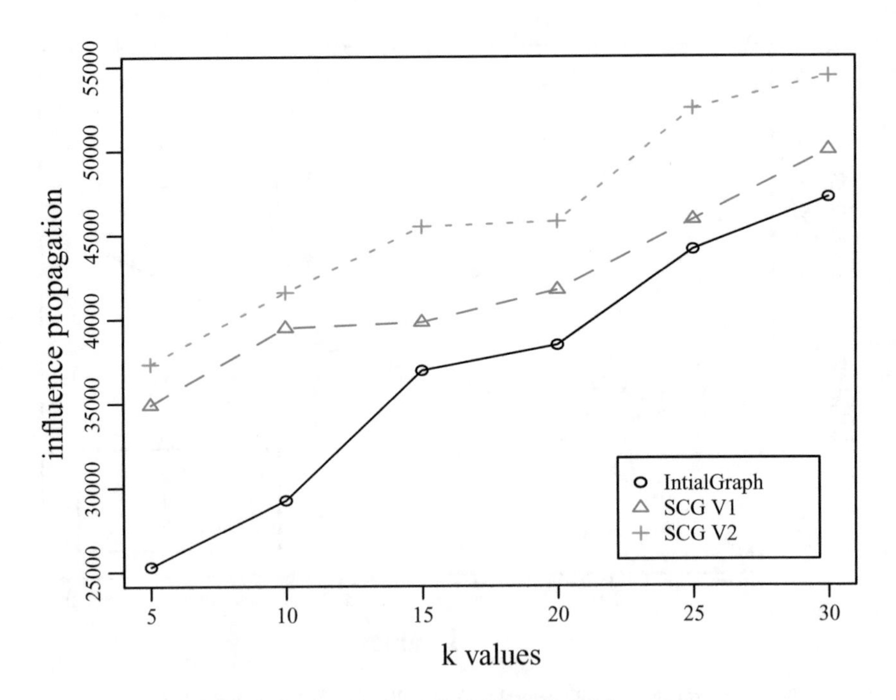

Fig. 13 Influence propagation with PageRank heuristic under *ICM*

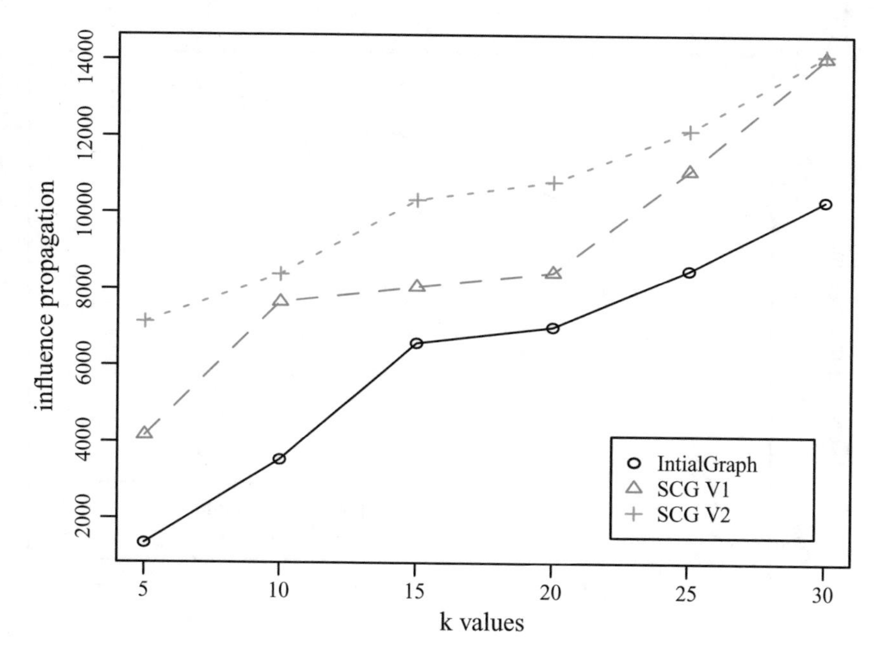

Fig. 14 Influence propagation with PageRank heuristic under *LTM*

In Figs. 15 and 16, we simulate the *SG*-algorithm. We show the number of nodes influenced by S_0^k and S_3^k in relation to the k values that vary between 1 and 6 by step of 1. The *ICM* is used as propagation model and the iteration number is fixed at 4. In Fig. 15, the *degree* heuristic is used to determine the seeds and in Fig. 16 the *degree discount* heuristic. The difference is not significant for this algorithm because the selected graph is not large. However, our approach gives the best results if the information feedback are not prevented.

In the process of maximization influence, the results of our simulation show that if we prevent the information feedback toward the seeds before to their selection, the results will be better than if they are ignored under the cascade models and the linear threshold models.

7 Conclusion

In this paper that is an extension of a previous work [13], we proposed an approach to determine the seeds that will start the information diffusion maximization process. Our approach prevents the information feedback toward the nodes that would be already considered as seeds. We proposed three algorithms: the two first algorithms use a connected graph and the third one is a generalization of the first them. It uses as input an arbitrary graph which is not necessary connected. We

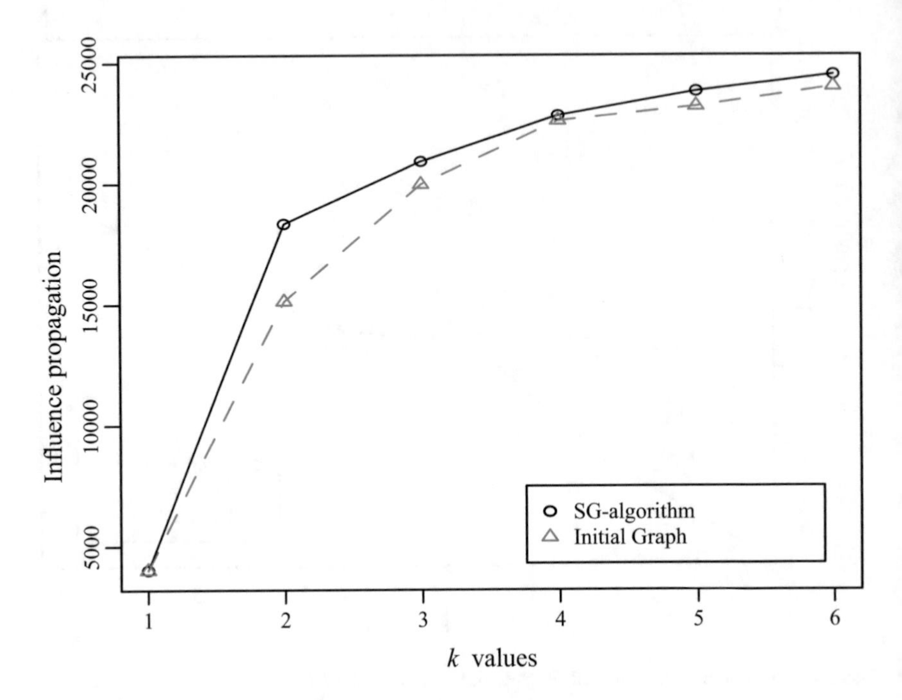

Fig. 15 Influence propagation with degree centrality heuristic under *ICM*

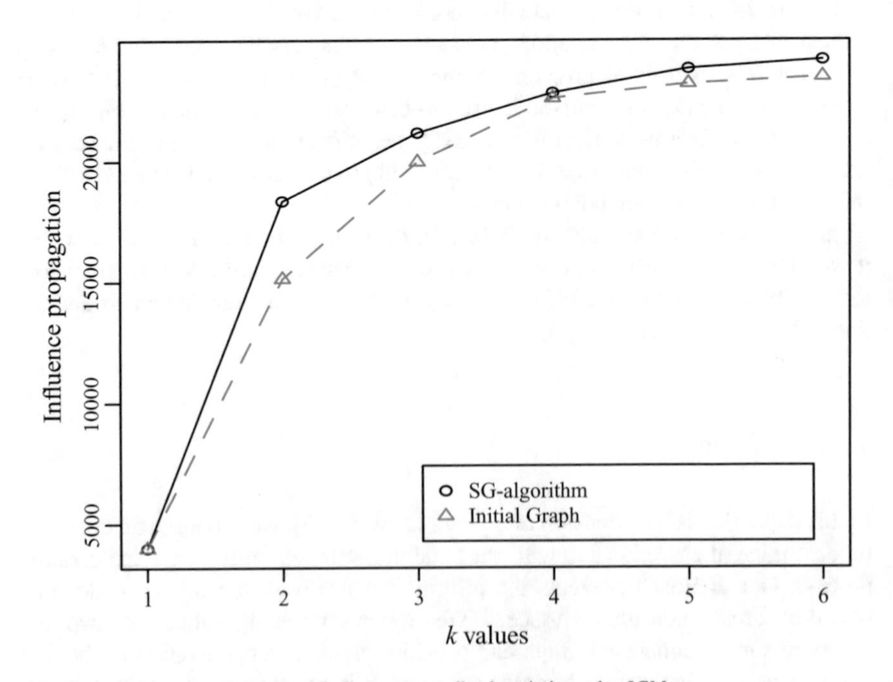

Fig. 16 Influence propagation with degree centrality heuristic under *ICM*

presented their motivations, and we discussed of the difference from the classical existing approaches. Finally to show the performance of our approach we used the *R* software and the *igraph* package for our simulation.

References

1. Domingos, P., Richardson, M.: Minimizing the expected complete influence time of a social network. In: 7-th ACM SIGKDD International Conference on Knowledge Discovery and Data Mining (2001)
2. Tsiporkova, E., Boeva, V.: Multi-step ranking of alternatives in a multi-criteria and multi-expert decision making environment. Inf. Sci. **176**(18), 2673–2697 (2006)
3. Domingos, P., Richardson, M.: Mining the network value of customers. In: 7-th ACM SIGKDD International Conference on Knowledge Discovery and Data Mining (2001)
4. Kempe, D., Kleinberg, J., Tardos, E.: Maximizing the spread of influence through a social network. In: Proceedings of the 9-th ACM SIGKDD International Conference on Knowledge Discovery and Data Mining (2003)
5. Freeman, L.C.: Centrality in social networks conceptual clarification. Soc. Netw. **1**(3), 215–239 (1979); Elsevier
6. Borgatti, S.P., Everett, M.G.: A graph-theoretic: perspective on centrality. Soc. Netw. **28**(4), 466–484 (2006); Elsevier
7. Brin, S.: Page: the anatomy of a large-scale hypertextual web search engine. Comput. Netw. ISDN Syst. **30**(1), 107–117 (1998)
8. Gaye, I., Mendy, G., Ouya, S., Seck, D: New centrality measure in Social Networks based on Independent Cascade (IC) model. In: 3rd International Conference on Future Internet of Things and Cloud, Rome, pp. 675–680 (2015)
9. Chen, W., Wang, Y., Yang, S.: Efficient influence maximization in social networks. In: Proceedings of the 15-th ACM SIGKDD International Conference on Knowledge Discovery and Data Mining, pp. 199–208 (2009)
10. Wei, C., Yajun, W., Siyu, Y.: Scalable influence maximization for independent cascade model in large-scale social networks. In: 16-th ACM SIGKDD International Conference on Knowledge Discovery and Data Mining (2012)
11. Chen, W., Wang, Y., Siyu, Y.: Scalable influence maximization in social networks under linear threshold model. In: 10-th IEEE Conference on Data Mining (2010)
12. Suman, K., Murthy, C.A., Pal, S.K.: New Centrality Measure for Influence Maximization in Social Networks. PReMI, Lecture Notes in Computer Science, vol. 6744, pp. 242–247. Springer, Berlin/Heidelberg (2011)
13. Gaye, I., Mendy, G., Ouya, S., Seck, D.: Spanning graph for maximizing the influence spread in Social Networks. In: Proceedings of the IEEE/ACM International Conference on Advances in Social Networks Analysis and Mining, Paris, pp. 1389–1394 (2015)
14. Enrico, B., Massimo, F.: Resistance distance, closeness, and betweenness. Soc. Netw. **35**(3) (2013); Elsevier
15. Klaus, W., Artur, Z.: DACCER: Distributed Assessment of the Closeness Centrality Ranking in complex networks. Soc. Netw. 43–48 (2013); Elsevier
16. Tore, O., Filip, A., John, S.: Node centrality in weighted networks: generalizing degree and shortest paths. Soc. Netw. **32**(3) (2010); Elsevier
17. Everetta, M.G., Borgattib, S.P.: Induced, endogenous and exogenous centrality. Soc. Netw. **32**(4) (2010); Elsevier
18. López-Pintado, D.: Diffusion in complex social networks. Games Econ. Behav. **62**(2), 573–590 (2008)

19. Missaoui, R., Sarr, I.: Social Network Analysis Community Detection and Evolution, pp. 243–267. Springer International Publishing, Switzerland (2014)
20. Jingyu, Z., Yunlong, Z., Cheng, J.: Preference-based mining of top-K influential nodes in social networks. Futur. Gener. Comput. Syst. **31**, 40–47 (2014); Elsevier
21. Zhu, T., Wang, B., Wub, B., Zhu, C.: Maximizing the spread of influence ranking in social networks. Inf. Sci. **278**, 535–544 (2014); Elsevier
22. Yang, J., Leskovec, J.: Defining and evaluating network communities based on ground-truth. International Conference on Data Mining (2012)
23. Leskovec, J., Lang, K., Dasgupta, A., Mahoney, M.: Community structure in large networks: natural cluster sizes and the absence of large well-defined clusters. Internet Math. **6**(1), 29–123 (2009)
24. Jure, L., Andrej, K.: SNAP datasets: stanford large network dataset collection. http://snap.stanford.edu/data (2014)

Energy Efficiency Analysis of the Very Fast Decision Tree Algorithm

Eva Garcia-Martin, Niklas Lavesson, and Håkan Grahn

1 Introduction

Data stream mining is gaining importance with the evolution of hardware, sensor systems, and technology. The rate at which data is generated is increasing day by day, challenging storage and computational efficiency [24]. Digital Universe Study [13] has predicted that by 2020, 40,000 exabytes of data will be processed, most of them originating from devices that automatically generate data. Many algorithms in data stream mining are designed to process fast and potentially infinite streams
[15, 35].

Traditionally, the machine learning community has considered accuracy as the main factor when building algorithms. With the appearance of big data analytics and data stream mining, scalability has also been a key factor to consider. In this context, scalability stands for how fast an algorithm can process the incoming data. The problem that we address in this study is the fact that few researchers in the data mining community consider energy consumption as an important measure.

It has been shown that energy consumption can be reduced in every layer of the Open Systems Interconnection (OSI) model [49, 56]. Hardware solutions to reduce energy consumption have been focused on, e.g., using the Dynamic Voltage Frequency Scaling (DVFS) technique and on parallel computing [41, 45]. During recent years, the interest in developing energy efficient software solutions has increased significantly, leading to a creation of applications to measure energy consumption in software [44].

This paper introduces energy consumption and energy efficiency as important factors to consider during data mining algorithm analysis and evaluation,

E. Garcia-Martin (✉) • N. Lavesson • H. Grahn
Blekinge Institute of Technology, SE-371 79 Karlskrona, Sweden
e-mail: eva.garcia.martin@bth.se; niklas.lavesson@bth.se; hakan.grahn@bth.se

R. Missaoui et al. (eds.), *Trends in Social Network Analysis*, Lecture Notes in Social Networks, DOI 10.1007/978-3-319-53420-6_10

and to demonstrate the use of these factors in a data stream mining context. The consideration of energy efficiency can help companies and researchers move towards green computing [33] while improving the business profits.

Social networks is a very good example of a domain in need of efficient data processing and analysis of algorithms. Companies such as Facebook, Twitter, and Instagram rely on large data clusters consuming vast amounts of energy. Facebook, for example, generates four million posts every minute [34], which creates interesting challenges for algorithms that are running continuously in their network. One of these challenges is to optimize such algorithms in terms of energy consumption. Even a small improvement will reduce energy on a large scale, due to the nature of the network.

We conducted an experiment to illustrate a possible scenario where energy consumption is relevant to study. More specifically, we studied how energy and accuracy are affected by changing the parameters of the VFDT (Very Fast Decision Tree) algorithm [15]. We make a comparison between the theoretical analysis on the algorithm and the experimental results, which indicate that it is possible to significantly reduce the energy consumption of the VFDT algorithm while maintaining similar levels of accuracy. The main contribution of this paper is the introduction of energy consumption as a key factor to consider in data mining algorithms. This is supported by a theoretical and empirical analysis that illustrate an example on how to build sustainable and efficient algorithms and the reasons behind energy consumption.

This paper is an extension of the paper titled: *Energy Efficiency in Data Stream Mining* [42]. While such publication centers on observing the energy consumption only from an empirical perspective, this study motivates the experimental setup and results by analyzing the behavior of the VFDT from a theoretical and empirical perspective. Therefore, we can compare how the algorithm behaves in reality from what we predicted theoretically. On top of that, more relevant parameters have been chosen and two real world datasets have been added to the experiment.

2 Background

In this section we first explain the importance of energy consumption in data mining. Then, we briefly explain data stream mining and why it is different from standard data mining, and finally we introduce some terminology related to power, energy, energy efficiency, and computational efficiency.

2.1 Energy Awareness

The demand for energy is increasing day by day [49]. World leaders and scientists focus on finding a solution towards this problem, centering on two key factors: developing new sources of clean energy and decreasing energy usage [1, 11], which would lead to a reduction in CO_2 emissions. The main reason why researchers and every citizen should be aware of energy consumption is because energy pollutes. Every device that we use in a daily bases that consumes energy produces CO_2. Nowadays, based on a study conducted by the World Health Organization, air pollution kills more people than malaria and AIDS combined [43]. This argument is based on what is known as ecological or environmental footprint [16], that measures how much impact a certain person or action has in relation to the environment. For instance, carbon footprint measures how many greenhouse gases are produced by an individual or event, expressed as CO_2 [10]. Therefore, if companies and individuals are aware of the footprint of their computations, their impact could be reduced by making them energy efficient.

There have been studies that measure the environmental impact of queries in search engines [50]. Considering that there are approximately 66 k Google queries per second [29], reducing the CO_2 emissions of search queries will significantly impact the environment. If we translate this example to data stream mining, we can picture the execution of data stream mining algorithms in servers running during 24 h a day, for a complete year. In this case, building energy-aware algorithms has the following consequences:

- Reduction of CO_2 emissions to the atmosphere.
- Reduction of air pollution, therefore reducing the number of deaths per year due to this matter.
- Reduction of the money spent on energy.
- Increase of the battery life of mobile devices and sensor networks, if the algorithm is implemented in such contexts.

2.2 *Data Stream Mining*

Data stream mining is the process of building models by exploring and extracting patterns from a stream of data.

The core assumption of data stream mining, in comparison to data mining, is that the examples are inspected only once, so we have to assume that if they are not processed immediately they are lost forever [7, 48]. Moreover, it is considered that the data arrives online, with no predefined order, at a high-speed and with time-changing characteristics. Data stream mining algorithms should be able to process potentially infinite streams while updating the model incrementally [25, 35].

2.3 *Terminology*

In this section we clarify several concepts related to energy, power, and efficiency. Energy is a measurement of the amount of *fuel* used for a specific application. It is measured in joules (J) or kWh. Power is a measurement of the rate at which energy is consumed. It is measured in joules/second, which is equal to watts (W). The following is an example that illustrates the relationship between power and energy: A process is running for 3.94 s consuming an estimate power of 1.81 W. The total energy consumed is: $3.94 \times 1.8 = 7.092\,\mathrm{J} = \mathrm{Ws} = 1.99 \times 10^{-3}$ Wh.

Energy efficiency has a specific definition at Green500 [30], being, *The amount of operations per watt a computer can perform.* This definition is related to hardware. In this study, whenever we mention energy efficiency we refer to reducing the energy consumption of some process or algorithm.

In theoretical machine learning, researchers introduced the computational learning theory [37], where they analyze the computational complexity of algorithms. They approach computational efficiency as a way of designing less computationally complex algorithms that can run in polynomial time.

3 Related Work

In this section we first review literature related to energy awareness in software and hardware. Then, we examine relevant work in the data stream mining field, focusing on the VFDT algorithm. Finally, we review papers that are related to both energy consumption and data stream mining.

Research in energy awareness at the software level started many years ago, when researchers began to realize the importance of the energy consumed by a software application. In 1994, the first systematic attempt to model the power consumption of the software components of a system was presented [53]. After that, in 1999, PowerScope was presented [21], a software tool for profiling the energy usage of applications. The novelty of this approach is that energy consumption can be mapped to program structure to analyze which procedures consume more energy. Companies such as Microsoft [18] and Intel [19] have invested in developing software tools to help developers reduce the energy consumption of their applications. During the past years, the Spiral research group [52] has gained interest in building energy efficient software. They show that energy consumption depends not only on the time and number of computations, but also on the stress of the processor, the I/O operations, and many other factors. They have developed a software tool, PowerAPI, where they show that they can get high accurate modeling of the power consumption of software applications. They have evaluated their model by comparing their results and method with hardware power meters obtaining promising results [9, 44].

In relation to energy efficiency at the hardware level, one of the most important techniques, implemented in most contemporary processors, is Dynamic Voltage Frequency Scaling (DVFS). DVFS is a power saving technique used and improved by many researchers. One improvement is Real Time DVFS, an implementation of DVFS for real time systems [45]. Another area that is gaining importance nowadays is parallel computing, where there are relevant energy savings by employing more cores on a processor [41]. Several energy-saving approaches, such as *Cost optimization for power-aware computing*, have been developed in the past years [49].

In relation to data stream mining, researchers have developed efficient approaches to mine data streams, as outlined below. There have been several reviews conducted in data stream mining since 2005. Two general reviews [3, 24], portray techniques and concepts such as data-based techniques, task-based techniques, data stream classification and frequent pattern mining. More specific reviews center on topics such as sensor networks [22] and knowledge discovery [25].

From the reviews explained above, we have extracted six main techniques and approaches in data stream mining: data stream clustering [31], data stream classification [15], frequent pattern mining [5], change detection in data streams [2, 38], sliding window techniques [12], and stream mining in sensor networks [22, 28]. We have decided to focus on data stream classification and change detection in data streams.

Concept drift refers to a change between the input data and the target variable on an online supervised learning scenario. The first framework that dealt with concept drift was proposed to also address efficiency and robustness [55]. Nowadays, researchers consider concept drift as an important aspect when building algorithms for other specific purposes. A survey on different methods that address concept drift has been conducted in 2014 [27].

Classification is considered a challenging problem in a data stream mining scenario[3]. The main reason is that many of the traditional classification techniques and algorithms were designed to build models from static data.

One of the key breakthroughs in supervised online learning was made with the development of the Hoeffding Tree algorithm and the Very Fast Decision Tree (VFDT) learner [15]. In contrast to previous algorithms, such as SPRINT [51] and ID5R [54], this new approach was able to deal with potential infinite streams, arriving at a fast pace and with low computational cost. The VFDT learner is able to process examples at a high rate in constant time. One year later, the same authors created a new version of the VFDT algorithm, CVFDT, that was able to adapt to concept drift [35]. Another extension on the VFDT algorithm appeared 2 years later, with a new decision tree learner that could efficiently process numerical attributes [36]. In the same line, a decision tree algorithm was created for spatial data streams [14]. We would like to mention relevant methods that address different classification problems, namely: On-Demand classification [2, 4], Online Information Network (OLIN) [39], LWClass [23], ANNCAD [40], and SCALLOP [20].

In relation to energy awareness in data stream mining, several researchers have conducted studies where they emphasize the importance of energy consumption [6, 24, 26]. While the first two are concerned on energy savings for sensor networks, the second one centers on examine the energy consumption of different data analysis techniques. To the best of our knowledge, the last work is the one most related to ours.

We can observe that there is no specific research on making energy consumption a key factor on data stream mining, since the research has been centered towards specific applications or hardware modifications. We would like to change this approach by proposing energy consumption as the new factor to consider when building, optimizing, or creating new algorithms in data stream mining. We believe that this is the next natural step to take, since other researchers in similar fields, hardware and software, have already taken that step.

4 Theoretical Analysis

This section aims to theoretically analyze the behavior of the Very Fast Decision Tree (VFDT) algorithm [15]. VFDT is an online decision tree algorithm able to build a decision tree from a stream of data by analyzing the data sequentially and only once.

The decision tree is built sequentially, where the tree waits until it gathers enough examples or instances from the stream. After those n instances arrive, the algorithm analyzes them and obtains the best attribute to split the tree on. The key feature is to obtain the optimal value of n that will split in the same attribute as if we had all examples available to analyze. To obtain the first best value of n, the authors make use of the Hoeffding Bound [32], represented by ϵ in Eq. (1).

$$\epsilon = \sqrt{\frac{R^2 \ln(1/\delta)}{2n}} \tag{1}$$

This bound states that with probability $1-\delta$, the chosen attribute at a specific node after seeing n number of examples will be the same attribute as if the algorithm had seen infinite number of examples. Therefore, δ represents one minus the probability of choosing the correct attribute to split on. The reason is that there will be no split on a certain attribute unless $\Delta\overline{G} > \epsilon$. $\Delta\overline{G}$ stands for the difference in information gain between the two best attributes. Thus, if the number of examples n is small, ϵ will be high, making it harder to split on an attribute unless $\Delta\overline{G}$ is big enough, meaning that there is a clear attribute that is the winner. Based on the equation, whenever we see more examples, n increases, making ϵ smaller and then making it easier to split on the top attribute. The reasoning behind this is that whenever we see more examples we are more confident on the split. In order to speed up the computations, some parameters are introduced, that will make the algorithm behave in a slightly different way that the one currently explained.

The next paragraphs theoretically analyze how accuracy and energy would differ when varying the different parameters of the VFDT algorithm. The chosen parameters to be varied are: *nmin*, τ, δ, *memory limits*, *memory management*, *split criterion*, and *poor attributes removal*. As a general observation, the theoretical analysis made about the parameters cannot be generalized to all cases, since it would vary depending on the input data. For that reason, the assumptions that are made in the following paragraphs are based on the reasoning and experiments from the original paper by the authors.

nmin parameter is the minimum number of examples that the algorithm must see before calculating ϵ to check if there are sufficient statistics for a good split.

The authors introduce this parameter to reduce execution time and computational effort when building the tree, since it is very unlikely that after just one instance the algorithm has a more convincing split. The default value of *nmin* is 200. If the value of *nmin* increases, then accuracy will be slightly reduced, since the tree will have a lower number of nodes. From the original paper, the difference in accuracy was of a merely 1.1%, but the execution time was 3.8 times faster when using and increasing *nmin*. Therefore, we predict that the energy would decrease when increasing *nmin*, since we would be decreasing the computational effort and time to analyze each instance.

τ parameter represents the tie breaking parameter. Whenever the difference between both attributes is small enough, that means both attributes are equally good, making no sense to wait a longer time for more examples to make a split. The absence of this parameter has been shown to decrease accuracy, since the decision tree contains fewer nodes in it. However, being able to make more splits on the data allows to obtain a finer grained decision tree. In an extreme situation where both attributes are exactly the same, the tree would stall, failing to grow. So increasing τ could, in an ideal scenario with an infinite stream of data, increase accuracy and decrease energy, since the tree is built before, reducing the time and the number of computations [7]. But if we analyze the same amount of examples, then increasing τ could increase the energy, due to the fact that with lower τ we make less number of computations.

δ parameter represents one minus the probability of choosing the correct attribute to split on. If δ increases, then the desired probability is smaller. Hence, the tree will grow faster, having more nodes. Since the difference on nodes between a higher and a lower δ will not be as high as when removing τ, and the probability of a correct split is lower, our assumption is that the accuracy will be lower when δ increases. At the same time, if δ decreases, then the probability of making a correct split increases, increasing accuracy. In terms of energy, we believe that the power and time consumed to build the tree will vary depending on the incoming data. We hypothesize that with a lower δ there will be more power spent on computing information gain rather than in building nodes. However, we currently do not have the knowledge of which consumes more power. At the same time, the tree could be built faster since we are spending less time on building the tree nodes.

In terms of *memory limits* and *memory management*, we predict that with a lower memory limit, the tree will induce fewer nodes, having less energy consumption and less accuracy. At the same time, the tree uses pruning techniques to reduce the memory spent on building the tree, removing less promising leaves, which in some cases could improve accuracy. It will depend on how different is the memory limit and how many nodes differ from each setting to correctly predict if the accuracy will be lower or higher. In a realistic case we assume that limiting the memory consumption and the tree growth will decrease accuracy and energy.

The default split criterion used by the authors in this algorithm is *information gain*. We have tested also *Gini index*. From a theoretical aspect, it has been shown [47] that it is not conclusive which of the two criterion will perform a better job in general, since they both have shown similar results, differing only by 2%.

Table 1 Summary of the theoretical behavior of the parameters

Parameter	Modification	ACC	Energy	Nodes
nmin	Increase	Lower	Lower	Fewer
	Decrease	Higher	Higher	More
τ	Increase	Higher	Higher	More
	Decrease	Lower	Lower	Fewer
δ	Increase	Lower	[a]	More
	Decrease	Higher	[a]	Fewer
MEM1	100 KB	Lower	Lower	Fewer
	2 GB	Higher	Higher	More
MEM2	ON	Lower	Lower	Fewer
SPLT CRIT	S2	[a]	[a]	[a]
RPA	ON	Higher	Lower	Fewer

[a]The variation depends on the input data

The last parameter that is going to be modified is *removing poor attributes*. This parameter aims to analyze attribute performance to find attributes that perform poorly and which are very unlikely to be chosen for a further split. The process that the authors follow is by analyzing the information gain of all attributes in every split, and when this value, for a specific attribute, is less than the information gain of the best attribute by more than a difference of ϵ, then the attribute is ignored for that leaf. In theory, this method should increase accuracy and decrease the amount of computations.

Table 1 represents a summary of all the predictions of the parameters variations explained above. It shows how energy, accuracy, and tree size will vary when increasing or decreasing the mentioned parameters.

5 Experimental Design

5.1 Problem Definition

In order to empirically study the different parameters of the VFDT algorithm we have created an experiment where we vary the parameters theoretically analyzed in Sect. 4. This experiment aims to evaluate the performance of the algorithm under different setups in terms of accuracy, energy, execution time, and size of the tree. An implicit goal is understanding why varying the parameters in a certain way increases or decreases accuracy and energy, and if it matches with the theoretical reasoning. The experiment has three phases. First, we obtain the datasets, then we input them into the algorithm under different setups, and we finally evaluate the performance of each model in terms of accuracy and energy consumption. The way to measure energy is explained in Sect. 5.3.

5.2 Data Gathering

We have gathered four different datasets to perform this experiment. Two datasets are synthetically generated and the other two are real world datasets. The main difference between synthetic and real world datasets is that the first ones are randomly generated based on a specific function and distribution, and the second ones are representations of some measure that exists in reality. The idea is to show that the solution proposed in this paper of analyzing energy consumption of algorithm applies to both real world and synthetic datasets. It improves the generalizability of the results.

The synthetic datasets have been generated with MOA (Massive Online Analysis) [8], and the functions: *Random Tree Generator*, *Hyperplane generator*. The random tree function generates a tree as explained by the authors of the VFDT algorithm [15]. We have chosen this dataset because it is the same dataset that the authors of the VFDT use in their experiments, so we consider it as a baseline of a standard behavior of the algorithm. Then we chose a more challenging synthetic dataset, since the hyperplane generator is often used to test algorithms that can handle concept drift, such as CVFDT. Even though this algorithm is not developed to handle concept drift, we wanted to test the different setups in a completely different dataset than the random tree. The hyperplane generator uses a function to generate data in the form of a plane in d dimensions [35]. The orientation of the hyperplane can easily be varied by adjusting its weights, creating different concepts. Depending on the coordinates of the plane, the examples are labeled as negative or positive. All synthetic generators have generated a total of one million instances, and we have chosen the number of numerical and nominal attributes based on the default settings in MOA. The tree-generated dataset is a binary classification dataset with 5 nominal and 5 numerical attributes. The hyperplane-generated dataset is also a binary classification dataset with 10 different attributes and 2 concept drift attributes.

Since usually synthetic datasets do not have the same properties as real world datasets, we have decided to add two real world datasets to the experiment. The first one represents instances that try to predict good poker hands based on a given hand [46]. There are a total of 1,025,010 instances and 11 attributes and have been normalized by the MOA researchers. The second real world dataset is the normalized airline dataset, created by Elena Ikonomovska [17]. This classification dataset classifies flights into delayed or not depending on the route of the flight, based on the departure and arrival airports. It contains a total of 8 attributes, being: Airline, flight number, origin, destination, day of the week, elapsed time, duration of the flight, and if there was a delay or not. The motivation behind these datasets is the amount of instances available, making them perfect candidates to test data stream mining algorithms. Table 2 summarizes the information from the different datasets, regarding the number of instances, attributes, and type of the dataset.

Table 2 Datasets summary

Dataset	Name	Type	Instances	Nominal attributes	Numeric attributes
1	Random tree	Synthetic	1,000,000	5	5
2	Hyperplane	Synthetic	1,000,000	1	9
3	Poker	Real world	1,025,010	6	5
4	Airlines	Real world	539,383	5	3

5.3 Methodology

This section explains the settings of the experiment, the parameters varied, and the tools used to perform it.

5.3.1 Parameter Choice

We have chosen to vary the parameters explained in Sect. 4, namely: *nmin*, τ, δ, *memory limits*, *memory management*, *split criterion*, and *removing poor attributes*. *nmin* was varied from the default value, 200, to a maximum value of 1,700 with steps of 500. τ was varied from the default value, 0.05, to a maximum value of 0.13, a minimum value of 0.01 and with steps of 0.04. δ was varied from the default value, 10^{-7}, to a maximum and minimum values of 10^{-1} and 10^{-10}, respectively. The step is of 10^{-3}. Memory limit varied from 100 KB to 30 MB (default value) until 2 GB, that was the maximum allowed by MOA, the tool that will be further explained. The memory management and removing poor attributes were tested by activating and deactivating them. Finally, Gini index was tested against Information Gain. Every parameter was varied while maintaining the other parameters constant, in their default value. The aim is to understand the behavior of each parameter on its own, without having external interference with the rest of the parameters. A summary of the parameters setup is shown in Table 3, where we have represented the varied parameters for each execution in bold.

5.3.2 Procedure

There will be a total of 15 parameter combinations for every dataset, indexed from A-O. Since there are a total of 4 datasets, the number of executions will be 60. Every execution represents the choice of applying one algorithm, with a specific parameter tuning, on one of the datasets. In parallel, we will be measuring how much energy is the execution consuming. Each combination of dataset and parameter tuning has been computed a total of ten times, to then obtain the average and standard deviation of all of them. The average of such computations are the results portrayed in the next section. A summary of these configurations is shown in Table 4. The experiment was carried out in a Linux machine with a 2.70 GHz Intel i7 processor (four cores), and with 8 GB of RAM. The models built from analyzing the synthetic generated

Table 3 Parameter configuration index. Bold parameters represent the parameters varied for each configuration

IDX	*nmin*	τ	δ	MEM1	MEM2	S.CRT	RPA
A	200	0.05	10^{-7}	30 MB	No	S1	No
B	**700**	0.05	10^{-7}	30 MB	No	S1	No
C	**1,200**	0.05	10^{-7}	30 MB	No	S1	No
D	**1,700**	0.05	10^{-7}	30 MB	No	S1	No
E	200	**0.01**	10^{-7}	30 MB	No	S1	No
F	200	**0.09**	10^{-7}	30 MB	No	S1	No
G	200	**0.13**	10^{-7}	30 MB	No	S1	No
H	200	0.05	$\mathbf{10^{-1}}$	30 MB	No	S1	No
I	200	0.05	$\mathbf{10^{-4}}$	30 MB	No	S1	No
J	200	0.05	$\mathbf{10^{-10}}$	30 MB	No	S1	No
K	200	0.05	10^{-7}	**100 KB**	No	S1	No
L	200	0.05	10^{-7}	**2 GB**	No	S1	No
M	200	0.05	10^{-7}	30 MB	**Yes**	S1	No
N	200	0.05	10^{-7}	30 MB	No	**S2**	No
O	200	0.05	10^{-7}	30 MB	No	S1	**Yes**

IDX Different configurations of the VFDT algorithm
MEM1 Memory limits, *MEM2* Memory management, *S.CRT* Split criterion, *S2* Gini index, *RPA* Removing poor attributes

Table 4 Design summary

	Quantity	Type
Datasets	4	Random tree generator, Hyperplane generator, poker, Airlines
Measures	6	Time, power, energy, accuracy, number of nodes, tree depth
Parameter configuration	15	Represented in Table 3, as: A, B, C, D, E, F, G, H, I, J, K, L, M, N, O
Executions	10	Ten executions for every parameter configuration on each dataset

data where trained and tested on one million instances. For the testing phase, new randomly generated data was used. On the other hand, for the real world datasets, the testing was performed on the same data as the training phase.

5.3.3 Tools

We need to differentiate between two tools. The first tool, MOA (Massive Online Analysis), is used to execute the VFDT with the different parameter settings. Running in parallel to MOA is PowerAPI [9], a tool developed by the Spirals Research team [52], which is able to measure how much power different processes are consuming. PowerAPI [9] has been successfully tested by the authors [44] to compute the differences in terms of energy between some software process in different laptops. The energy is calculated by integrating the power consumed from the process during the execution time.

5.4 Evaluation

The last step of the experiment is the evaluation process. In order to evaluate the different settings on the different datasets, we have chosen four measures. The first two are accuracy and energy. We want to discover if there is a trade-off between energy and accuracy, i.e., we will only obtain a lower energy consumption by reducing the accuracy. Or, on the other hand, if there are specific setups where we can reduce energy consumption without loosing accuracy, i.e., smart setups. From the theoretical analysis on the algorithm we have observed that a possible relationship with accuracy is with the number of nodes of the tree. Therefore, the last two evaluation measures considered are the number of nodes (size) and the depth of the tree.

6 Results and Analysis

6.1 General Analyses

Table 5 shows the results obtained from the experiment for each dataset and measuring energy, time, power, number of nodes, depth of the tree, and accuracy. First, we compare energy and accuracy to understand if there is a visible trade-off between the increase of energy and the increase of accuracy. We can observe from Fig. 1 how there seems to be a linear relationship between the increase of energy and the decrease of accuracy. Apparently, based on datasets 1, 3, and 4, whenever the energy increases the accuracy decreases. This result is promising in terms of our ultimate goal, trying to make energy efficient algorithms. Since there is no trade-off between energy and accuracy, a sacrifice of accuracy is not needed to develop energy efficient algorithms. In all datasets there is one outlier, that presents a lower accuracy than the rest, this is the parameter split criterion set to Gini index. This configuration presents a significantly lower levels of accuracy without reducing energy consumption in comparison with the other parameters.

Figure 2 shows how energy and accuracy vary for the different parameter configurations. Although this energy variation between parameters is analyzed in depth in the next subsection, we believe it is relevant to mention that energy differs significantly between specific configurations. For instance, for the third dataset, energy and accuracy vary for every single parameter, suggesting the relevancy of measuring energy consumption and not leaving the development choices to pure chance.

Table 5 Experimental results. Bold values represent lower energy and higher accuracy values for each setup

	1						2					
S	T	P	E	A	N	D	T	P	E	A	N	D
A	4.44	8.59	38.10	96.91	1,134	8	5.85	7.39	43.19	90.93	655	12
B	3.90	7.32	28.57	96.31	661	8	5.65	7.12	40.22	90.99	607	11
C	3.90	7.10	27.71	96.24	570	7	5.66	7.75	43.90	91.22	575	10
D	3.82	6.99	**26.67**	95.91	495	7	5.57	7.77	43.31	90.98	515	11
E	4.46	8.44	37.63	95.57	699	8	5.36	9.15	49.08	90.57	57	7
F	4.74	7.35	34.84	97.93	1,541	9	7.42	6.89	51.10	90.38	2,071	19
G	5.09	7.30	37.11	97.94	2,074	11	9.25	6.87	63.55	89.87	3,863	18
H	5.20	7.55	39.21	**98.27**	2,181	12	9.51	6.78	64.45	89.75	3,971	19
I	4.58	7.90	36.22	97.54	1,403	9	6.32	7.01	44.32	90.82	1,129	12
J	4.29	8.18	35.10	96.35	888	8	5.73	7.58	43.48	**91.24**	471	11
K	3.98	7.37	29.31	95.43	1,134	8	5.03	7.60	**38.20**	84.89	655	12
L	4.40	8.32	36.60	96.91	1,134	8	5.92	7.17	42.48	90.93	655	12
M	4.39	7.95	34.91	96.91	1,134	8	5.91	7.12	42.11	90.93	655	12
N	6.12	6.67	40.79	83.11	1,735	117	5.77	7.60	43.90	90.72	619	11
O	4.34	7.85	34.08	96.91	1,134	8	5.83	7.24	42.23	90.93	655	12
	3						4					
S	T	P	E	A	N	D	T	P	E	A	N	D
A	7.29	7.33	53.48	76.63	297	16	6.54	9.17	59.96	67.31	8,582	4
B	7.03	8.29	58.27	70.67	181	11	6.09	8.41	51.20	67.01	7,984	3
C	7.25	8.57	62.16	79.44	195	16	6.06	8.40	50.91	67.01	8,228	4
D	7.70	8.76	67.45	73.66	149	16	5.99	8.19	**49.01**	66.65	7,895	3
E	8.36	7.49	62.60	73.74	149	13	6.19	8.81	54.50	67.09	5,127	3
F	7.14	7.46	53.26	82.56	791	17	6.94	8.46	58.74	67.77	12,307	3
G	7.46	7.53	56.25	84.88	1,285	22	7.32	8.41	61.56	**67.92**	14,137	3
H	6.87	7.06	**48.49**	**93.06**	1,991	20	7.29	8.43	61.46	67.90	14,001	4
I	7.32	7.36	53.88	78.93	575	18	6.75	8.61	58.14	67.70	10,871	3
J	7.89	8.98	70.89	68.71	119	13	6.41	9.14	58.58	66.97	6,618	3
K	7.26	7.43	53.92	76.63	297	16	6.56	8.96	58.74	67.31	8,582	4
L	7.19	7.57	54.41	76.63	297	16	6.51	8.88	57.82	67.31	8,582	4
M	7.34	7.61	55.85	76.63	297	16	6.60	9.47	62.48	67.31	8,582	4
N	8.99	7.25	65.17	37.83	1,380	102	7.62	9.24	70.37	58.77	1490	53
O	7.18	7.37	52.95	76.63	297	16	6.52	9.06	59.07	67.31	8,582	4

The best accuracy and energy results for each dataset are highlighted

T(s) Time in seconds, *P(W)* Power in watts, *E(J)* Energy in joules, *A(%)* Percentage of correctly classified instances, *N* Nodes, *D* Depth

In this paragraph we analyze the relationship between the number of nodes and accuracy, portrayed in Fig. 3. As has been explained in the theoretical analysis section, some parameter configuration can increase accuracy although intuitively it should be decreased. This is the case of the parameter τ. When τ is increased, intuitively the accuracy should decrease, since you are allowing splits on *not so*

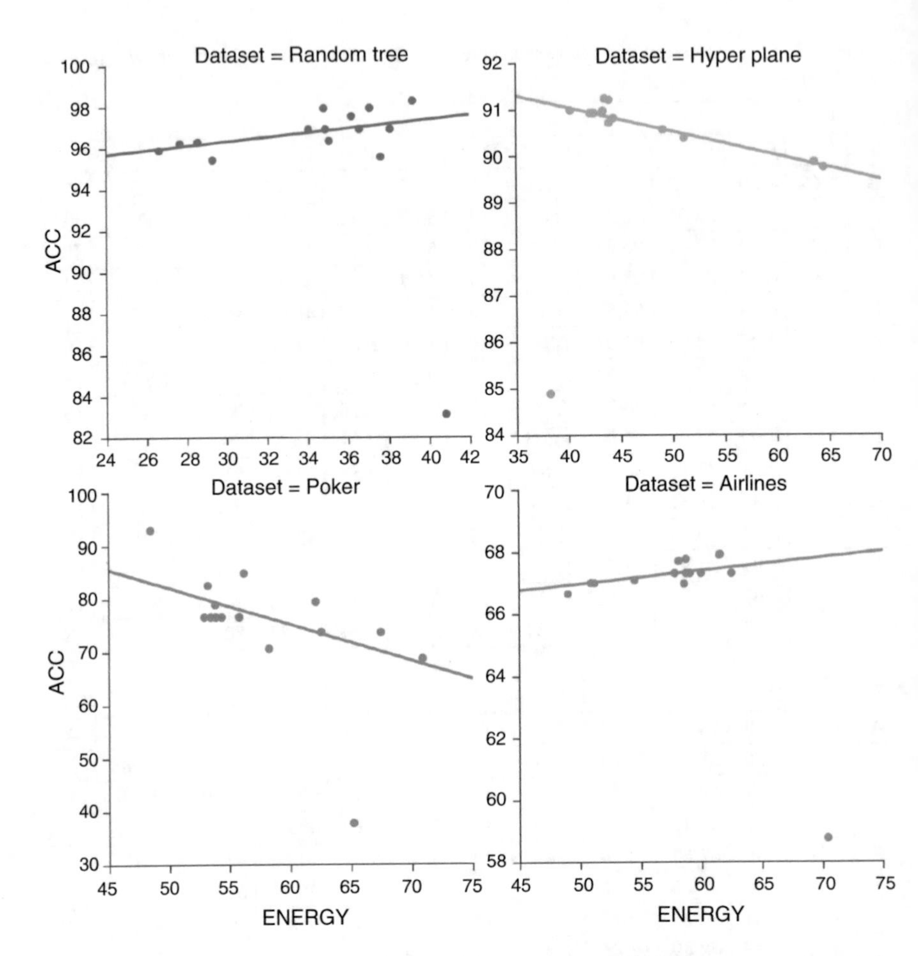

Fig. 1 Scatter plot with a linear relationship between accuracy (ACC) and energy (Joules) for every dataset

good attributes. However, although at first we predicted a decrease in accuracy, since there are significantly more nodes in the tree, the accuracy increases. This can be observed from Tables 5 and 6. The main reason is that with a significant increase of nodes, the tree is able to represent the data in a more fine-grained way, thus increasing accuracy. If we zoom in the highest values for each dataset, we can observe that for datasets 1, 3, and 4, there seems to be a higher accuracy whenever the number of nodes is higher.

A final analysis is related to the number of nodes and the energy consumed. When we look at the data from Table 5, whenever energy increases, time or power increases (since it is the product of both). The interesting measure is to see whether is the power or the time the one causing this energy increase. We have observed from Fig. 4 that whenever the number of nodes of a tree increases, so does the execution

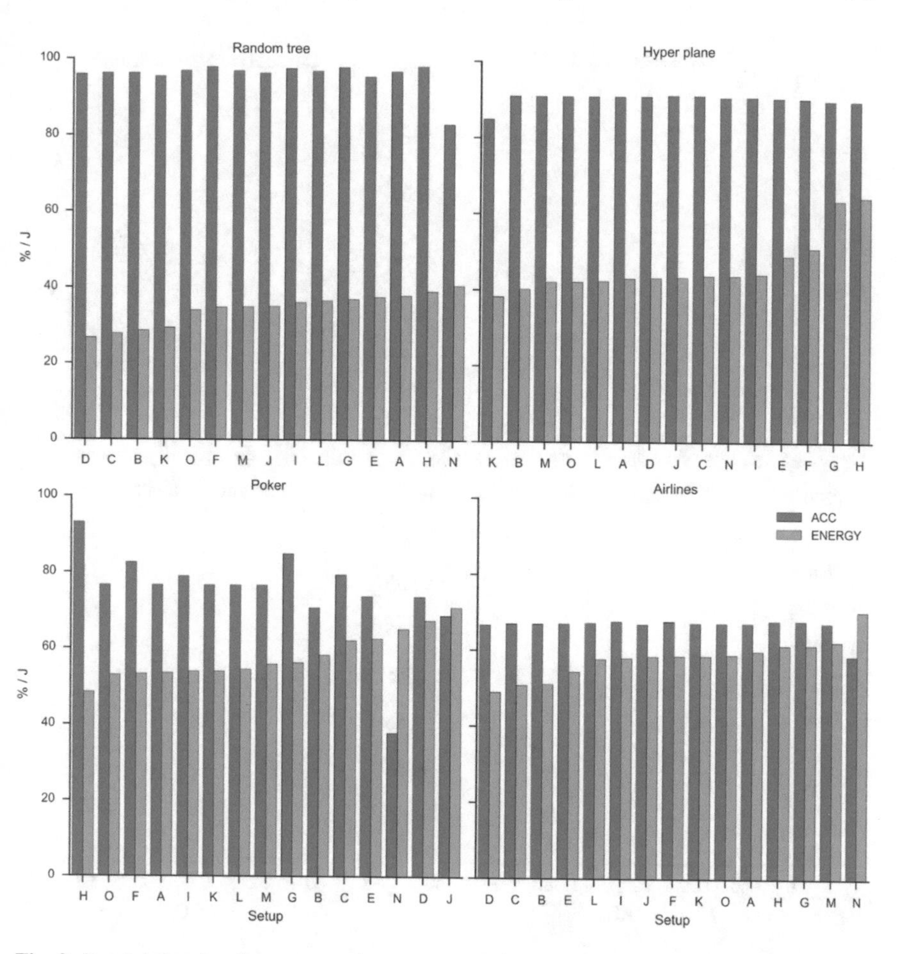

Fig. 2 Barplot showing the energy and accuracy variations for every parameter configuration on each dataset. Accuracy (ACC) is measured in percentage of correctly classified instances and energy is measured in Joules

time. This phenomenon occurs more clearly for datasets 1, 2, and 4, suggesting that there is a relationship between time and nodes. Therefore, a conclusion is the increase of energy is related to an increase on time, probably due to an increase in the number of nodes.

A second measurement is power. We have observed that the power is not linearly correlated with the number of nodes. Figure 5 suggests that whenever the number of nodes is higher, the power is lower, higher, or the same. We have not encountered a variable that is directly correlated with the increase of power, so we will analyze its behavior for each parameter in the next subsection.

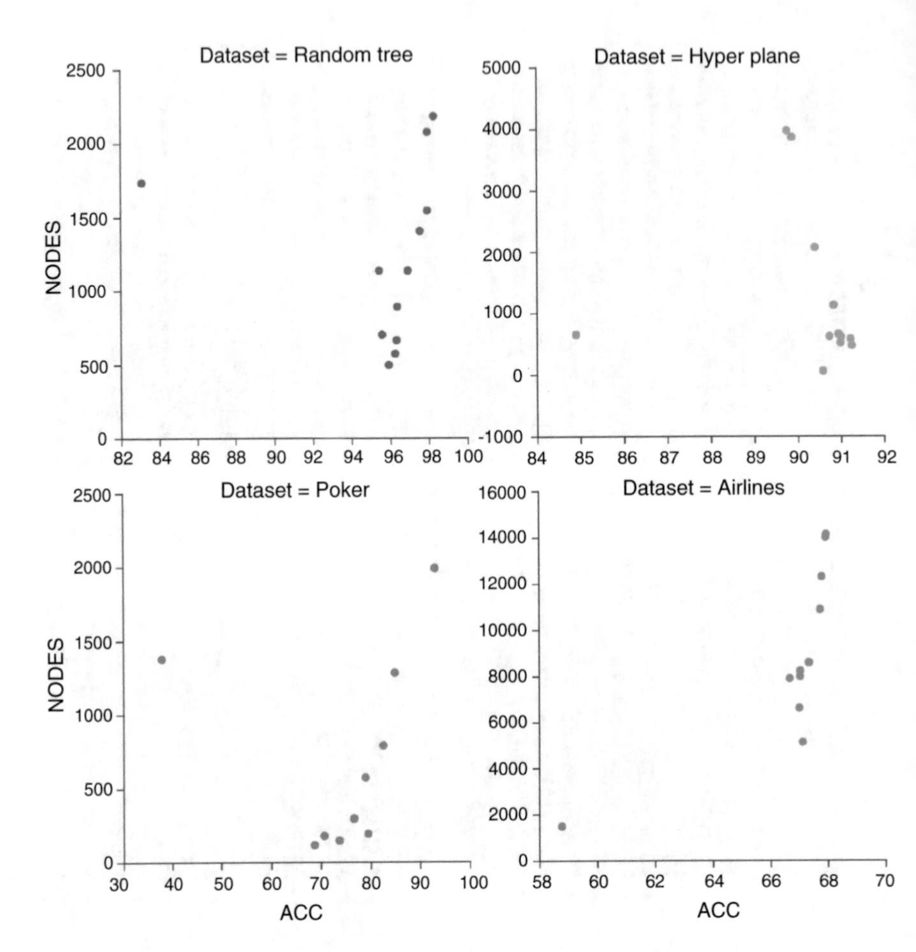

Fig. 3 Scatter plot with a linear relationship between accuracy and nodes for every dataset

6.2 Parameter Analysis

This section aims to compare the parameters' behavior from the theoretical predictions with the empirical results. For that, we have created Table 6 that shows the summary of the real behavior of the parameters obtained in the experiment.

6.2.1 Parameter *nmin*

Observing the behavior of the *nmin* parameter across all datasets, we see that the accuracy is not significantly affected by this parameter, there is some decrease in dataset 1, but only of a 1%. In terms of power, however, we can see an important variation of watts in datasets 1, 3, and 4. Both datasets 1 and 4 have an increase of

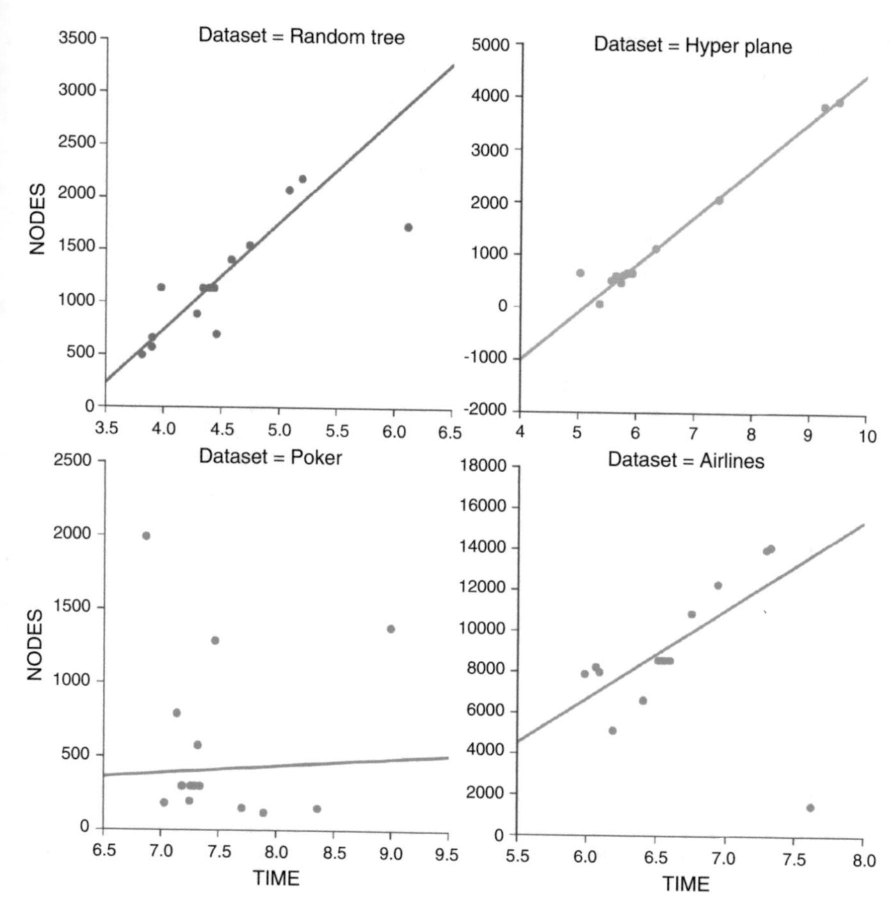

Fig. 4 Scatter plot with a linear relationship between time (seconds) and nodes for every dataset

power when *nmin* is increased. We checked what both datasets have in common to see the reasons behind this increase. The only characteristic that we find in common is that when power decreases, the depth of the tree is lower and also small in comparison to the other datasets. Another reason for this decrease in power when *nmin* increases is because we are computing less times the value of ΔG, therefore saving power. In terms of energy, when *nmin* increases, energy decreases and increases depending on the dataset. For datasets 1 and 4 it decreases, as was predicted in the theoretical section, for dataset 2 it is stable and for dataset 3 it increases. In terms of nodes, we predicted that with higher *nmin* the number of nodes will decrease, and it is exactly what happened in all datasets. Finally, looking at time, in general for all datasets except for the third one, time decreases when *nmin* increases, which is reasonable since we are looking into less batches of data, therefore the tree is computed faster.

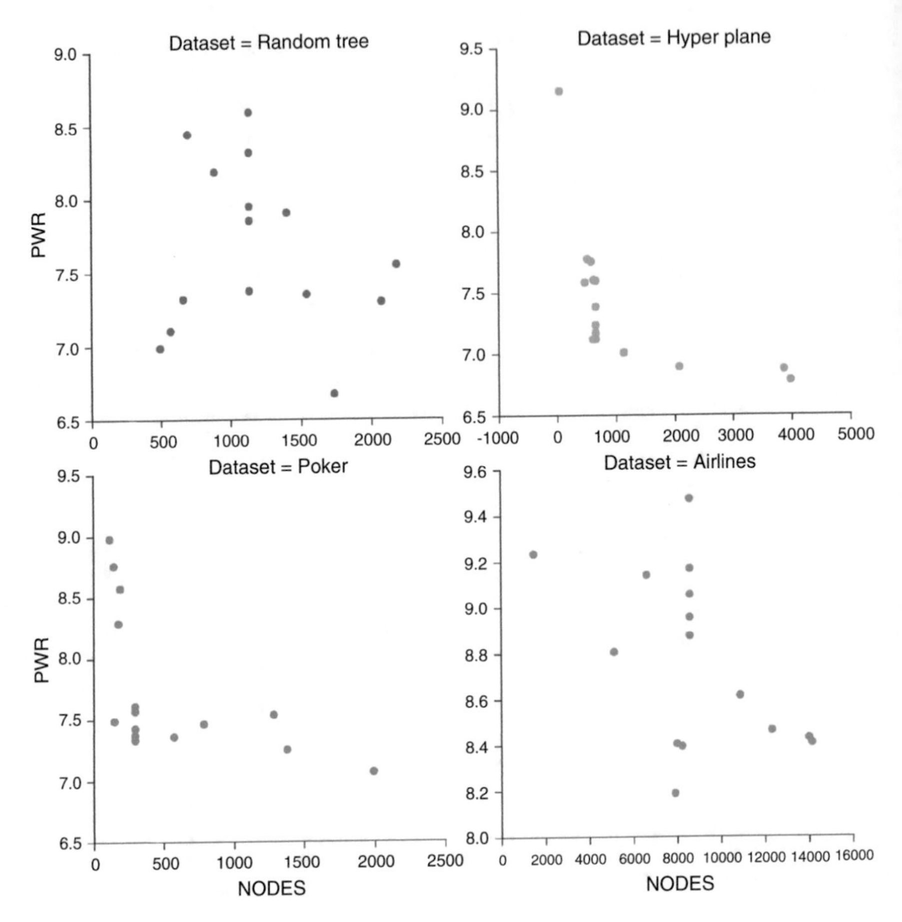

Fig. 5 Scatter plot with a linear relationship between power (W) and nodes for every dataset

6.2.2 Parameter τ

When the value of τ increases, so does the accuracy for datasets 1, 3, and 4. Dataset 2 experiences a non-stable accuracy value when increasing τ. Dataset 3 has an increase in accuracy of 11%, mainly due to the increase in the number of nodes, increasing around 2,000 nodes. That is also the case for all datasets, they significantly increase their number of nodes when τ increases, making this the main reason, from our understanding, for the accuracy increase. It matches with what we theoretically predicted, and with the results shown in the original paper of the authors [15]. In terms of energy, it varies depending on the parameter value. For datasets 2 and 4, there is a significant increase in energy. Dataset 3 experiences an important decrease in energy and dataset 1 does not vary energy significantly. Datasets' 2 and 4 increase of energy is due to a significant increase in time, and a slight reduction in power. Maybe this increase in time is due to the time that it

Table 6 Summary of the real behavior of the parameters in the experiment

		nmin	τ	δ	MEM1				
		↑	↑	↓	2 GB	100 KB	MEM2	S2	RPA
D1	ACC	Dec	Inc	Dec	[a]	Dec	[a]	Dec	[a]
	ENG	Dec	[a]	Dec	Dec	Dec	Dec	Inc	Dec
	NDS	Few	More	Few	[a]	[a]	[a]	More	[a]
D2	ACC	[a]	Dec[b]	Inc	[a]	Dec	[a]	[a]	[a]
	ENG	[a]	Inc	Dec	Dec	Dec	Dec	Inc	Dec
	NDS	Few	More	Few	[a]	[a]	[a]	Few	[a]
D3	ACC	[b]	Inc	Dec	[a]	[a]	[a]	Dec	[a]
	ENG	Inc	Dec[b]	Inc	Inc	[a]	Inc	Inc	Dec
	NDS	Few	More	Few	[a]	[a]	[a]	More	[a]
D4	ACC	[a]	Inc	Dec	[a]	[a]	[a]	Dec	[a]
	ENG	Dec	Inc	Dec	Dec	[a]	Inc	Inc	Dec
	NDS	Few[b]	More	Few	[a]	[a]	[a]	Fewer	[a]

This table portrays how energy, accuracy, and the number of nodes vary when increasing, decreasing, or modifying certain parameters of the VFDT algorithm

D Datasets, *Inc* Increase, *Dec* Decrease, *Few* Fewer nodes, *More* More nodes, *ACC* Accuracy, *ENG* Energy, *NDS* Nodes

[a] There is no variation

[b] It is not a constant decrease or increase throughout the configurations

takes to build more nodes. Dataset 3 decreases energy because it decreases time, while power varies between the setups. We predicted that energy would increase, matching with the behavior of datasets 2 and 4.

6.2.3 Parameter δ

When δ decreases, the probability of making a correct split increases. In this case, datasets 1 and 3 experience a significant decrease in accuracy. Especially dataset 3 that varies its accuracy a 25% (from $\delta = 10^{-1}$ to $\delta = 10^{-10}$). Comparing δ from the default value to the highest value ($\delta = 10^{-1}$), accuracy increases from 76.6% to 93%. Intuitively, it should be the opposite, that increasing δ would decrease accuracy, since there would be less confidence of making a correct split. However, since the number of nodes increases significantly, being almost seven times more nodes for the value $\delta = 10^{-1}$, then accuracy increases by that much. In all datasets, when δ decreases, the number of nodes decreases also, as we predicted, since there are less splits. Taking a look into the energy aspect, when δ increases energy significantly decreases for datasets 1 and 2. Dataset 4 does not vary energy significantly and dataset 3 increases energy when δ decreases. We discovered that when $\delta = 10^{-1}$ for dataset 3, not only do we get a 25% increase in accuracy, as shown before, but we also get a decrease in energy of 22 J.

6.2.4 Parameter MEM1

When the memory for the tree is restricted to 100 KB, accuracy decreases by 1% for the first dataset and by 6% for the second dataset. For the other two datasets the accuracy stays the same, suggesting that the tree did not need more memory than 100 KB. The number of nodes does not vary either in this case. The same happens when the memory is set to 2 GB, the accuracy is the same across all datasets and the number of nodes does not vary at all. The reason the number of nodes is the same is because the implementation deactivates nodes when the memory is limited, but it does not remove them. In terms of power, time, and energy, for the first two datasets there is a significant decrease of energy when the memory limit is set to 100 KB. In the first dataset, the decrease is due to both a decrease in time and a decrease in power, which makes sense since the algorithm needs to analyze less number of nodes, consuming less power, and taking less time. This is not the case for the two last datasets, since we already mentioned that probably the tree is not making use of such parameter, keeping energy at similar levels. When this parameter is set to the value of 2 GB, the energy decreases slightly across all datasets except for the third one.

6.2.5 Parameter MEM2

When the parameter memory management (MEM2) is active, the tree stops growing when the memory limit is hit. The achieved accuracy and the number of nodes are exactly the same across all datasets. Energy does vary across all datasets, although it is not a big difference. Datasets 1 and 2 have a decrease in energy, while datasets 3 and 4 have an increase in energy. These variations are both due to the change of power and time, except for the second dataset, where there is a decrease on power but an increase on time.

6.2.6 Parameter S2

When the split criterion is set to *Gini index*, there is a significant decrease of accuracy across all datasets except for the second one. It is interesting to notice that even though it creates more nodes for the first and third datasets, the accuracy is still significantly lower. In the second dataset the accuracy is maintained and the number of nodes decreased. In terms of energy, there is an increase of energy across all datasets, although power and time vary in a different way in all of them, in most of them when this parameter is chosen the time to build the tree increases. Based on these results, we would not recommend to choose this splitting criterion. However, we have not investigated the reasons behind these results, therefore we suggest to do that in future work studies.

6.2.7 Parameter RPA

When the parameter *removing poor attributes* is chosen, we expect to have a higher accuracy and a decrease in energy. From the experiment we can observe that accuracy is maintained across all datasets, and energy is decreased across also all datasets, although the decrease is not significant in all of them. This increase is mostly due to a slightly decrease in time and a decrease in power. The nodes are maintained for all datasets, which is not what we predicted. The reason could be that even though the tree is using less attributes, the amount of instances to analyze is the same.

6.2.8 Summary of the Parameter Analysis

In general, we can observe that tuning the parameters in a different way outputs different values of energy consumption and accuracy. At the same time, it depends on the type of dataset which parameters will give better results in terms of energy consumption. For instance, while for the first dataset, the set of *nmin* to 1700 will give the best results, for the third dataset it gives very poor results in terms of energy consumption. The reason behind this is that the VFDT algorithm will create different decision trees, depending on the input data. A general observation is that the third dataset, the poker dataset, behaves in a different way in comparison to the other three datasets. On the same line, setting *nmin* to 700 seems to give good results across all datasets except the third one, what suggests that is a good option for future researchers. From looking at Table 1, we conclude that, in general, *removing poor attributes* has a positive impact on energy consumption without affecting accuracy. Also, decreasing the value of δ, decreases the energy consumption and accuracy, but the accuracy decrease is not significant.

7 Conclusions and Future Work

The aim of this paper is to introduce energy consumption as an important factor during data mining algorithm evaluation and analysis. While performance and computational effort are factors usually considered in data mining, energy consumption is seldom evaluated. Energy awareness leads to reducing CO_2 emissions, increasing battery life of mobile devices, and reducing air pollution.

In order to understand the impact of taking energy consumption into consideration, we have analyzed the behavior in terms of energy and accuracy of the VFDT (Very Fast Decision Tree) algorithm when modifying certain parameters. First, we theoretically analyzed how increasing or decreasing certain parameters of such algorithm would affect the tree structure, the accuracy, and the energy consumed. Then, we created an experiment where we empirically vary the same parameters of the VFDT algorithm under four different datasets. We have compared the empirical with the theoretical results, and found that there is indeed a significant variation in terms of energy consumption that depends on how the algorithmic setup is designed. The results also indicate that it is possible to significantly reduce the energy consumption of an algorithm without reducing accuracy by varying correctly the parameters of the algorithm.

Future work is to investigate why certain parameter choices consume more energy than others. For this purpose, we aim to break down data stream mining algorithms into generic subtasks to allow a more fine-grained comparison of energy consumption across various algorithms and algorithm configurations. Finally, we plan to obtain more challenging real world datasets to test how energy can vary on these type of datasets.

Acknowledgements This work is part of the research project "Scalable resource-efficient systems for big data analytics" funded by the Knowledge Foundation (grant: 20140032) in Sweden.

References

1. Address of the president, Lord Rees Of Ludlow om kt prs, given at the anniversary meeting on 1 December 2008. Notes Rec. R. Soc. Lond. **63**(2), 183–190 (2009)
2. Aggarwal, C.C.: A framework for diagnosing changes in evolving data streams. In: Proceedings of the 2003 ACM SIGMOD International Conference on Management of Data, pp. 575–586. Association for Computing Machinery, New York (2003)
3. Aggarwal, C.C.: Data Streams: Models and Algorithms, vol. 31. Springer Science and Business Media, New York (2007)
4. Aggarwal, C.C., Han, J., Wang, J., Yu, P.S.: On demand classification of data streams. In: Proceedings of the Tenth ACM SIGKDD International Conference on Knowledge Discovery and Data Mining, pp. 503–508 (2004)
5. Agrawal, R., Imieliński, T., Swami, A.: Mining association rules between sets of items in large databases. In: ACM SIGMOD Record, vol. 22, pp. 207–216. Association for Computing Machinery, New York (1993)
6. Bhargava, R., Kargupta, H., Powers, M.: Energy consumption in data analysis for on-board and distributed applications. In: Proceedings of the ICML, vol. 3, p. 47 (2003)
7. Bifet, A., Kirkby, R.: Data stream mining a practical approach (2009) doi:10.1.1.192.1957
8. Bifet, A., Holmes, G., Kirkby, R., Pfahringer, B.: MOA: massive online analysis. J. Mach. Learn. Res. **11**, 1601–1604 (2010); http://portal.acm.org/citation.cfm?id=1859903
9. Bourdon, A., Noureddine, A., Rouvoy, R., Seinturier, L.: Powerapi: a software library to monitor the energy consumed at the process-level. ERCIM News (2013)
10. Carbon footprint: (2017). https://en.wikipedia.org/wiki/Carbon_footprint
11. Chu, S.: The energy problem and Lawrence Berkeley National Laboratory. Talk given to the California Air Resources Board (2008)
12. Datar, M., Gionis, A., Indyk, P., Motwani, R.: Maintaining stream statistics over sliding windows. SIAM J. Comput. **31**(6), 1794–1813 (2002)
13. Digital universe study: (2013). https://www.emc.com/collateral/analyst-reports/idc-digital-universe-united-states.pdf
14. Ding, Q., Ding, Q., Perrizo, W.: Decision tree classification of spatial data streams using peano count trees. In: Proceedings of the 2002 ACM Symposium on Applied Computing, SAC '02, pp. 413–417. Association for Computing Machinery, New York (2002)

15. Domingos, P., Hulten, G.: Mining high-speed data streams. In: Proceedings of the sixth ACM SIGKDD International Conference on Knowledge Discovery and Data Mining, pp. 71–80. Association for Computing Machinery, New York (2000)
16. Ecological footprint: (2017). https://en.wikipedia.org/wiki/Ecological_footprint
17. Elena Ikonomovska airline dataset: (2017). http://kt.ijs.si/elena_ikonomovska/data.html
18. Empowering developers to estimate app energy consumption: (2012). http://research.microsoft.com/apps/pubs/default.aspx?id=166288
19. Energy efficient software development: (2015). https://software.intel.com/en-us/energy-efficient-software
20. Ferrer-Troyano, F., Aguilar-Ruiz, J.S., Riquelme, J.C.: Discovering decision rules from numerical data streams. In: Proceedings of the 2004 ACM Symposium on Applied Computing, pp. 649–653. Association for Computing Machinery, New York (2004)
21. Flinn, J., Satyanarayanan, M.: PowerScope: a tool for profiling the energy usage of mobile applications. In: WMCSA '99 Proceedings of the Second IEEE Workshop on Mobile Computer Systems and Applications, pp. 2–10. IEEE, Los Alamitos (1999)
22. Gaber, M.M.: Data stream processing in sensor networks. In: Learning from Data Streams, pp. 41–48. Springer, New York (2007)
23. Gaber, M.M., Krishnaswamy, S., Zaslavsky, A.: On-board mining of data streams in sensor networks. In: Advanced Methods for Knowledge Discovery from Complex Data, pp. 307–335. Springer, New York (2005)
24. Gaber, M.M., Zaslavsky, A., Krishnaswamy, S.: Mining data streams: a review. ACM Sigmod Rec. **34**(2), 18–26 (2005)
25. Gama, J.: Knowledge Discovery from Data Streams. CRC Press, Boca Raton (2010)
26. Gama, J., Gaber, M.M.: Learning from data streams. Springer, New York (2007)
27. Gama, J., Žliobaitė, I., Bifet, A., Pechenizkiy, M., Bouchachia, A.: A survey on concept drift adaptation. ACM Comput. Surv. **46**(4), 44 (2014)
28. Garofalakis, M., Gehrke, J., Rastogi, R.: Querying and mining data streams: you only get one look a tutorial. In: SIGMOD Conference, p. 635 (2002)
29. Google search statistics: (2009). http://searchengineland.com/calculating-the-carbon-footprint-of-a-google-search-16105
30. Green 500: (2017). www.green500.org
31. Guha, S., Mishra, N., Motwani, R., O'Callaghan, L.: Clustering data streams. In: Proceedings 41st Annual Symposium on Foundations of Computer Science, pp. 359–366 (IEEE, 2000)
32. Hoeffding, W.: Probability inequalities for sums of bounded random variables. J. Am. Stat. Assoc. **58**(301), 13–30 (1963)
33. Hooper, A.: Green computing. Commun. ACM **51**(10), 11–13 (2008)
34. How much data is generated every minute on social media? http://wersm.com/how-much-data-is-generated-every-minute-on-social-media/. Published 19 Aug 2015
35. Hulten, G., Spencer, L., Domingos, P.: Mining time-changing data streams. In: Proceedings of the Seventh ACM SIGKDD International Conference on Knowledge Discovery and Data Mining, pp. 97–106 (2001)
36. Jin, R., Agrawal, G.: Efficient decision tree construction on streaming data. In: Proceedings of the Ninth ACM SIGKDD International Conference on Knowledge Discovery and Data Mining, pp. 571–576. Association for Computing Machinery, New York (2003)
37. Kearns, M.J., Vazirani, U.V.: An Introduction to Computational Learning Theory. MIT press, Cambridge, MA (1994)
38. Kifer, D., Ben-David, S., Gehrke, J.: Detecting change in data streams. In: Proceedings of the Thirtieth International Conference on Very Large Data Bases, vol. 30, pp. 180–191 (2004)
39. Last, M.: Online classification of nonstationary data streams. Intelligent Data Anal. **6**(2), 129–147 (2002)
40. Law, Y.N., Zaniolo, C.: An adaptive nearest neighbor classification algorithm for data streams. In: Knowledge Discovery in Databases: PKDD 2005, pp. 108–120. Springer, New York (2005)
41. Li, J., Martínez, J.F.: Power-performance considerations of parallel computing on chip multiprocessors. ACM Trans. Arch. Code Optim. **2**(4), 397–422 (2005)

42. Martín, E.G., Lavesson, N., Grahn, H.: Energy efficiency in data stream mining. In: Proceedings of the 2015 IEEE/ACM International Conference on Advances in Social Networks Analysis and Mining 2015, pp. 1125–1132. Association for Computing Machinery, New York (2015)
43. Naghavi, M., Wang, H., Lozano, R., Davis, A., Liang, X., Zhou, M., Vollset, S.E.V., Abbasoglu Ozgoren, A., Norman, R.E., Vos, T., et al.: Global, regional, and national age sex specific all-cause and cause-specific mortality for 240 causes of death: a systematic analysis for the global burden of disease study 2013. The Lancet **385**(9963), 117–171 (2015)
44. Noureddine, A., Rouvoy, R., Seinturier, L.: Monitoring energy hotspots in software. Autom. Softw. Eng. **22**(3), 291–332 (2015)
45. Pillai, P., Shin, K.G.: Real-time dynamic voltage scaling for low-power embedded operating systems. In: ACM SIGOPS Operating Systems Review, vol. 35, pp. 89–102. Association for Computing Machinery, New York (2001)
46. Poker dataset: (2017). http://moa.cms.waikato.ac.nz/datasets/
47. Raileanu, L.E., Stoffel, K.: Theoretical comparison between the gini index and information gain criteria. Ann. Math. Artif. Intell. **41**(1), 77–93 (2004)
48. Rajaraman, A., Ullman, J.D.: Mining of Massive Datasets. Cambridge University Press, Cambridge (2011)
49. Reams, C.: Modelling energy efficiency for computation. Ph.D. thesis, University of Cambridge (2012)
50. Search queries: (2009). http://searchengineland.com/calculating-the-carbon-footprint-of-a-google-search-16105
51. Shafer, J., Agrawal, R., Mehta, M.: Sprint: a scalable parallel classifier for data mining. In: Proceedings of the International Conference on Very Large Data Bases, pp. 544–555 (1996); Citeseer
52. Spirals research group: (2017). https://team.inria.fr/spirals/
53. Tiwari, V., Malik, S., Wolfe, A.: Power analysis of embedded software: a first step towards software power minimization. IEEE Trans. Very Large Scale Integr. VLSI Syst. **2**(4), 437–445 (1994)
54. Utgoff, P.E.: Incremental induction of decision trees. Mach. Learn. **4**(2), 161–186 (1989)
55. Wang, H., Fan, W., Yu, P.S., Han, J.: Mining concept-drifting data streams using ensemble classifiers. In: Proceedings of the Ninth ACM SIGKDD International Conference on Knowledge Discovery and Data Mining, pp. 226–235. Association for Computing Machinery, New York (2003)
56. Zimmermann, H.: OSI reference model–the iso model of architecture for open systems interconnection. IEEE Trans. Commun. **28**(4), 425–432 (1980)

Glossary

Assortativity The tendency for individuals ("nodes") within a network to be connected to other nodes possessing a similar characteristic to themselves. Also used as short-hand to refer to "degree assortativity", a particular class of assortativity.

Big data analytics Analysis of massively large data sets to discover hidden patterns, trends, correlations, user preferences as well as other relevant business information and knowledge.

BPR **B**ipartite **P**rojection via **R**andom Walks is a similarity measure between objects based on their object-attribute associations as links on a bipartite graph.

Citation network A social network which contains paper references linked by co-citation relationships.

Computational mechanics The study of how a stochastic process evolves according to its ϵ-machine.

Degree assortativity The correlation between the number of connections ("degree") of a node in a network and the degree of those it is connected to. Can range from -1 to 1.

Degree correlation Commonly used synonym of degree assortativity.

Delurking The process of encouraging silent members of an online community to more actively be involved in it, i.e., by creating original content and tangibly contributing to knowledge sharing.

Egocentric network A special type of social network that involves a focal user called "ego". The egocentric network of a user (ego) consists of the individuals called "alters" having a direct relationship with the user (ego) together with the relationships between these individuals.

Energy Efficiency The amount of operations per watt a computer can perform.

ϵ-Machine The minimally complex and optimally predictive representation of a stochastic process constructed via equivalence classes over pasts induced by statistically equivalent futures.

R. Missaoui et al. (eds.), *Trends in Social Network Analysis*, Lecture Notes in Social Networks, DOI 10.1007/978-3-319-53420-6

Graph-based models The abstraction of the architecture and/or the behavioral patterns of a system using mathematical objects from graph theory (edges, nodes, layers, node attributes, and so on).

Green Computing Environmentally responsible use of computers and their applications.

LDA Latent Dirichlet Allocation is a generative model for a collection of documents. This model represents documents as a combination of multiple topics. Each topic is comprised of words with certain probabilities.

Linear Threshold model Classical graph-based information diffusion model whose key idea is that exposure to multiple sources (i.e., friends) is needed for a user before taking a decision. To enable the unfolding of the influence propagation process (from "activated" to "non-activated" users), it holds the assumption that users are assigned activation thresholds, which are distributed uniformly at random. If the sum of incoming influence from activated neighbors is above the activation threshold, then the user becomes active.

Link prediction Given a disjoint node pair (x, y), the purpose is to predict if the node pair has a relationship, or in the case of dynamic interactions, will form one in the near future.

Lurker Silent member of a social network who gains benefit from the observation of information produced by other users, while rarely visibly participating in the community life and providing information themselves. Unlike inactive users, lurkers are effective members who actually correspond to the crowd of the social network. Lurking behavior has been recognized as strictly related to legitimate peripheral participation to learn the netiquette and social norms, individual information strategy of micro-learning and knowledge sharing barriers, as well as individual motivation for interpersonal surveillance.

LurkerRank A family of eigenvector centrality based algorithms designed to identify and rank users showing a lurking behavior in a social network.

Multi-layered graph A graph that allows different types of interactions and possibly different types of nodes.

***N*-Gram** A contiguous sequence of n items ($N = 1, 2, 3, \ldots, n$) from a given sequence of text or speech. An N-gram model is a type of probabilistic language model that is widely used in numerous disciplines, such as computational linguistics, computational biology, probability, and communication theory.

Null model A model generating a result where the effect of interest is not present, so representing the expectation if the null hypothesis is true.

Opinion mining A type of natural language processing task for tracking the mood of the public about a particular product.

Predictive modeling A modeling approach where the statistical model is optimized for the prediction, rather than the description, of a process.

Sarcasm A type of sentiment where people express their negative feelings using positive or intensified positive words in the text.

Sentiment One's opinion, attitude, and emotion towards any specific target such as individuals, events, topics, products, organizations, services, etc.

Social engineering The psychological manipulation of people into executing actions and revealing confidential information. An example of social engineering would be that an attacker befriends a victim and gives him a USB key that contains malicious software. The victim would then use the key infecting his company's information system.

Social media forecasting Prediction of user's activity on social media.

Social profile A particular user profile in which the interests are extracted from the information about user social network members.

Targeted influence maximization A special case of the problem of influence maximization in networks, whereby the influence is meant to be propagated toward a specific target (i.e., subset of nodes in the network) rather than to any node in the network.

User profile A record of user information (personnel data, preferences, interests). In information systems, adaptive information mechanisms (e.g. personalization, information access, recommendation) rely on user profiles to propose relevant content according to the user specific needs.

Very Fast Decision Tree An online decision tree algorithm that is able to handle potentially infinite streams of data and produce similar results compared to offline algorithms.

Vulnerability assessment The identification of weaknesses in a system and the estimation of the likelihood and the cost of their exploitation.

Printed in the United States
By Bookmasters